D1528840

Bioarrays

Bioarrays

From Basics to Diagnostics

Edited by

Krishnarao Appasani, PhD, MBA

Founder and CEO
GeneExpression Systems, Inc.
Waltham, MA

Foreword by

Sir Edwin M. Southern, PhD, FRS

Emeritus Professor, Department of Biochemistry
University of Oxford
Founder and Chairman, Oxford Gene Techology
Oxford, United Kingdom

HUMANA PRESS ✳ TOTOWA, NEW JERSEY

Library of Congress Cataloging-in-Publication Data

Bioarrays : from basics to diagnostics / edited by Krishnarao Appasani
  ; foreword by Edwin M. Southern.
    p. ; cm.
  Includes bibliographical references and index.
  ISBN: 978-1-58829-476-0 (alk. paper)
  1. DNA microarrays.   I. Appasani, Krishnarao, 1959-   .
  [DNLM: 1. Oligonucleotide Array Sequence Analysis.   QU 450
B615 2007]
  RB43.8.D62B5622 2007
  616.9'11--dc22

                                        2006024162

To my mentors,

Maharani Chakravorty-Burma

and

the late Debi P. Burma

Foreword

It seems a while since it was possible for one person to keep abreast of all aspects of microarrays. Early in the development of the technology there were few fabrication methods, the amount of data generated was small enough that a human could comprehend it without too much help from computers, and there was a handful of publications on the topic. But that changed around ten years ago with the advent of the application of arrays to gene expression analysis. And now the field is so wide that it embraces a number of quite narrow specialisations; there are people whose careers are founded on array bioinformatics or whose working day is largely filled in sample preparation for microarrays. The technology continues to advance along many paths. New modes of use, such as array CGH, appear frequently and spawn a range of applications. There can be few areas of biology that do not use the technology.

For those of us who have been in the field since its beginning, it is satisfying to see the technology producing research data that could not be obtained in any other way, and which is advancing so many fields. None is more satisfying than the advances in disease research, where the molecular characterisation of genetic and infectious diseases is laying the foundations for new diagnostics and treatments. *Bioarrays: From Basics to Diagnostics* is a timely review of applications of arrays to human biology and to human disease. The articles illustrate the flexibility of the technology and its great potential. But reading these reviews makes it clear that array technology is only a part of the picture. Some of the biological systems addressed by the technology, interacting networks of gene expression, for example, are complex and several chapters stress the need in such cases for careful experimental design and data interpretation. Others share the authors' experience with difficulties such as poorly defined samples, which also produce poor results. These are extremely important issues. These chapters are a timely reminder that the technology, if carelessly applied, can lead to poor data that may be difficult to correct.

Most of the experience with arrays has been with nucleic acids. The main reason for this is the simplicity of the relationship between analyte and probe molecules, which relies on the rules of Watson-Crick base pairing. There are easy ways to make probes by replicating clones, cDNA or genomic DNA, or by synthesising oligonucleotides using the efficient chemistry developed in the 1980s: when the sequence of the target analyte is known, it is possible to make a probe. No simple relationship exists for any other type of analyte. It is, of course, possible to make arrays from antibodies to proteins and examples are described in this volume, but the process is an empirical one and the rules of interaction between antibody and epitope are not so well understood. However, it is also clear that a description of proteins and their modifications is a necessary complement to the description of the genes and their RNA products. It is testament to the power and effectiveness of the array platform that this large investment in reagents for protein analysis is seen to be worthwhile.

The articles in this volume address these issues and make clear that microarrays have proven their value as research tools. Their future as diagnostic devices is also addressed; much effort is going into discovering reagents that can be used in routine testing. However, it seems unlikely that the tests that will emerge from this research will be carried out on the discovery platform. Migration to diagnostics is well under way, but simpler, quicker and cheaper methods and devices will be needed to translate the technology to the point of care or to the testing laboratory. The regulatory authorities are engaged with the technology and their interaction, for example, the FDA's Micro Array Quality Control (MAQC) project, is having a salutary effect in introducing quality standards. This, in turn, will feed back to improve the standard of work done in basic research settings.

The technology has come a long way in the past decade. It still has far to go. The chapters in *Bioarrays: From Basics to Diagnostics* present a fascinating view of the field at this interesting point in the journey.

Edwin Southern
Oxford, December 2006

Preface

Molecular Biology was once considered to be a completely useless subject, remote from medical applications, and of only academic interest. But now molecular biology may help create a new public health paradigm.

SYDNEY BRENNER, NOBEL LAUREATE
SALK INSTITUTE, LA JOLLA, CA
(From *Science* 2003, 302, p. 533.)

Gene expression or genomics is the branch of molecular biology that describes the functional architecture of genes. The wealth of genetic information revealed from genome sequencing projects, in combination with the discovery of array technology, has paved the way for a new discipline—molecular diagnostics—within which scientists can dissect and understand the molecular pathogenesis of the biological cell. The use of biological arrays became a routine practice in several molecular, cell biological, and toxicological laboratories around the globe, and the technology is collectively referred as BioArrays or BioChips. Although *Bioarrays: From Basics to Diagnostics* is focused primarily on applications in molecular diagnostics, bioarrays have potential applications in biology, medicine, and agriculture.

Bioarrays: From Basics to Diagnostics is mainly intended for readers in the molecular cell biology, genomics, and molecular diagnostics fields. It will also be useful in advanced graduate level courses, as well as interesting to those in biotechnology and molecular medicine. Although a number of books already cover array technology, *Bioarrays: From Basics to Diagnostics* differs from its forefunners: it is the first text completely devoted to applications diagnostic of human diseases. The book focuses on the concepts of oligonucleotide, cDNA, protein, antibody, and carbohydrate arrays, in 17 chapters, grouped into four parts. The chapters are written by experts in the field, from both academic and industrial backgrounds. This book will serve as a reference for graduate students, postdoctoral researchers, and professors from academia and as an explanatory analysis for executives, and scientists in biotechnology and pharmaceutical companies. My hope is that *Bioarrays: From Basics to Diagnostics* will provide both a prolog to bioarrays for newcomers and insight to those already active in the field.

Edwin Southern at the University of Oxford first developed a way to use inkjet printing to show oligonucleotide sequences on glass slides in the late 1980s. Subsequently, his group demonstrated the first array of all 256 octapurines in a simple eight-step process by using in-house combinatorial chemistry approaches. During the early 1990s Stephan Fodor at Affymetrix, using photolithography-based methods, developed the *miniaturized oligonucleotide array* of eight nucleotides. The work of both pioneers became the foundation for today's *in situ* oligonucleotide arrays. At the same time, Patrick Brown at Stanford University devised an essentially different array fabrication method. Brown proposed and demonstrated for the first time an arrangement of already-synthesized snippets of DNA placed at regular intervals on a glass surface by a robot. The successful collaboration between the Brown laboratory and Ron Davis's laboratory at Stanford produced the groundbreaking paper published in 1995 in *Science* that used the word *microarray* for the first time. Since then, the buzz words "DNA arrays" and "chips" have become common parlance in the scientific community. The same technology platform was extended later to development of protein arrays, glycoarrays, and tissue arrays, jointly referred to as bioarrays or biochips.

Before the discovery of this bioarray technology, only a limited number of techniques were available (such as, differential display or serial analysis of gene expression) for investigating gene expression and regulation. However, use of microarrays paved new avenues to study *expression profiling* (the process of measuring simultaneously the expression of thousands of individual genes in a given biological sample). Use of microarrays became a routine practice for studying gene expression. Today, with the availability of data from various genome projects, it is now feasible to get "genomes-on-a-chip" from several vendors. Current bioarray platforms also allow the analysis of the function, expression, and disease involvement of several thousands of genes. However, in reality, the discovery of only a few good genes is enough for the routine clinical molecular diagnostics that have emerged recently. To develop 100% accuracy of disease diagnosis, a combination of two or more methods is pivotal.

Use of bioarrays falls under the new subject of *Systemomics*, or the holistic study of expression profiling (gene, protein, lipid, and drug), function, physiological circuits, and developmental networks in human (animal) body systems. Several innovators like Fodor envisioned these developments almost 15 years ago. Despite their proof-of-concept papers, which are reviewed here, microarray-based diagnostics had to wait several more years to become a routine tool in the clinical diagnostics laboratories, a development hastened by studies of cancer prognosis, prediction, and classification.

In the last decade, DNA microarray research has provided a wealth of information on gene expression. From a simple technique to analyze gene expression patterns of relatively few genes, microarrays have evolved into indispensable tools for scientists in biomedicine. Characterizing cancer-gene expression patterns from a systems perspective will also involve understanding the protein–protein networks and transcriptional regulatory programs. Protein chip technologies are helping academicians to better understand the basic research. Proteomics approaches are uniquely useful in biomarker discovery and subsequent immunoassay development. Therefore, proteomics can be seen as a platform for identifying new analytes for the diagnostic industry, rather than just a tool.

Antibody microarrays are also useful because they can measure protein abundance independently of gene expression, unlike the DNA microarrays. It is known that changes in gene expression do not always correlate with protein expression. Patterns of protein expression can be used to diagnose disease or determine appropriate treatment for patients. In addition, antibody arrays can be used to study protein expression in tumor cells, phosphorylation states of cellular proteins, and posttranslational modifications of proteins. In contrast, tissue microarrays will allow researchers to validate the targets identified from the DNA and protein arrays in a high-throughput manner by using immunohistochemistry and *in situ* hybridization methods. The advantage conferred by tissue microarray validation is pivotal to rapid screening of several hundreds of different types of tumors, where a specific protein is studied in different tumor histogenesis pathways. This approach is particularly useful in the screening of new genes and antibodies, and it is ideal for target-specific therapeutic development. Many of the applications mentioned here are described in the four parts of *Bioarrays: From Basics to Diagnostics,* and a summary of each part can be found in the intro-duction on each part title page.

In the near future, we will be able to use molecular signatures or panels of multi-plexed biomarkers to identify whether a tumor is malignant or benign, its site of origin, its prognostic subtype, and even predict its response to therapy. Integrative analyses of gene–drug pairs from therapeutic drug databases along with their gene expression signatures may lead to personalized multidrug regimens based on an individual's tumor gene expression signature. Clearly, high-throughput approaches and bioinformatics will have a primary role in future clinical oncology and pathology. To fully realize the potential of biomarkers to aid in drug development, industry must implement best practices for biomarker development, and promote translational research strategies. However, the biggest obstacle to translating discovery from bench to bedside is not only technological but regulatory.

It is possible that no single type of assay, DNA based or protein based, will be able to completely replace the other type. However, in a nutshell, array technology indeed has the potential to accommodate all required assay formats on one testing platform, and to provide better reagents for pathological diagnosis in the future.

Many people have contributed to making my involvement in this project possible. I thank my teachers throughout my life for their excellent teaching, guidance, and mentoring, which led me to become a scientist, and to bring this educational enterprise of editing to readers. I am thankful to all of the contributors to this book. Without their commitment this book may not have emerged. Many people have had a hand in the launch of *Bioarrays: From Basics to Diagnostics.* Each chapter has been reviewed and revised so that it represents a joint composition. Thanks are also due to the readers, who make my hours putting together this volume worthwhile if they find value in the hours they spend with our book. I am indebted to the capable staff of Humana Press for their generosity during development of the manuscript and their efficiency in bringing it to print. I gratefully acknowledge Drs. Patrick O. Brown and Michael J. Fero of Stanford Univesity for providing the images of a droplet from the finger and oliogarray background that were used in the cover design. I am grateful to Sir Edwin M. Southern for his generous foreword. Last, but not least, I thank my wife, Shyamala, and two wonderful sons, Raakish and Raghu, for their under-standing and cooperation during the entire endeavor.

A portion of the royalties will be contributed to the Dr. Appasani Foundation, a nonprofit organization devoted to bringing social change in developing countries through the education of youth.

Krishnarao Appasani

Contents

Contributors

MELANIE ABONNENC, PhD • *Center of Excellence on Electronic Systems (ARCES), University of Bologna, Bologna, Italy*

LUIGI ALTOMARE, PhD • *Center of Excellence on Electronic Systems (ARCES), University of Bologna, Bologna, Italy*

KRISHNARAO APPASANI, PhD, MBA • *GeneExpression Systems, Inc., Waltham, MA*

VICTORIA ARANGO, PhD • *New York State Psychiatric Institute, Department of Neuroscience, and Department of Psychiatry, Columbia University, New York, NY*

MURALI D. BASHYAM, PhD • *Laboratory of Molecular Oncology and National Genomics and Transcriptomics Facility, Centre for DNA Fingerprinting and Diagnostics, Hyderabad, India*

CARL A. K. BORREBAECK, PhD • *Department of Immunotechnology, Lund University, Lund, Sweden*

MONICA BORGATTI, PhD • *Biotechnology Center and ER-GenTech, Department of Biochemistry and Molecular Biology, University of Ferrara, Ferrara, Italy*

HELENA PAULA BRENTANI • *Hospital do Câncer A.C. Camargo, São Paulo, Brazil*

RICARDO RENZO BRENTANI, MD, PhD • *Hospital do Câncer A.C. Camargo; Faculdade de Medicina, Universidade de São Paulo, São Paulo, Brazil*

KONRAD BÜSSOW, PhD • *Max Planck Institute for Molecular Genetics, Berlin, Germany*

DIRCE MARIA CARRARO, PhD • *Hospital do Câncer A. C. Camargo, São Paulo, Brazil*

PAULO COSTA CARVALHO, PhD • *Laboratory for Functional Genomics and Bioinformatics, Oswaldo Cruz Institute, Rio de Janeiro, Brazil*

MARIA DA GLORIA COSTA CARVALHO, PhD • *Instituto de Biofisica Carlos Chagas Filho, Universidade Federal do Rio de Janeiro, Rio de Janeiro, Brazil*

SABINE CEPOK, PhD • *Department of Neurology, Heinrich-Heine-Universität, Düsseldorf, Germany*

GIRIRAJ RATAN CHANDAK, MD, DNB • *Genome Research Group, Centre for Cellular and Molecular Biology, Hyderabad, India*

VIKAS CHANDHOKE, PhD • *Center for Biomedical Genomics and Informatics, George Mason University, Manassas, VA*

RONG CHENG, PhD • *Columbia Genome Center, Columbia University College of Physicians and Surgeons, New York, NY*

JEN-TSAN ASHLEY CHI, MD, PhD • *Institute of Genome Sciences and Policy, Department of Molecular Genetics and Microbiology, Duke University Medical Center, Durham, NC*

ROLAND CONTRERAS, PhD • *Department for Molecular Biomedical Research, VIB; Department of Molecular Biology, Ghent University, Ghent, Belgium*

STEVE G. CULP • *Target Biology, Astra Zeneca Pharmaceuticals, Wilmington, DE*

TOM R. DEFAY, PhD • *Target Biology, Astra Zeneca Pharmaceuticals, Wilmington, DE*

CHAO DENG, PhD • *Columbia Genome Center, Columbia University College of Physicians and Surgeons, New York, NY*

LOUBNA ERRAJI-BENCHEKROUN, PhD • *New York State Psychiatric Institute, Department of Neuroscience, and Department of Psychiatry, Columbia University, New York, NY*

ENRICA FABBRI, PhD • *Biotechnology Center and ER-GenTech, Department of Biochemistry and Molecular Biology, University of Ferrara, Ferrara, Italy*

CINZIA FORTINI, PhD • *ER-GenTech, Department of Biochemistry and Molecular Biology, University of Ferrara, Ferrara, Italy*

MARCIA V. FOURNIER, PhD • *Oncology Center of Excellence for Drug Discovery, Glaxo SmithKline, Collegeville, PA*

ROBERTO GAMBARI, PhD • *Professor, Biotechnology Center and ER-GenTech, Department of Biochemistry and Molecular Biology, University of Ferrara, Ferrara, Italy*

RICCARDO GAVIOLI, PhD • *Professor, ER-GenTech, Department of Biochemistry and Molecular Biology, University of Ferrara, Ferrara, Italy*

FRANCESCO GORRETA • *Center for Biomedical Genomics and Informatics, George Mason University, Manassas, VA*

GERALDINE GRANT, PhD • *Center for Biomedical Genomics and Informatics, George Mason University, Manassas, VA*

ROBERTO GUERRIERI, PhD • *Professor, Center of Excellence on Electronic Systems (ARCES), University of Bologna, Bologna, Italy*

KEVIN W. HAGAN • *Target Biology, Astra Zeneca Pharmaceuticals, Wilmington, DE*

SEYED E. HASNAIN, PhD • *University of Hyderabad, Hyderabad, India*

BERNHARD HEMMER, MD • *Department of Neurology, Heinrich-Heine-Universität, Düsseldorf, Germany*

HEMAVATHI JAYARAM • *Genome Research Group, Centre for Cellular and Molecular Biology, Hyderabad, India*

WOUTER LAROY, PhD • *Department for Molecular Biomedical Research, VIB; Department of Molecular Biology, Ghent University, Ghent, Belgium*

TOWIA A. LIBERMANN, PhD • *Associate Professor of Medicine, Director, BIDMC Genomics Center, Beth Israel Deaconess Medical Center, Boston, MA*

KE LIU, MD, PhD • *Columbia Genome Center, Columbia University College of Physicians and Surgeons, New York, NY*

SHAOYI LIU, MD • *Columbia Genome Center, Columbia University College of Physicians and Surgeons, New York, NY*

DAVID D. MAGEE • *Life Science Division, Lawrence Berkeley National Laboratory, Berkeley, CA*

MICHAEL A. MALLAMACI, PhD • *Target Biology, Astra Zeneca Pharmaceuticals, Wilmington, DE*

NICOLÒ MANARESI, PhD • *Silicon Biosystems, Bologna, Italy*

J. JOHN MANN, MD • *New York State Psychiatric Institute, Department of Neuroscience, and Department of Psychiatry, Columbia University, New York, NY*

AMY MEDD • *Neuroscience, Astra Zeneca Pharmaceuticals, Wilmington, DE*

GIANNI MEDORO, PhD • *Silicon Biosystems, Bologna, Italy*

SUSAN MCDONNELL, PhD • *School of Biotechnology, National Institute for Cellular Biotechnology, Dublin City University, Dublin, Ireland*

LADISLAV MRZLJAK • *Neuroscience, Astra Zeneca Pharmaceuticals, Wilmington, DE*

DAVID MURRAY, PhD • *School of Biotechnology, National Institute for Cellular Biotechnology, Dublin City University, Dublin, Ireland*

CLAUDIO NASTRUZZI, PhD • *Professor, Center of Excellence on Electronic Systems (ARCES), University of Bologna, Bologna, Italy*

HASAN H. OTU, PhD • *BIDMC Genomics Center, Beth Israel Deaconess Medical Center, Boston, MA*

JURAJ PETRIK, PhD • *Microbiology Research and Development, Scottish National Blood Transfusion Service, TTI Department, LCMV, Royal Veterinary College, Summerhall, Edinburgh, UK*

ANIL POTTI, MD • *Department of Medicine, Duke University Medical Center, Durham, NC*

LUIZ FERNANDO LIMA REIS • *Ludwig Institute for Cancer Research; Hospital do Câncer A. C. Camargo, São Paulo, Brazil*

JANINE SCOTT ROBB • *Alba Bioscience, Edinburgh, UK*

ALDO ROMANI, PhD • *Center of Excellence on Electronic Systems (ARCES), University of Bologna, Bologna, Italy*

VANI SANTOSH, MD, DBM • *National Institute of Mental Health and Neurological Sciences, Bangalore, India*

DHAVAL SHAH, MD • *Columbia Genome Center, Columbia University College of Physicians and Surgeons, New York, NY*

RAVI SIRDESHMUKH, PhD • *Center for Cellular and Molecular Biology, Hyderabad, India*

FERNANDO AUGUSTO SOARES, MD, PhD • *Hospital do Câncer A.C. Camargo, São Paulo, Brazil*

SIR EDWIN M. SOUTHERN, PhD, FRS • *Chairman, Oxford Gene Technology, Oxford, United Kingdom*

ANUSHA SRIKANTH • *Center for Cellular and Molecular Biology, Hyderabad, India*

MARCO TARTAGNI, PhD • *Center of Excellence on Electronic Systems (ARCES), University of Bologna, Bologna, Italy*

BRIAN J. TRUMMER • *Columbia Genome Center, College of Physicians and Surgeons, Columbia University, New York, NY*

KAZUHIKO UCHIDA, MD, PhD • *Department of Molecular Biology and Oncology, Institute of Basic Medical Sciences, Graduate School of Comprehensive Human Sciences, University of Tsukuba; Clinical Bioinfomatics Research Initiative, AIST, Tsukuba, Ibaraki, Japan*

MARK D. UNDERWOOD, PhD • *New York State Psychiatric Institute, Department of Neuroscience, and Department of Psychiatry, Columbia University, New York, NY*

AILI WANG, MD • *Columbia Genome Center, Columbia University College of Physicians and Surgeons, New York, NY*

DENONG WANG, MD, PhD • *Carbohydrate Microarray Laboratory, Departments of Genetics, Neurology and Neurological Sciences, Stanford University School of Medicine, Stanford, CA*

RUOBING WANG • *Carbohydrate Microarray Laboratory, Departments of Genetics, Neurology and Neurological Sciences, Stanford University School of Medicine, Stanford, CA*

ZHEN WANG, MD • *Department of Surgery, Stanford University School of Medicine, Stanford, CA*

AMANDA J. WILLIAMS • *Target Biology, Astra Zeneca Pharmaceuticals, Wilmington, DE*

CHRISTER WINGREN, PhD • *Department of Immunotechnology, Lund University, Lund, Sweden*

XIAOYUAN XU, PhD • *Columbia Genome Center, Columbia University College of Physicians and Surgeons, New York, NY*

PART I

BIOARRAY TECHNOLOGY PLATFORMS

Krishnarao Appasani

The seminal work of Edwin Southern paved the way for the development of "oligoarrays" as a tool for studying gene expression and regulation. Subsequently, gene expression profiling by using DNA microarrays has become an integral part of basic and applied research in both academic and industrial scientific communities. This technology has been successfully used for several distinct applications ranging from disease classification and functional genomics to pharmacogenomics, biomarker identification, and single-nucleotide polymorphism analyses. In Part I, Chapter 1, Murray et al. summarize the applicability of cDNA microarrays in the study of gene expression profiles in colorectal cancer, with a particular interest in identifying biomarkers for metastasis. Metastasis represents the most lethal aspect of cancer and is the prime cause of cancer deaths. Thus, new insight into the mechanisms of metastasis is needed for the development of successful therapies. One mechanism is elucidated by the microarray approach in this chapter. In addition, Murray et al. present a thoughtful overview of the applicability of microarrays in cancer research, especially to predict response of cancer cells to various chemotherapies.

The applicability of microarrays to tumor classification, as demonstrated by several investigators in the past few years, opened new avenues for the development of more efficient strategies for prognosis, diagnosis, and treatment. Molecular taxonomy of cancers is well described in Chapter 2 by Carraro et al. They provide a microarray-based experimental strategy that addresses cancer tissue collection through gene expression profiling and final validation by using immunohisto-chemistry, in situ hybridization by using tissue, or both. Human tumor samples are heterogeneous mixtures of diverse cell types, including malignant cells, stromal elements, blood vessels, and inflammatory cells. Because of this heterogeneity, the interpretation of gene expression studies is not always simple. To harvest a homogeneous population of cells, a laser capture microdissection approach is an alternative that is used in combination with microarrays. In Chapter 3, Fournier et al. summarize the experimental flow of performing microarray experiments and the cautionary steps that should be considered in developing expression profile data. In addition, they provide a comprehensive overview of the application of proteomics in the discovery of new cancer biomarkers, protein signatures, and protein networks.

Following the advances in sequencing of genetic information, biology has become an increasingly information-intensive discipline. Although microarray technology has occupied a central role in functional genomics, the functionality and interrelationships of genes thus far identified have not yet come into focus. Microarrays give a snapshot of the enormous interaction network that defines the working mechanisms in the cell by measuring thousands of data points at a given time. This approach opens up the possibility and creates the challenge to reverse engineer biological networks by using high-throughput systems. Using gene expression data to infer genetic regulatory networks is just one example and is the subject of the Chapter 4, in which Otu and Libermann detail how the mathematical concepts-based network theory provides a promising framework for studying biological systems. Network theory-based biological systems study becomes especially important as biology undergoes paradigm shifts, particularly with the rapid advancements in technology.

1

Investigation of Tumor Metastasis by Using cDNA Microarrays

David Murray, Francesco Gorreta, Geraldine Grant, Vikas Chandhoke, and Susan McDonnell

Summary

Microarray-based technologies are powerful and widely used genomic techniques for the study of gene expression patterns on a genome-wide scale. The applications of microarrays as research tools in all areas of biology are immense, and modern approaches using these technologies to understand tumor metastasis are described in this chapter. Attention is placed on the steps involved in analysis from start to finish. We also have highlighted our own work, in which gene expression profiles in colorectal metastasis were monitored using cDNA microarrays.

Key Words: cDNA microarray; colorectal cancer; metastasis.

1. Introduction

1.1. Introduction to Tumor Metastasis

Metastasis represents the most lethal aspect of cancer and is the cause of 90% of cancer deaths, usually because it is difficult to treat patients when the cancer is at this late stage *(1)*. The metastatic spread of cancer cells from their initial site of origin enables them to escape the primary tumor mass and colonize new terrain in the body. Metastasis is a multistep process that involves local area invasion at the primary site, followed by intravasation of tumor cells to either the blood or lymphatic vessels, thereby accessing the general circulation (**Fig. 1**). After circulating in the vascular systems, tumor cells arrest and extravasate at distant organs where they establish secondary tumors. The process of metastasis, therefore, involves the participation of numerous biomolecules in a variety of intricate cellular functions, including altered cell adhesion, proteolysis, and migration *(2)*. Advanced metastasis represents a major hurdle in cancer treatment, mainly because these molecular mechanisms remain unclear. New insight into the mechanisms of metastasis is needed for the development of successful therapies. Modern therapeutic approaches to eliminate or control metastasis are based on defining its critical events and the specific targets that regulate these events. Therefore, the identification and understanding of novel cellular and molecular determinants of metastasis are crucial.

1.2. DNA Microarrays

Microarray-based technologies have become the most widely used analytical techniques for the study of gene expression patterns on a genome-wide scale *(3,4)*. In addi-

From: *Bioarrays: From Basics to Diagnostics*
Edited by: K. Appasani © Humana Press Inc., Totowa, NJ

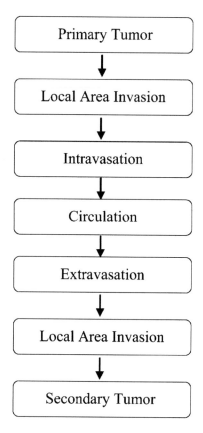

Fig. 1. Metastasis is a multistage process. This simplified representation shows that metastasis involves local area invasion at the primary site, followed by intravasation of tumor cells to either the blood or lymphatic vessels, thereby accessing the general circulation. After circulating in the vascular systems, tumor cells arrest and extravasate at distant organs where they establish secondary tumors.

tion to other levels of regulation, cellular processes are governed by the repertoire of expressed genes and the levels and timing of their expression. It is therefore important to have tools to monitor a large number of messenger RNAs (mRNAs) in parallel.

One of the first decisions to be made while planning a microarray experiment involves choosing a platform. Several choices are available. Microarrays are collections of gene-specific sequences represented on a solid support. The two major methods of microarray production are 1) *in situ* synthesis of oligonucleotides (oligos) and 2) spotting oligos or cDNA on glass slides. With *in situ* synthesis, several sequences of short oligos (~25 base pairs [bp]) are obtained using a combination of photolithography and oligo chemistry *(6,7)*. Because shorter oligos do not allow high specificity, each gene is represented at least five times with equally as many mismatch negative controls, thereby providing reliable and reproducible data. One disadvantage of this platform is that it allows only one color hybridization. The alternative microarray-manufacturing method consists of spotting DNA sequences on glass slides by using robotic systems. Spotted microarrays can be divided into oligo arrays and cDNA arrays. If spotted, the oligos are usually 50mers or 70mers, thereby allowing higher specificity. cDNA microarrays can be manufactured by PCR amplification of cDNA libraries. Amplicons can be purified and spotted on glass. The major advantage of cDNA microarrays is the reduction in cost of production. In contrast, amplicon sizes vary between clones, and it can therefore be difficult setting stringent hybridization conditions for all spots, thereby increasing the changes of cross-hybridization.

On a microarray, each spotted sequence represents a gene. High-density microarrays can represent whole genomes. These sequences offer a solid support for the investigation of unknown samples. Moreover, a query RNA population can be reverse transcribed and fluorescently tagged and hybridized to the microarray. For each gene, specific signals are obtained in proportion to the messenger abundance in the RNA under analysis. Typical samples for analysis include cell lines and both clinical and laboratory animal-derived tissue biopsies. In a two-color experiment, it is possible to compare two RNA populations (e.g., "experimental: cell state and a "control" cell state) on the same microarray.

Using DNA microarray technologies to characterize expression profiles from different biological states and to identify alterations in gene modulations has become a powerful and valuable genomic tool in elucidating the molecular basis of almost every aspect of cancer biology as well as many other diseases, because DNA array-based technologies allow expression levels of many genes to be studied in parallel, thereby providing static (i.e., in which sample a gene of interest is expressed) and dynamic (i.e., how the expression pattern of one gene relates to that of other genes) information about gene expression *(5,6)*. Essentially, the expression level of either the same gene in different samples or of different genes in the same sample can be compared, thus providing insight into gene function and regulation.

Most of this chapter concerns cDNA microarrays; however, other array technologies, including oligo arrays, have been developed. cDNA microarrays are discussed here because they allow direct comparison of two samples on the same slide.

1.3. Application of Array Technologies to Cancer Research

Cancer is characterized by alterations in molecular cellular processes, and as a powerful genomic tool, microarray analysis holds great promise in the molecular medicine of cancer research, because cancer is a complex polygenic and multifactorial disease, resulting from successive changes in the genome of cells and from the accumulation of molecular alterations in both tumor and host cells *(1)*. Such genetic alterations include expression, suppression, or enhancement as well as deletions, mutations, insertions, and rearrangements of genes controlling regulatory pathways and cellular processes such as proliferation, differentiation, cell cycle, DNA repair, and apoptosis. These genetic changes lead to genetic instability, tumorigenesis, malignancy, and an invasive and often drug-resistant phenotype. Therefore, an understanding of the molecular behavior of tumors would aid their molecular classification and in therapeutic decisions *(8)*. New tools are required to predict the clinical behavior and outcome of individual tumors and to ideally prescribe individual tailored treatments.

Microarray techniques have been used to investigate the underlying biology of almost all cancers, including acute myelogenous leukemia *(9)*, breast *(10)*, ovarian *(11,12)*, lung *(13)*, and colon cancer *(14)*. Array analysis is applicable to any area or stage (e.g., initiation, progression, and drug-resistance) of cancer biology where it is used to elucidate the underlying molecular mechanisms *(15)*. Because most current cancer medications are relatively nonspecific cytotoxic agents that exert their effect on both normal and tumor cells, new selective agents are needed to further improve cure rates and reduce host toxicity. Therefore, several studies have investigated the use of DNA microarrays to predict the response of cancer cells to chemotherapies *(16–18)*. A better understanding of the disease at a molecular level would therefore allow the design of individualized therapies targeted to the specific molecular cause of cancer as well as the development of logical and effective drug combinations.

Array technology provides an ideal tool for the investigation of metastasis as well as all areas of cancer, because the activity of many genes can be simultaneously monitored, thereby giving insight into the complexity of cancer genetics. The control of metastasis is an appealing target in cancer treatment, because it is the main cause of mortality in cancer and is commonly the reason for the failure of chemotherapy to eradicate cancer. The metastatic process has been described as an obstacle course in which rare cancer cells with appropriate combinations of attributes survive to form a metastatic colony *(19,20)*. The challenge, therefore, is to discover the minimal set of genes that are functionally necessary for metastasis. Consequently, several studies using microarrays have focused on metastasis by comparing the expression profiles of highly metastatic cells with less or nonmetastatic paired cells *(21–24)*. By using paired or genetically related cell lines, it can be presumed that the expression of the majority of genes is shared, with the exception of those genes promoting or related to metastasis. This application of microarray technology is therefore ideal for the identification of biomarkers of metastasis.

2. Experimental Outline

In our study, cDNA microarray technology was used to compare the expression profiles of two colorectal cancer cell lines. Together, the SW480 and SW620 cell lines represented a validated model of colorectal metastasis *(25)*. Both cell lines were derived from the same patient at different stages of tumor progression. The SW480 cell line was cultured from a primary lesion (Dukes' stage B colon carcinoma) isolated from a 50-yr-old Caucasian male patient, and the SW620 cell line was derived from a lymph node metastasis in the same patient a year later *(26)*. These cell lines represented an ideal model for studying the later stages of colorectal cancer progression and for comparing the expression profiles of these isogenic cell lines, thereby giving insight into tumor metastasis. Because these cell lines share a common genetic background, studying their genetic differences was simplified because background genetic variation was minimized and expression changes most likely represented metastasis-specific as opposed to individual-specific changes.

The aim of this array study was to identify genes and pathways of importance in the metastatic process or in genes that conferred a metastatic potential or survival advantage on these cells. There was also the potential of identifying novel genes as possible therapeutic targets or prognostic markers.

2.1. Experimental Design

Proper planning of microarray experimental design was required to ensure that the questions of interest could be answered accurately. Inadequate experimental design can result in either the loss of real biological data or in the acquisition of false and

Table 1
Experimental Design for the Comparison of Gene Expression in the SW480 and SW620 Colorectal Cell Lines by cDNA Microarray Analysis [a]

Experiment	Cy3	Cy5
1	$SW480_1$	$SW620_1$
2	$SW480_2$	$SW620_2$
3	$SW480_3$	$SW620_3$
4	$SW620_3$	$SW480_3$
5	$SW480_1$	$SW480_1$

[a] Experiments 1–3 represent the analysis of biological replicates, where the subscripted number indicates the separate RNA preparation batch. Experiment 4 is a technical replicate of experiment 3, where the dye assignments have been reversed.

untrustworthy information. Biological variance is unavoidable and is intrinsic to all biological systems; therefore, biological replication, by using multiple and separate RNA preparations, was used to ensure the end results were statistically significant, real, and reproducible *(27,28)*. To reduce the technical variation introduced by the processes of RNA extraction, labeling, and hybridization, a dye-swap experiment was carried out (**Table 1**), an effective step for the comparison of two samples, because it accounts for dye bias with the dye assignments being reversed in the second experiment *(29)*. The fluorescent cyanine (Cy) dyes Cy3-dUTP and Cy5-dUTP were used, because they have good photostability and are widely separated in their excitation (550 and 650 nm, respectively) and emission spectra (570 and 670 nm, respectively), which allows highly selective optical filtration. A "self" *vs* "self" experiment also was carried out to control the amount of false positives that the technique and subsequent data analyses generated.

Another important part of experimental design concerns the array itself. Considerations here include the selection of genes, the arrangement of these elements on the array, and how many replicates of these genes should be spotted and where these replicates should be distributed on the slide *(30)*. The cDNA arrays used in this study were constructed from 39,360 human cDNA clones (Research Genetics, Huntsville, AL). cDNA inserts were amplified, purified, and printed on poly-L-lysine–coated glass slides. Poly-L-lysine treatment enhances the hydrophobicity of the slide and the adherence of the printed DNA, thereby limiting the spread of the spotted DNA droplet on the slide. This treatment also offers a support for covalent binding after ultraviolet irradiation. Negative controls consisting of no-template PCR amplifications also were printed on the microarray slide as well as blank controls (i.e., areas on which nothing was printed but were quantified as normal spots). Blanks and negative controls were distributed in several subarrays to monitor different areas on the slide surface. The complete list of genes with accession numbers and spot locations is published at http://www.dcu.ie/~biotech/dmurray/COMBINED-GENE-KEY.xls.

2.2. Hybridization, Image Acquisition, and Image Processing

Labeled samples were then hybridized to the printed slides. The main goal in the hybridization step was to obtain high specificity while minimizing background. The Institute of Genomic Research (TIGR) has developed protocols that yield reproducible, quality hybridizations while maximizing the measured fluorescence on the array (http://www.tigr.org/; *31*). These protocols were used to block nonspecific interactions and to remove any unbound cDNA probe that might have escaped the crosslinking step from the slide before the hybridization step. This free DNA would otherwise compete with bound DNA for hybridization.

After stringency washing, hybridized slides were then scanned using a confocal laser scanner capable of detecting both the Cy3- and Cy5-labeled probes and producing separate 16-bit gray scale TIFF images for each when scanned using 75% laser power and 75% PMT power. QuantArray 3.0 software (PerkinElmer Life and Analytical Sciences, Boston, MA) was then used to analyze the images and calculate the relative expression level of each individual gene and hence identify differentially regulated genes. The hypothesis underlying microarray analysis is that the measured intensity for each arrayed gene represents its relative expression level.

2.3. Data Normalization

Before biologically relevant patterns of expression were identified by comparing expression levels between samples on a gene-by-gene basis, it was appropriate to carry out normalization on the data to eliminate questionable, low-quality measurements and to adjust the measured intensities to make them comparable. Obtaining actual readings of expression levels from a cDNA microarray is subject to several sources of variability, including biological variability and mechanical variability (e.g., printing of slides and pipetting errors, cDNA labeling, and measurement of the fluorescent light intensity), that are represented in the random fluctuations of observed expression levels *(32)*. True differentially expressed genes need to be distinguished from differences generated by those erroneous fluctuations mentioned above. The many sources of systematic variation in microarray experiments that affect the measured expression levels are best demonstrated in a "yellow test," an experiment where two identical mRNA samples are labeled with different dyes and hybridized to the same slide *(33)*. In this example, it was rare to have the dye intensities equal across all spots. Although such systematic differences may be small, they also may have a confounding effect when assaying for subtle biological differences. Normalization is a process that attempts to remove this variation. Normalization of the individual hybridization intensities in each of the two scanned channels is essential to adjust for label-specific differences (dye bias) such as incorporation and detection efficiencies, thus facilitating channel-to-channel comparisons *(34)*. Normalization is also necessary to adjust for differences in the starting RNA in the samples. All these issues can shift the average Cy3/Cy5 ratio if these intensities are not rescaled. Only by balancing these individual intensities appropriately can meaningful biological comparisons be made.

Locally weighted linear regression (LOWESS) analysis is a nonlinear intensity-dependent normalization procedure that is suited to two-color experiments where there are more than 100 elements on the chip *(30,35)*. LOWESS normalization was used to eliminate dye-related artefacts in these two-color experiments that can cause the Cy5/Cy3 ratio to be affected by the total intensity of the spot. The artefacts that LOWESS attempts to correct for include nonlinear rates of dye incorporation as well as inconsistencies in the relative fluorescence intensity between the two dyes used and bias because of low-intensity spots, which give rise to a higher grade of variability.

Other normalization approaches include linear total intensity normalization, which considers all the genes in an array experiment together. Linear total intensity normalization is based on the simple assumption that the total amount of RNA labeled with either Cy3 or Cy5 is equal. This form of global normalization is a useful tool in instances of closely related samples where the transcription level of many genes remains unchanged. Although the intensity for one spot may be higher in one channel than in the other channel, when averaged over thousands of spots in the array, these differences average out. Therefore, the total fluorescence across all the spots in the array will be equal for both channels, and a scatter plot of the measured Cy5 versus Cy3 intensities should have a slope of 1. A normalization method that makes use of a specific subset of "housekeeping" or control genes as opposed to all elements on the slide also has been developed *(36)*. This method is useful for more divergent samples and for arrays with low gene numbers where the total intensity model would give a poorer estimation of normalization. This method assumes that the distribution of transcription levels for this set of housekeeping genes remains unchanged and has a mean value and standard deviation that are independent of the sample. In this case, the ratio of measured Cy5-to-Cy3 ratios for these genes is used as a model to adjust the mean of ratios to 1.

2.4. Data Mining

The extraction of biologically relevant information from large array generated gene lists still proves to be a serious bottleneck *(37)*. Regardless of the nature of the experiment, the major interest in performing microarray analysis is in the identification of genes that are differentially expressed between samples in the data set. Therefore, simply generating the data is not enough; extracting meaningful information about the system being studied also is necessary. This approach currently requires the combined efforts of biologists, bioinformaticians, computer scientists, statisticians, and software engineers.

After scanning, the raw data generated was analyzed using GeneSpring 6.0 software (Silicon Genetics, Redwood City, CA), a powerful commercially available expression data visualization and analysis tool. Data were imported and normalized, and replicate and dye-swap experiments were identified. A twofold cutoff was used to identify genes that were variable between samples *(4,5,38)*. Therefore, genes were taken to be differentially expressed if the expression under one condition was more than twofold greater or less than that under the other condition.

Because the "fold" change test is not a statistical test and does not provide information on the level of confidence in the designation of genes as differentially expressed or not differentially expressed, *t*-tests were carried out. For statistical purposes, biological replicates, i.e., RNA samples obtained from independent biological sources are preferable to technical replicates, i.e., RNA samples from one biological source, especially if conclusions are to be made regarding the significance of expression changes. Ideally, each biological condition for comparison should be represented by at least three independent biological samples *(28)*. This sample size depends on the expense of the microarray experiment and on the availability of samples and therefore RNA. Because independent microarray assays were conducted starting from independent mRNA isolations, they could be used to define differential expression based on their statistical consensus. Therefore, genes that were significantly differentially expressed could be distinguished from random changes. The *t*-test is one of many statistical methods suitable for confirming differentially expressed genes *(39)*. The *t*-test uses the error variance for a given gene over replicated experiments to determine whether that gene is differentially expressed and whether this difference is significant *(40)*. A *t*-value for a given gene is computed using **Eq. 1** , where R_g is the mean ratio of expression levels for that gene and SE_g is its standard error. After a *t*-value is calculated, it is converted to a *p* value. Genes with *p* values falling below a prescribed "nominal" level of 0.005 were regarded as significant. In our experiments with the SW480 and SW620 cell lines, 441 genes in total were upregulated in the SW620 cells compared with the SW480 cells. After *t*-tests, 233 genes were identified as being significantly twofold upregulated and 208 genes were identified as being significantly twofold downregulated in the SW620 cell line compared with the SW480 cell line. A selection of genes that upregulated and downregulated in the SW620 cells are listed in **Tables 2** and **3**, respectively.

$$t = \frac{R_g}{SE_g} \tag{1}$$

Several problems exist with this twofold approach to identifying significantly differentially expressed genes as genes of importance in the biology under investigation. The first is that not only changing genes but also the genes that are unchanged from one state to another (i.e., from normal to tumor or from primary to metastatic) may understandably have an important role in both states. The rudimentary analysis of picking changing genes at the top of a list may draw attention away from the biological importance of maintaining transcription of a gene from normal to disease state. Also, real biological data might be lost when alterations in gene expression are more modest than twofold. Modern approaches to data analysis aim to capture the real biology underlining the distinction of interest as apposed to subjectively "cherry picking" top-ranking genes and creating hypotheses about pathway membership *(41)*. Nevertheless, -fold change and statistical significance are different from biological relevance.

2.5. Postanalysis Follow-Up: Validation of Microarray Data

After genes that were significantly up- or down-regulated were identified, the next step was to carry out detailed literature searches to help select genes for further validation and investigation. Postarray verification of results using an independent laboratory approach provides experimental verification of gene expression levels and ideally begins with the same biological samples that were compared in the array experiment. Common mRNA and protein methods used to validate array data include RT-PCR *(42)*, Northern blot analysis *(43)*, ribonuclease protection assay *(44)*, *in situ* hybridization *(45)*, immunoblot analysis *(46)*, and immunohistochemistry *(45)*.

Table 2
Selection of Genes With Significantly (*p* < 0.005) Twofold or More Increased Expression in SW620 Cells Compared With SW480 Cells

Accession no.	Fold up	Gene description
AA916325	22	Aldo-keto reductase family 1, member C3
AA608575	17	Propionyl coenzyme A carboxylase, α polypeptide
AA135152	12	Glutathione peroxidase 2 (gastrointestinal)
R37743	5	T54 protein
R16134	4	Transmembrane 4 superfamily member 11 (plasmolipin)
N70463	4	B-cell translocation gene 1, antiproliferative
H79534	4	Hemoglobin, epsilon 1
W46900	3	Chemokine (C-X-C motif) ligand 1
AA779165	3	ADP-ribosylation factor-like 4
T66320	3	Glutathione *S*-transferase
AA460463	3	Cytokine-like protein C17
R60170	3	Guanine deaminase
R98936	3	Membrane metallo-endopeptidase
AA478585	3	Butyrophilin, subfamily 3, member A3
AA486220	3	Lysyl-tRNA synthetase

Table 3
Selection of Genes With Significantly (*p* < 0.005) Twofold or More Decreased Expression in SW620 Cells Compared With SW480 Cells

Accession no.	Fold down	Gene description
AA598601	7	Insulin-like growth factor binding protein 3
AA406552	6	Solute carrier family 2 (facilitated glucose transporter), member 3
AA160507	3	K5
AA487425	3	Zinc finger protein 463
AA291163	3	Glutaredoxin (thioltransferase)
AA460286	3	Guanine nucleotide binding protein
AA486556	3	CD81 antigen (target of antiproliferative antibody 1)
AI439571	3	Apoptosis antagonizing transcription factor
N67766	2	Acetyl-coenzyme A synthetase 2 (AMP forming)-like
AA017526	2	Collagen, type IX, α3
AA487486	2	Cyclin D1 (PRAD1: parathyroid adenomatosis 1)
AA827551	2	Notch homolog 2 (*Drosophila*)
AI688757	2	Cytochrome-*c* oxidase subunit Vb

Before genes of interest were pursued for further investigation and validation, sequencing was required as a quality control measure. Because many array experiments are carried out on the genome-wide scale, the probability of error associated with spot identity and printing also is increased. Array production is a high-throughput operation involving multiple steps of sample handling and processing, and it is therefore understandably subject to error. Consequently, sequence verification is required to identify and correct for such errors *(47)*. The sequences of the genes chosen for further analysis were positively identified by their sequence using an ABI 377 DNA sequencing platform (Applied Biosystems Inc., Foster City, CA). The BLAST tool on the National Center for Biotechnology Information website (http://www.ncbi.nlm.nih.gov/blast/) was used to confirm the identity of the resulting sequence for the clones of interest.

In our study, quantitative real-time PCR was used to confirm the -fold changes in expression. In comparison with conventional PCR, real-time PCR provides a much more accurate and quantitative representation of the differences in the expression of a gene from one state to another, because it is more sensitive and has a larger range of detection. For postarray validation, real-time RT-PCR is the most popular method for quantitatively measuring specific mRNA transcripts, because it is reproducible, rapid, inexpensive, and requires little starting template *(48,49)*. However, it has been shown that for many genes studied there were significant quantitative differences between array and RT-based data *(42)*. As well as validating array results at the mRNA level, it is also worth investigating the changes in expression at the level of the corresponding protein, because the observed changes at the mRNA level may not always translate into the protein *(50)*. This discrepancy may be a factor of either the laboratory test used to determine protein expression or of the biology of the protein, because protein function in the cell is affected by several factors besides abundance.

In this study, the importance of the genes of interest in colorectal cancer was further investigated and validated by examining their expression in a pilot study of paired normal and tumor colorectal biopsy specimens. RNA was extracted from samples and analyzed for expression of these genes by using quantitative real-time RT-PCR.

Recently, the same SW480 and SW620 cell model has been used in an array study into the mechanisms of drug resistance *(51)*. Although different array technology to that described here was used and the array used had only 12,625 elements in comparison with the 39,365 used in this study, there are several differentially expressed genes that are common to both studies, including glutathione transferase, guanine nucleotide-binding protein, and cyclin D. In our study, glutathione transferase expression was 2.1-fold higher in the SW620 cells compared with the SW480 cells. Glutathione transferase levels are elevated in patients with a high risk of developing colon cancer, and their overexpression leads to chemotherapeutic resistance in ovarian cancer cell lines *(52,53)*. Serial analysis of gene expression (SAGE) also has been used to study the same cell model to identify changes in gene expression during colorectal cancer progression *(54)*. Five genes with differential expression were reported in this SAGE study where 5000 tagged clones were studied. SAGE is an effective technique, but it does not allow for the same statistical interpretation and analysis as cDNA microarray analysis. Only one of these five differentially expressed genes was common to our study, keratin 5 (K5). K5 was threefold downregulated in the SW620 cells compared with the SW480 cells, whereas K5 was expressed in SW480 cells and undetected in SW620 cells by SAGE analysis *(54)*. K5 is a cytoskeletal filament protein known to play a major role in cell motility, invasion, proliferation, and differentiation and in the transduction of extracellular signals *(55)*. The same cell model was used, along with two other cell models of colorectal metastasis, in an elegant 19,000-element array study *(22)*. Hegde et al. showed that 1569 genes were differentially regulated between the SW480 and SW620 cells and that 176 of these genes were differentially regulated in all three models of

colorectal metastasis. Of the 176 reported genes, very few were common to those identified in our study.

3. Future of Microarray Analysis and Conclusions

The applications of microarrays to all areas of investigative biology are endless.
Microarrays have become an accessible and common oncology research tool that will
have a great impact on the evolution of molecular medicine of cancer. More and more
cancers will be classified into new previously unrecognized subclasses based on their
molecular signatures, thus facilitating tailored treatments *(8,16,56,57)*. For a relatively
young technique, microarray analysis is already a key element in the investigation of
the complexity of cancer. As well as having great impact on the development of novel
therapeutics, the recent completion of the human genome sequence also spurs a desire
to carry out microarray analysis on the whole genome level *(58,59)*. However, the
current approach of users selecting and printing sets of genes to produce customized
arrays is likely to remain attractive. Further advances and improvements in array analysis will concern the care of the biological samples under investigation. In particular,
and of clinical relevance, is the care of tissue samples, which is important in avoiding
RNA degradation. Stricter rules are needed for the collection, storage, and processing
of tissues. Another major issue in need of improvement is the purity of tissue samples.
Most biopsies are heterogeneous, containing many different cell types, including normal surrounding cells such as epithelial, endothelial, adipose, and stromal cells; infiltrating lymphocytes; and tumor cells *(60)*. A pure homogenous sample is desirable to
obtain representative results. One solution to obtaining a pure sample is to use laser
capture microdissection. Laser capture microdissection is used to harvest subpopulations of cells; however, obtaining sufficient RNA from homogenous material for
microarray analysis usually requires an amplification step *(61)*. Although it introduces
another variable into the process, RNA amplification allows array analysis to be carried out on 1–50 ng of mRNA *(62)*.

This chapter focused primarily on the practicalities involved in the identification of
metastasis-associated genes by using microarray analysis to compare the expression
profiles of a nonmetastatic cell line, SW480, with its metastatic derivative SW620.
Four hundred and forty-one genes were shown to be differentially expressed between
these two cell lines. Analysis was carried out in triplicate, allowing these expression
changes to be identified as statistically significant. Several genes were chosen for further validation analysis. As quality control steps, sequencing of the spotted cDNA and
real-time PCR of the genes of interest confirmed their identity and changes in expression, respectively. The expression of these potential markers of colorectal metastasis
was investigated in paired normal and tumor colorectal specimens. This work validates
the further investigation of these targets as players in the metastatic process.

The future of arrays and the interpretation of the vast amount of data generated
revolve around the synergy between different scientific cultures. Multidisciplinary col-
laborations between medical researchers, mathematicians, computer scientists, and
bioinformaticians are required to carry studies forward to the clinic. Synergy also is
required between different types of information, such as the transcriptional and translational data for the same samples. Such data should be comparable from one laboratory
to another, thus providing a platform for connectivity.

References

1. Hanahan, D., and Weinberg, R. A. (2000) The hallmarks of cancer. *Cell* **100**, 57–70.
2. Murray, D., Morrin, M., and McDonnell, S. (2004) Increased invasion and expression of MMP-9 in human colorectal cell lines by a CD44-dependent mechanism. *Anticancer Res.* **24**, 489–494.
3. Leung, Y. F., and Cavalieri, D. (2003) Fundamentals of cDNA microarray data analysis. *Trends Genet.* **19**, 649–659.
4. Schena, M., Shalon, D., Davis, R. W., and Brown, P. O. (1995) Quantitative monitoring of gene expression patterns with a complementary DNA microarray. *Science* **270**, 467–470.
5. Schena, M., Shalon, D., Heller R., Chai, A., Brown, P. O., and Davis, R. W. (1996) Parallel human genome analysis: microarray-based expression monitoring of 1000 genes. *Proc. Natl. Acad. Sci. USA* **93**, 10,614–10,619.
6. Lockhart, D. J., Dong, H., Byrne, M. C., et al. (1996) Expression monitoring by hybridization to high-density oligonucleotide arrays. *Nat. Biotechnol.* **14**, 1675–1680.
7. Lipshutz, R. J., Fodor, S. P., Gingeras, T. R., and Lockhart, D. J. (1999) High density synthetic oligonucleotide arrays. *Nat. Genet.* **21**, 20–24.
8. Khan, J., Simon, R., Bittner, M., et al. (1998) Gene expression profiling of alveolar rhabdomyosarcoma with cDNA microarrays. *Cancer Res.* **58**, 5009–5013.
9. Virtaneva, K., Wright, F. A., Tanner, S. M., et al. (2001) Expression profiling reveals fundamental biological differences in acute myeloid leukemia with isolated trisomy 8 and normal cytogenetics. *Proc. Natl. Acad. Sci. USA* **98**, 1124–1129.
10. Sgroi, D. C., Teng, S., Robinson, G., LeVangie, R., Hudson, J. R., Jr., and Elkahloun, A. G. (1999) In vivo gene expression profile analysis of human breast cancer progression. *Cancer Res.* **59**, 5656–5661.
11. Ono, K., Tanaka, T., Tsunoda, T., et al. (2000) Identification by cDNA microarray of genes involved in ovarian carcinogenesis. *Cancer Res.* **60**, 5007–5011.
12. Welsh, J. B., Zarrinkar, P. P., Sapinoso, L. M., et al. (2001) Analysis of gene expression profiles in normal and neoplastic ovarian tissue samples identifies candidate molecular markers of epithelial ovarian cancer. *Proc. Natl. Acad. Sci. USA* **98**, 1176–1181.
13. Anbazhagan, R., et al. (1999) Classification of small cell lung cancer and pulmonary carcinoid by gene expression profiles. *Cancer Res.* **59**, 5119–5122.
14. Alon, U., Barkai, N., Notterman, D. A., et al. (1999) Broad patterns of gene expression revealed by clustering analysis of tumor and normal colon tissues probed by oligonucleotide arrays. *Proc. Natl. Acad. Sci. USA* **96**, 6745–6750.
15. Bertucci, F., Houlgatte, R., Nguyen, C., Viens, P., Jordan, B. R., and Birnbaum, D. (2001) Gene expression profiling of cancer by use of DNA arrays: how far from the clinic? *Lancet Oncol.* **2**, 674–682.
16. Scherf, U., Ross, D. T., Waltham, M., et al. (2000) A gene expression database for the molecular pharmacology of cancer. *Nat. Genet.* **24**, 236–244.
17. Zembutsu, H., Ohnishi, Y., Tsunoda, T., et al. (2002) Genome-wide cDNA microarray screening to correlate gene expression profiles with sensitivity of 85 human cancer xenografts to anticancer drugs. *Cancer Res.* **62**, 518–527.
18. Blower, P. E., Yang, C., Fligner, M. A., et al. (2002) Pharmacogenomic analysis: correlating molecular substructure classes with microarray gene expression data. *Pharmacogenomics J.* **2**, 259–271.
19. Poste, G., and Fidler, I. J. (1980) The pathogenesis of cancer metastasis. *Nature* **283**, 139–146.
20. Fidler, I. J., and Kripke, M. L. (1977) Metastasis results from preexisting variant cells within a malignant tumor. *Science* **197**, 893–895.
21. Clark, E. A., Golub, T. R., Lander, E. S., and Hynes, R. O. (2000) Genomic analysis of metastasis reveals an essential role for RhoC. *Nature* **406**, 532–535.

22. Hegde, P., Qi, R., Gaspard, R., et al. (2001) Identification of tumor markers in models of human colorectal cancer using a 19,200-element complementary DNA microarray. *Cancer Res.* **61,** 7792–7797.
23. Bittner, M., Meltzer, P., Chen, Y., et al. (2000) Molecular classification of cutaneous malignant melanoma by gene expression profiling. *Nature* **406,** 536–540.
24. Maniotis, A. J., Folberg, R., Hess, A., et al. (1999) Vascular channel formation by human melanoma cells in vivo and in vitro: vasculogenic mimicry. *Am. J. Pathol.* **155,** 739–752.
25. Hewitt, R. E., McMarlin, A., Kleiner, D., et al. (2000) Validation of a model of colon cancer progression. *J Pathol.* **192,** 446–454.
26. Leibovitz, A., Stinson, J. C., McCombs, W. B., 3rd., McCoy, C. E., Mazur, K. C., and Mabry, N. D. (1976) Classification of human colorectal adenocarcinoma cell lines. *Cancer Res.* **36,** 4562–4569.
27. Rosenbaum, P. R. (2001) Replicating effects and biases. *Am. Stat.* **55,** 223–227.
28. Lee, M. L., Kuo, F. C., Whitmore, G. A., and Sklar, J. (2000) Importance of replication in microarray gene expression studies: statistical methods and evidence from repetitive cDNA hybridizations. *Proc. Natl. Acad. Sci. USA.* **97,** 9834–9839.
29. Kerr, M. K., and Churchill, G. A. (2001) Statistical design and the analysis of gene expression microarray data. *Genet Res.* **77,** 123–128.
30. Yang, Y. H., Dudoit, S., Luu, P., et al. (2002) Normalization for cDNA microarray data: a robust composite method addressing single and multiple slide systematic variation. *Nucleic Acids Res.* **30,** e15.
31. Hegde, P., Qi, R., Abernathy, K., et al. (2000) A concise guide to cDNA microarray analysis. *Biotechniques* **20,** 548–562.
32. Schuchhardt, J., Beule, D., Malik, A., et al. (2000) Normalization strategies for cDNA microarrays. *Nucleic Acids Res.* **28,** e47.
33. Yang, Y. H., Dudoit, S., Speed, T. P., and Callow, M. J. (2002) Statistical methods for identifying differentially expressed genes in replicated cDNA microarray experiments. *Stat. Sin.* **12,** 111–139.
34. Quackenbush, J. (2000) Microarray data normalization and transformation. *Nat. Genet.* **32,** 496–501.
35. Cleveland, W. S. (1979) Robust locally weighted regression and smoothing scatterplots. *J Am. Stat. Assoc.* **74,** 829–835.
36. Chen, Y., Dougherty, E. R., and Bittner, M. L. (1997) Ratio-based decisions and the quantitative analysis of cDNA microarray images. *J Biomed. Opt.* **2,** 364–374.
37. Butte, A. (2002) The use and analysis of microarray data. *Nat. Rev. Drug Discov.* **1,** 951–960.
38. Yang, I. V., Chen, E., Hasseman, J. P., et al. (2002) Within the fold: assessing differential expression measures and reproducibility in microarray assays. *Genome Biol.* **3,** 62.1–62.13.
39. Cui, X., and Churchill, G. A. (2003) Statistical tests for differential expression in cDNA microarray experiments. *Genome Biol.* **4,** 210.
40. Callow, M. J., Dudoit, S., Gong, E. L., Speed, T. P., and Rubin, E. M. (2000) Microarray expression profiling identifies genes with altered expression in HDL-deficient mice. *Genome Res.* **10,** 2022–2029.
41. Mootha, V. K., Lindgren, C. M., Eriksson, K. F., et al. (2003) PGC-1alpha-responsive genes involved in oxidative phosphorylation are coordinately downregulated in human diabetes. *Nat. Genet.* **34,** 267–273.
42. Rajeevan, M. S., Vernon, S. D., Taysavang, N., and Unger, E. R. (2001) Validation of array-based gene expression profiles by real-time (kinetic) RT-PCR. *J. Mol. Diagn.* **3,** 26–31.
43. Chaib, H., Cockrell, E. K., Rubin, M. A., and Macoska, J. A. (2001) Profiling and verification of gene expression patterns in normal and malignant human prostate tissues by cDNA microarray analysis. *Neoplasia* **3,** 43–52.

44. Taniguchi, M., Miura, K., Iwao, H., and Yamanaka, S. (2001) Quantitative assessment of DNA microarrays-comparison with Northern blot analyses. *Genomics* **71,** 34–39.
45. Mousses, S., Bubendorf, L., Wagner, U., et al. (2002) Clinical validation of candidate genes associated with prostate cancer progression in the CWR22 model system using tissue microarrays. *Cancer Res.* **62,** 1256–1260.
46. Al Moustafa, A. E., Alaoui-Jamali, M. A., Batist, G., et al. (2002) Identification of genes associated with head and neck carcinogenesis by cDNA microarray comparison between matched primary normal epithelial and squamous carcinoma cells. *Oncogene* **21,** 2634–2640.
47. Taylor, E., Cogdell, D., Coombes, K., et al. (2001) Sequence verification as quality-control step for production of cDNA microarrays. *Biotechniques* **31,** 62–65.
48. Giulietti, A., Overbergh, L., Valckx, D., Decallonne, B., Bouillon, R., and Mathieu, C. (2001) An overview of real-time quantitative PCR: applications to quantify cytokine gene expression. *Methods* **25,** 386–401.
49. Walker, N. J. (2002) Tech.Sight. A technique whose time has come. *Science* **296,** 557–558.
50. Chuaqui, R. F., Bonner, R. F., Best, C. J., et al. (2002) Post-analysis follow-up and validation of microarray experiments. *Nat. Genet.* **32,** 509–514.
51. Huerta, S., Harris, D. M., Jazirehi, A., et al. (2003) Gene expression profile of metastatic colon cancer cells resistant to cisplatin-induced apoptosis. *Int. J. Oncol.* **22,** 663–670.
52. Grubben, M. J., Nagengast, F. M., Katan, M. B., and Peters, W. H. (2001) The glutathione biotransformation system and colorectal cancer risk in humans. *Scand. J. Gastroenterol. Suppl.* 234, **68–76.**
53. Zhang, F., Qi, L., and Chen, H. (2001) Value of P-glycoprotein and glutathione S-transferase-pi as chemoresistant indicators in ovarian cancers. *Zhonghua Zhong Liu Za Zhi,* 313–316.
54. Parle-McDermott, A., McWilliam, P., Tighe, O., Dunican, D., and Croke, D. T. (2000) Serial analysis of gene expression identifies putative metastasis-associated transcripts in colon tumour cell lines. *Br. J. Cancer* **83,** 725–728.
55. Fuchs, E., and Weber, K. (1994) Intermediate filaments: structure, dynamics, function, and disease. *Annu. Rev. Biochem.* **63,** 345–382
56. Golub, T. R., Slonim, D. K., Tamayo, P., et al. (1999) Molecular classification of cancer: class discovery and class prediction by gene expression monitoring. *Science* **286,** 531–537.
57. Su, A. I., Welsh, J. B., Sapinoso, L. M., et al. (2001) Molecular classification of human carcinomas by use of gene expression signatures. *Cancer Res.* **61,** 7388–7393.
58. The Genome International Sequencing Consortium (2001) Initial sequencing and analysis of the human genome. *Nature* **409,** 860–921.
59. Austin, C. P. (2004) The impact of the completed human genome sequence on the development of novel therapeutics for human disease. *Annu. Rev. Med.* **55,** 1–13.
60. Ronnov-Jessen, L., Petersen, O. W., and Bissell, M. J. (1996) Cellular changes involved in conversion of normal to malignant breast: importance of the stromal reaction. *Physiol. Rev.* **76,** 69–125.
61. Emmert-Buck, M. R., Bonner, R. F., Smith, P. D., et al. (1996) Laser capture microdissection. *Science* **274,** 998–1001.
62. Wang, E., Miller, L. D., Ohnmacht, G. A., Liu, E. T., and Marincola, F. M. (2000) High-fidelity mRNA amplification for gene profiling. *Nat. Biotechnol.* **18,** 457–459.

2

From Tissue Samples to Tumor Markers

Dirce Maria Carraro, Helena Paula Brentani, Fernando Augusto Soares, Luiz Fernando Lima Reis, and Ricardo Renzo Brentani

Summary

Changes in the general transcription profile have been observed through silencing and activating at the transcriptional level of genes in tumor cells. After the genomic era, molecular biology has changed, and new technologies have allowed assessing transcriptional alterations among different tissue types in a high-throughput manner. Microarray technology is one of the technologies that has contributed to improving our understanding about the defective molecular processes in cancer cells. In this chapter, we report issues about the selection of cDNA clones spotted on the platform, the purity of the tumor samples used in the microarray experiments, and the integrated database with important clinical information of patients that can be associated to a specific molecular portrait. Finally, we focus on the validation of candidate genes selected from microarray experiments through real-time RT-PCR and high-throughput tissue microarray analyses that dramatically facilitate testing of the potential molecular markers.

Key Words: Cancer; cDNA microarray; integrated database; laser microdissection; tissue microarray; tumor markers.

1. Introduction

The major characteristic of a normal tissue is to have growth and division of its cells highly regulated. The cancer cell is an exception, because it has the property to lose growth regulation and propagate indefinitely. The malignant phenotype is the result of a range of genetic and epigenetic alterations (1). The molecular mechanisms of cellular transformation involve many DNA disorders, such as point mutations, deletions, insertions, and chromosomal translocations and also alteration in methylation status (2,3). These DNA disorders can arise in different regions of chromosomes, including intergenic and genic regions such as transcribed and coding sequences, regions responsible for chromosome stability, and also gene promoters. Consequently, changes in the general transcription profile can be observed through silencing and activating at the transcriptional level of genes. Therefore, tumor cells differ from their normal cell counterparts in their general expression pattern, and this difference is focused mainly in the mechanisms related to cell proliferation, differentiation, and survival.

Recently, with the completion of the human genome sequence (4,5) and the identification of many human genes (6–8), molecular biology has changed. Functional studies have applied new technologies that allow assessing genomic, transcriptional, and proteomic alterations among different tissue types in a high-throughput manner. This

From: *Bioarrays: From Basics to Diagnostics*
Edited by: K. Appasani © Humana Press Inc., Totowa, NJ

new way of investigation has improved our understanding about the molecular basis of cancer.

In transcriptional assessment, methods for detecting differences in messenger RNA (mRNA) level have been proved to be very useful. Many sequence-based approaches can be cited, such as serial analysis of gene expression *(9,10)*, massive parallel signature sequencing *(11)*, and expressed sequence tags *(12)*. These methodologies produce quantitative measurements of gene expression and present great analytic potential and sensitivity. However, they can be hampered by individual genetic differences, because they are not suited for large sample numbers; consequently, only a few samples of a given tumor class can be assessed. In contrast, microarray technology *(13)* has the capacity for assessing multiple samples, thereby generating specific gene expression profiles characteristic of a given tumor class. In short, this technology consists of an ordered collection of DNA fragments (cDNA or oligonucleotides) immobilized on a solid surface. The immobilized sequences represent genes, and each microspot is composed of a single species of sequences. The technology is based on the hybridization between DNA sequences immobilized on the slide and labeled free sequences generated from the mRNA population from a given biological sample. The methodology allows the quantitative detection of thousands of genes simultaneously, and its application has been contributing to the characterization of aberrant events that enable cells to bypass normal control and generate tumors *(14–16)*. Additionally, microarray technology is improving the ability to subclassify tumors according to their clinical behavior *(17–20)*, and it is uncovering novel possibilities for the development of new and more efficient strategies for prognosis, diagnosis, and treatment.

2. Technical Issues: Array Platform, Samples, and Bioinformatics Tools

We designed a probe as the known entity that is immobilized on the solid support and a target as the labeled molecule generated from the mRNA population of a given cell culture or tissue *(21)*.

In general terms, integrity and purity of total RNA from each used sample and also proper selection of the DNA sequence that represent the genes in the platform are very important issues to guarantee reliability and reproducibility of the experimental data.

2.1. cDNA Microarray Platform

Advances in microarray technology concerning the probe that better represent a single gene, avoiding cross-hybridization, have been investigated by several groups *(22–24)*. Cross-hybridization involves multiple sequences belonging to distinct genes hybridizing to the same spot, interfering in the measuring accuracy of the real expression level of a given gene in a given sample.

Regarding probes, our group has been using a customized cDNA platform with open reading frame expressed sequence tags (ORESTES) clones selected from a collection of 1,200,000 clones that represent 4608 different human genes. ORESTES clones are partial cDNA clones that represent different regions of the full-lengths gene, but they are biased toward the central portion of the coding region *(25,26)*. The selection of ORESTES clones that are used as probes follows several criteria that allow stringent hybridization conditions and avoids cross-hybridization. In short, probes should have similar size and identity, while corresponding to only one full-length mRNA *(27)*. Specifically, the clone is excluded if, in a 100-base pair sequence, a hit equal to or higher than 85% with any other sequence other than the full-length sequence in question is found. This criterion is important to ensure that the intensity measurement is related to a unique gene, even if it belongs to a gene family. A decrease in intensity level when the probe location is distant from the 3¥ extremity also has been observed by us, probably because of low efficiency of the reverse transcriptase to reach the 5¥ extremity of the gene, especially for the long genes. Therefore, the probe should be mapped as proximal as possible to the 3¥ extremity considering the first poly-A signal mapped on the sequence. This criterion ensures that the probe can detect all possible poly-A variants of a given gene present in the sample under investigation, because a very large fraction of human genes contain several polyadenylation signals *(28)*. The last selection criterion is to filter out repeat elements, such as Alu signal-related sequences or other less abundant repetitive DNA elements, because they may influence signal intensity. It is also very important to spot positive and negative controls in the chip. We strongly recommend that spots should be replicated. Clone repeatability can be used to adapt thresholds for filtering criteria while focusing on data quality and maximization of the number of observations amenable for further analysis. Furthermore, during array fabrication, spots that represent the same gene should not be located in the same subarray.

2.2. RNA Samples: Quality Checking and Biodiversity

The preparation of target RNA is as important as correct probe selection in microarray technology. In cancer studies, the most useful kind of target used in microarray experiments is solid tissue, which is composed not only of tumor cells but also normal cells, stromal cells, infiltrating immunological system cells, and endothelial cells from lymphatic and blood vessels. This heterogeneity interferes with the gene expression profile. Microdissection, manual and laser capture microdissection (LCM), allow homogeneity of the tissue. In manual microdissection, depending on the kind of tissue, there can be approx 80% homogeneity. The process is guided by stained histopathological slides, which allow planning of microdissection, such as removal of necrotic areas and infiltrating or undesirable tissue. The LCM station basically incorporates an inverted microscope with a low-power infrared laser and a photocamera that sends the image to a video monitor. The image provided by the video monitor allows selection of the desirable cells from the whole tissue sections. A low-power infrared laser activates the transparent thermoplastic film attached to an optical grade plastic cap. The optical grade plastic cap is positioned over the tissue section and with a shoot; the thermoplastic film adheres to the selected cells. Once the selected cells have been captured, the transfer film cap, with the cells attached to the film surface, is lifted (LCM with PixCell II LCM instrument, Arcturus, Molecular Devices, Sunnyvale, CA). Because the thermoplastic film used for RNA extraction is supplied with a green pigment that prevents warming of the sample, the RNA integrity is preserved. This methodology can increase the accuracy and sensibility of molecular studies, because a more homogenous population of tumor cells is recovered (approx 99%), leading to a more precise correlation of tumor morphology with its gene expression profile.

A second important issue is the limiting amounts of RNA obtained from freshly isolated samples; especially with LCM-dissected samples and fine needle aspiration biopsies. For microarray experiments, in a first-strand cDNA reaction in the presence of fluorescent dye, approx 30–100 μg of total RNA is required, and this amount of total RNA is normally not recovered from such tissue types. Therefore, expression profile

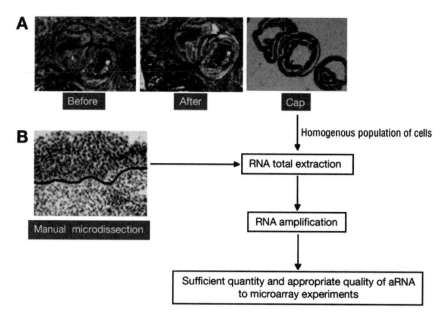

Fig. 1. Tumor tissue microdissection. (**A**) Laser capture microdissection in breast cancer. The three pictures showed the tissue before the capture, the selected cells and the captured cell adhered to the plastic cap. (**B**) Manual microdissection of blastema component of Wilm's tumor.

assessment of this kind of sample requires methodologies for mRNA amplification in order to properly delineate the experiment considering the necessary number of replicates. Several mRNA amplification protocols are available *(29)*, and in general, they are based on two rounds of in vitro transcription that may lead to 1300-fold linear mRNA amplification *(29)*. In short, total RNA is reverse transcribed into cDNA by using an oligonucleotide that consists of a T7 RNA polymerase promoter sequence at the 5¥ extremity of the dT(15) sequence. The cDNA single strand is converted to double strand and submitted to an in vitro transcription reaction. The procedure is repeated twice to provide the necessary amount of RNA for the microarray experiments (**Fig. 1**). Several analyses have shown the maintenance of the relative mRNA concentrations compared with the results obtained from microarray experiments by using total RNA *(30–32)*.

Another important point concerning the target is the correct selection of samples that will be representative of the investigated classes. The complex and heterogeneous origins of most cancer types require careful sample selection, and several key issues have to be considered to render meaningful the long list of different expressed genes normally obtained from comparative analyses between two different investigated classes. It has been suggested that to have more reliable data, samples that are representative of each class should be as closely matched as possible *(21)*. In this aspect, the critical point in cancer research is the great biological diversity among samples of the same class. This problem can be bypassed using cell lines as model systems, which can be the best experimental model for some specific questions where they can pose as an inducible

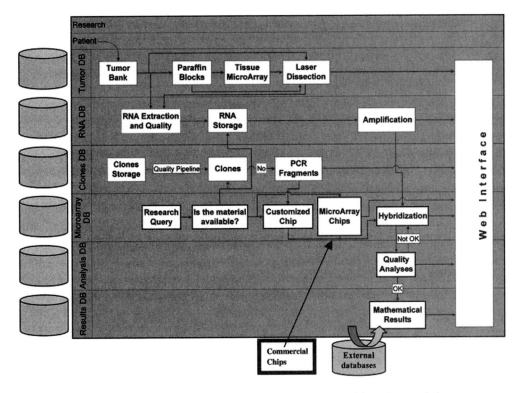

Fig. 2. Schematic fluxogram of the array pipeline. In this scheme, it is easy to see all the steps from tumors to arrays and the necessity of integration between all databases as well as a friendly interface to access the data.

system, for example, comparison of a knockout gene cell line compared with its wild-type counterpart and/or by addition of an extracellular factor. Nevertheless, for clinically oriented questions, it is not recommended to use cell lines because the model probably differs from the fresh tissue characteristic in its gene expression profile. In this context, to generate data in a meaningful manner, large collections of tissue samples with a minimal follow-up and clinical data are required.

2.3. Bioinformatics Necessary to Organize a Microarray Infrastructure and Data Analysis

An important issue related to the organization of a microarray infrastructure is the necessity of a solid bioinformatics unit able to: select and track clones to be printed on slides, maintain a relational database with clinical and sample information, and also analyze all the microarray data.

In cDNA clone selection with the criteria discussed in Subheading 2.1., bioinformatics also allows choosing the best-suited clone representing a specific gene to be spotted on a slide. The same requirement is afforded by the selection of tissue samples, RNA samples, or both. **Figure 2** presents all the relational databases available in our infrastructure and how they are interrelated. It is very important to have an interactive graphical visualization system to help the biologist to make some decisions about quality control issues during the experiments. Without the aid of an appropriate data visualizer, it would be almost impossible to interpret data and make decisions in real time.

Concerning time and space required for recording raw data, two arguments should be considered. First, microarray data analyses are in a progressive learning process. Consequently, with improvement of the process, data are required to be reanalyzed in different ways. Thus, raw data should be available. Second, it is increasingly clearer that the number of samples used in a given microarray project is critical and must be representative of the class under examination. Thus, it is important to be able to make some kind of analysis with data provided by a lot of different projects and also with publicly available data from other groups. In this situation, all the data should be reanalyzed together, and the raw data are required.

In general, two conceptually different issues can be addressed by the array technology. One issue is the identification of candidate genes that would help in understanding the biology of the problem under investigation. The second issue involves building diagnostic or prognostic tools in which the identity of genes is not the critical point, but rather the information associated with a given expression profile or expression signature is critical. For these applications, different strategies can be used to build an array that would favor a specific issue or question.

We are studying molecular classification, or the identification of novel patient subgroups, of new markers for diagnosis and of novel targets for therapeutic intervention in cancer. The literature reveals that an approach based on single-gene analysis is not sufficiently robust for diagnostic applications. In contrast, a combination of genes provides much more useful information.

Working with microarrays, we are not measuring expression levels directly but rather signal intensity levels, represented by the amount of phosphorescent dye that is recorded by a scanning device. Different concentrations of mRNA, or the brightness, relative binding affinity, and concentration of the dye labels, are some of the factors that could produce artifacts in the "measurement of the expression levels." After considering corrections in microarray experiments such as image acquisition, background correction or spot quality assessment and trimming, to correct differences in intensity levels we need to perform normalization. There are many approaches for normalization, such as total intensity normalization, linear regression analysis, log centering, and rank invariant methods. Most normalization algorithms can be applied globally (to the entire data set) or locally (to some physical subset of the data, for example, within printing tip groups). Local normalization has the advantage that it can help correct for systematic spatial variations in the array *(33,34)*.

After normalization, data for each gene is reported as an "expression ratio" or as the logarithm of the expression ratio. The expression ratio is the ratio of the normalized value of the expression level for a gene in one sample by its normalized level for the control. By using the logarithm of the expression ratio, it is easier to visualize downregulated genes that are squashed between 1 and 0 if the logarithm scale is not applied.

The next obvious step is to search for differentially expressed genes among the tested samples. To find these differentially expressed genes, it is possible to use a parametric test (e.g., *t*-test) or a nonparametric test (e.g., Mann–Whitney). By using any statistical test, the result will be a list of genes that are more likely to be indeed differentially expressed and their respective *p* values. Because in a typical microarray experiment the *p* value for thousands of genes is determined, it is very common to have a sizeable set of false-positive results when considering only individual *p* values. There are several methods to adjust the *p* value for multiple testing; however, all of the methods have drawbacks.

Cluster analysis has been applied to microarray data to cluster both genes with similar expression patterns and samples with similar expression profiles. There are several methods for performing cluster analysis, and many of these have already been applied to microarray data as hierarchical clustering, K-means clustering, and self-organizing maps *(33,34)*. To perform a clustering analysis, we use some distance measure to assess the similarity of two samples. The goal of cluster analysis is to group objects that have similar distance measures leading to clusters where the distance measure is small within clusters and large between clusters. Bootstrap analysis, as a robustness measure is well suited to be taken into account when making clusters *(35)*. Cluster analysis is the standard procedure for performing unsupervised pattern recognition. Supervised pattern recognition contrasts with unsupervised pattern recognition in that, for the latter recognition, we ask the computer to discover categories without any previous training cases. Supervised pattern recognition has been widely applied to identify patterns of gene expression and for sample classification. One of the goals of supervised expression data analysis is to construct classifiers, such as linear discriminants, decisions trees, or support vector machines, which assign predefined class to a given expression profile.

3. Validation of Candidate Genes Through Real-Time Reverse Transcription-PCR and Tissue Microarray

Cancer research has benefited from several advances in technology, and this advance has resulted in an exponential increase in the number of novel gene candidates to be implicated in tumorigenesis, diagnosis, prognosis, and new therapeutic targets. The large amount of new information has created a new challenge in the identification of an individual gene, and mainly in its validation as a clinically useful diagnostic or prognostic marker.

There are two useful methodologies to validate candidate genes shown to be differentially expressed in microarray experiments: real-time reverse transcription-PCR and immunohistochemistry or *in situ* hybridization. The difference between these technical approaches is the kind of molecule that is quantitatively being measured and the number of samples that can be analyzed at the same time. In real-time reverse transcription-PCR the level of a given mRNA molecule is being measured, which is exactly the same measured molecule as in the microarray technology. In immunohistochemistry, the molecule whose level is being measured is a protein. It has been reported that the level of mRNA does not always correspond to the protein level *(36)*, because of a variety of mechanisms beyond the transcriptional level involved in regulation of protein expression of a specific gene. However, the latter methodology has the advantage to monitor *in situ* reaction, with determination of the cell compartment involved.

Real-time PCR is a molecular biological technique used for detection of small changes in the expression level of specific genes. Real-time PCR using cDNA as a template, which corresponds to the mRNA population, has been used as technical validation (when the same set of samples assessed by microarray technology is used) and also as biological validation (when a significant greater number of the samples representative of a given class is used) of candidates for molecular markers identified by microarray experiments. Real-time PCR is based on an amplification process that can

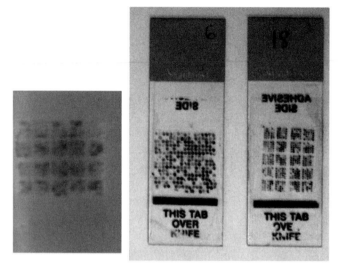

Fig. 3. (Left) TMA paraffin block with 500 cores. **(Right)** Two slides obtained from TMA with 270 and 500 cores, respectively.

be monitored by detection of either directly or indirectly accumulated fluorescence, resultant of a fluorescent molecule that is associated with newly amplified DNA. The fluorescence measurement corresponds to the quantity of newly synthesized DNA molecule in the reaction; thus, the more copies of a specific cDNA molecule that are present in the sample, the fewer cycles of amplification in the PCR reaction that are needed to make a specific number of amplified molecules detected by the florescent measuring, providing a relatively quantitative measurement *(37)*.

Another classical approach to test candidate genes is by either immunohistochemistry or *in situ* hybridization, a laborious process based on the observation of individual tissue sections placed on a single microscope slide. Konoken and collaborators *(38)* described a method for a high-throughput molecular test in archival paraffined tissues, in which a large number of tumor samples can be arrayed in a single paraffin block, called a high-density tissue microarray (TMA). With this technique, more than 500 samples can be tested simultaneously, facilitating the quick assessment of several molecular markers. Thus, TMA enables pathologists to perform large-scale analyses by using immunohistochemistry, fluorescence in situ hybridization, or RNA *in situ* hybridization substantially faster and at markedly lower costs compared with the conventional approach. The construction of TMA is based in placement of hundreds of cases of a determined tumor or several different tumors in a single paraffin block. The histological aspects of the tumor are reviewed and classified, and the donor paraffin block is identified. Core tissue biopsies ranging from 0.6 to 2 mm are taken and transferred to a recipient paraffin block. A mechanical arrayer is used to place the cores in an organized manner, and the coordinates of the core biopsies are recorded on a grid in which one corner position called zero corresponds to a special marker. As marker, we use a normal tissue from a different organ compared with the tumor that will be placed as study subject. Precise x-y positioning of the specimens provides a basis for automation of TMA construction. The resulting paraffin block and respective slide can be seen in **Fig. 3**.

Several reports have shown the validity of the TMA approach compared with analysis of conventional tissue sections *(39–43)*. Because of the known heterogeneity of tumors, the limited sampling represented by the TMA may not fully represent the "donor" tumor. This heterogeneity can be avoided by using multiple sections and spotting the same tumor more than one time. An interesting approach is to put cases in duplicate or triplicate and perform immunohistochemistry in deeper sections obtained from the TMA block. Therefore, the assay can be repeated several times, and the results can be expressed as mean or median of the experiments, a new approach in terms of immunohistochemistry. Preservation of the donor block is another advantage in using TMA. Multiple cores can be collected from the same paraffin block without damage to the original tissue.

To construct a TMA, case selection is essential *(44,45)*. The cases should be carefully screened, to select cases appropriate to the research object. Thus, clinical and pathological information should be available for each primary site or histological type. A different approach is a construction of a multitumor TMA. In this case, several different types of tumors are collected in a same TMA to study a specific protein in different tumor histogenesis *(46)*. This approach is particularly useful in the process of screening new genes and new antibodies. Also, with the emerging field of pharmaco-immunostaining, a large-scale test will be necessary to address target-specific therapeutics *(47)*.

4. Perspectives

We believe that microarray technology can help cancer scientists by contributing to a better and more comprehensive understanding of the establishment and progression of cancer and with reliable molecular markers. Additionally, its contribution can be increased in cancer research, especially if a better platform design and a larger number of samples that adequately represent the investigated tumor class are used. Taken together, microarray technology can significantly increase the statistical power of the correlations with clinical information, help pathologists to better classify the different subtypes of more heterogeneous tumors, and address the uncountable questions about this complex disease.

Acknowledgments

We thank the researchers, technicians, and students of Microarray infrastructure's laboratories of São Paulo branch of the Ludwig Institute for Cancer Research and the AC Camargo Cancer Hospital. We also thank the team of AC Camargo Cancer Hospital for tissue collecting. This work was supported by Fundação de Amparo à Pesquisa do Estado de São Paulo Grant 98/14335-2 CEPID.

References

1. Balmain, A., Gray, J., and Ponder B. (2003) The genetics and genomics of cancer. *Nat. Genet.* **33,** 238–244.
2. Popescu, N. C. (2000) Comprehensive genetic analysis of cancer cells. *J Cell Mol. Med* **4,** 151–163.
3. Vogelstein, B., and Kinzler, K. W. (2004) Cancer genes and the pathways they control. *Nat Med.* **10,** 789–799.
4. Lander, E. S., Linton, L. M., Birren, B., et al. (2001). Initial sequencing and analysis of the human genome. *Nature* **409,** 860–921.

5. Venter, J. C., Adams, M. D., Myers, E. W., et al. (2001) The sequence of the human genome. *Science* **291,** 1304–1351.

6. de Souza, S. J., Camargo, A. A., Briones, M. R., et al. *Proc. Natl. Acad. Sci. USA* **97,** 12,690–12,693.

7. Camargo, A. A., Samaia, H. P., Dias-Neto, E., et al. (2004) The contribution of 700,000 ORF sequence tags to the definition of the human transcriptome *Proc. Natl. Acad. Sci. USA* **98,** 12,103–12,108.

8. Strausberg, R. L., Feingold, E. A., Grouse, L. H., et al. (2002) Generation and initial analysis of more than 15,000 full-length human and mouse cDNA sequences. *Proc. Natl. Acad. Sci. USA* **99,** 16,899–16,903.

9. Velculescu, V. E., Zhang, L., Vogelstein, B., and Kinzler, K. W. (1995) Serial analysis of gene expression. *Science* **270,** 484–487.

10. Boon, K., Osorio, E. C., Greenhut, S. F., et al. (2002) An anatomy of normal and malignant gene expression *Proc. Natl. Acad. Sci. USA* **99,** 11,287–11,292.

11. Jongeneel, C. V., Iseli, C., Stevenson, B. J., et al. (2003) Comprehensive sampling of gene expression in human cell lines with massively parallel signature sequencing. *Proc. Natl. Acad. Sci. USA* **100,** 4702–4705.

12. Jia, L., Young, M. F., Powell, J., et al. (2003) Gene expression profile of human bone marrow stromal cells: high-throughput expressed sequence tag sequencing analysis. *Genomics* **79,** 7–17.

13. Schena, M., Shalon, D., Davis, R. W., and Brown, P. O. (1995) Quantitative monitoring of gene expression patterns with a complementary DNA microarray. Science **270,** 467–470.

14. Wu, Y. M., Robinson, D. R., and Kung, H. J. (2004) Signal pathways in up-regulation of chemokines by tyrosine kinase MER/NYK in prostate cancer cells. *Cancer Res.* **64,** 7311–7320.

15. Scian, M. J., Stagliano, K. E., Ellis, M. A., et al. (2004) Modulation of gene expression by tumor-derived p53 mutants *Cancer Res.* **64,** 7447–7454.

16. Thompson, H. J., Zhu, Z., and Jiang, W (2004) Identification of the apoptosis activation cascade induced in mammary carcinomas by energy restriction. *Cancer Res.* **4,** 1541–1545.

17. Sotiriou, C., Neo, S. Y., McShane, L. M., et al. (2003) Breast cancer classification and prognosis based on gene expression profiles from a population-based study. *Proc. Natl. Acad. Sci. USA* **100,** 10,393–10,398.

18. Sorlie, T., Tibshirani, R., Parker, J., et al. (2003) Repeated observation of breast tumor subtypes in independent gene expression data sets. *Proc. Natl. Acad. Sci. USA* **100,** 8418–8422.

19. van't Veer, L. J., Dai, H., van de Vijver, M. J., et al. (2002) Gene expression profiling predicts clinical outcome of breast cancer. *Nature* **415,** 530–536.

20. Ma, X. J., Salunga, R., Tuggle, J. T., et al. (2000) Gene expression profiles of human breast cancer progression *Proc. Natl. Acad. Sci. USA* **100,** 5974–5979.

21. Schulze, A., and Downward, J. (2001) Navigating gene expression using microarrays: a technology review *Nat. Cell Biol.* **3,** E190–E195.

22. Handley, D., Serban, N., Peters, D., et al. (2004) Evidence of systematic expressed sequence tag IMAGE clone cross-hybridization on cDNA microarrays *Genomics* **83,** 1169–1175.

23. Flikka, K., Yadetie, F., Laegreid, A., and Jonassen, I. (2004) XHM, a system for detection of potential cross hybridizations in DNA microarrays BMC *Bioinformatics* **5,** 117.

24. Chou, C. C., Chen, C. H., Lee, T. T., and Peck, K. (2004) Optimization of probe length and the number of probes per gene for optimal microarray analysis of gene expression. *Nucleic Acids Res.* **32,** e99.

25. Dias, N. E., Correa, R. G., Verjovski-Almeida, S., et al. (2000) Shotgun sequencing of the human transcriptome with ORF expressed sequence tags. *Proc. Natl. Acad. Sci. USA* **97,** 3491–3496.

26. Brentani, H., Caballero, O. L., Camargo, A. A., et al. (2003) The generation and utilization of a cancer-oriented representation of the human transcriptome by using expressed sequence tags. *Proc. Natl. Acad. Sci. USA* **100,** 418–423.

27. Brentani, R. R., Carraro, D. M., Verjovski-Almeida, S., et al. (2005) Gene expression arrays in cancer research: methods and applications. *Crit. Rev. Oncol. Hematol.* **54,** 95–105.

28. Iseli, C., Stevenson, B. J., de Souza, S. J., et al. (2002) Long-range heterogeneity at the 3′ ends of human mRNAs. *Genome Res.* **12,** 1068–1074.

29. Zhao, H., Hastie, T., Whitfield, M. L., Borresen-Dale, A. L., and Jeffrey, S. S. (2002) Optimization and evaluation of T7 based RNA linear amplification protocols for cDNA microarray analysis. *BMC Genomics* **3,** 31.

30. Gomes, L. I., Silva, R. L., Stolf, B. S., et al. (2003) Comparative analysis of amplified and nonamplified RNA for hybridization in cDNA microarray. Anal. Biochem. **321,** 244–251.

31. Feldman, A. L., Costouros, N. G., Wang, E., et al. (2002) Advantages of mRNA amplification for microarray analysis. *Biotechniques* **33,** 906–912.

32. Wang, E., Miller, L. D., Ohnmacht, G. A., Liu, E. T., and Marincola, F. M. (2000) High-fidelity mRNA amplification for gene profiling. *Nat. Biotechnol.* **18,** 457–459.

33. Yang, Y. H., Dudoit, S., Luu, P., et al. (2002) Normalization for cDNA microarray data: a robust composite method addressing single and multiple slide systematic variation. *Nucleic Acids Res.* **30,** 1–10.

34. Quackenbush, J. (2001) Computational analysis of microarray data. *Nat. Rev. Genet.* **2,** 418–427.

35. Kerr, M. K., and Churchill, G. A. (2001) Bootstrapping cluster analysis: assessing the reliability of conclusions from microarray experiments. *Proc. Natl. Acad. Sci. USA* **98,** 8961–8965.

36. Anderson, L., and Seilhamer, J. (1997) A comparison of selected mRNA and protein abundances in human liver. *Electrophoresis* **18,** 533–537.

37. Walker, N. J. (2002) Tech.Sight. A technique whose time has come. *Science* **296,** 557–559.

38. Kononen, J., Bubendorf, L., and Kallioniemi, A. (1998) Tissue miroarrays for high-troughput molecular profiling of tumor specimens. *Nat. Med.* **4,** 844–847.

39. Camp, R. L., Charette, L. A., and Rimm, D. L. (2000) Validation of tissue microarray technology in breast carcinoma. *Lab Invest.* **80,** 1943–1949.

40. Engellau, J., Akerman, M., Anderson, H., et al. (2001) Tissue microarray technique in soft tissue sarcoma: immunohistochemical Ki-67 expression in malignant fibrous histiocytoma. *Appl. Immunohistochem. Mol. Morphol.* **9,** 358–363

41. Fernebro, E., Dictor, M., Bendahl, P. O., Ferno, M., and Nilbert, M. (2002) Evaluation of the tissue microarray technique for immunohistochemical analysis in rectal cancer. *Arch. Pathol. Lab. Med.* **126,** 702–705.

42. Gancberg, D., Di Leo, A., Rouas, G., et al. (2002) Reliability of the tissue microarray based FISH for evaluation of the HER-2 oncogene in breast carcinoma. *J. Clin. Pathol.* **55,** 315–317.

43. Rubin, M. A., Dunn, R., Strawderman, M., and Pienta, K. J. (2002) Tissue microarray sampling strategy for prostate cancer biomarker analysis. *Am. J. Surg. Pathol.* **26,** 312–319.

44. Skacel, M., Skilton, B., Pettay, J. D., and Tubbs, R. R. (2002) Tissue microarrays: a powerful tool for high-throughput analysis of clinical specimens: a review of the method with validation data. *Appl. Immunohistochem. Mol. Morphol.* **10,** 1–6.

45. Hoos, A., and Cordon-Cardo, C. (2002) Tissue microarray profiling of cancer specimens and cell lines: opportunities and limitations. *Lab Invest.* **81,**1331–1338.

46. Terris, B., and Bralet, M. P. (2002) Tissue microarrays, or the advent of chips in pathology. *Ann. Pathol.* **22,** 69–72.
47. Wang, H., Zhang, W., and Fuller, G. N. (2002) Tissue microarrays: applications in neuropathology research, diagnosis, and education. *Brain Pathol.* **12,** 95–107.

3

Experimental Design for Gene Expression Analysis

Answers Are Easy, Is Asking the Right Question Difficult?

Marcia V. Fournier, Paulo Costa Carvalho, David D. Magee,
Maria da Gloria Costa Carvalho, and Krishnarao Appasani

Summary

More and more, array platforms are being used to assess gene expression in a wide range of biological and clinical models. Technologies using arrays have proven to be reliable and affordable for most of the scientific community worldwide. By typing microarrays or proteomics into a search engine such as PubMed, thousands of references can be viewed. Nevertheless, almost everyone in life science research has a story to tell about array experiments that were expensive, did not generate reproducible data, or generated meaningless data. Because considerable resources are required for any experiment using arrays, it is desirable to evaluate the best method and the best design to ask a certain question. Multiple levels of technical problems, such as sample preparation, array spotting, signal acquisition, dye intensity bias, normalization, or sample-contamination, can generate inconsistent results or misleading conclusions. Technical recommendations that offer alternatives and solutions for the most common problems have been discussed extensively in previous work. Less often discussed is the experimental design. A poor design can make array data analysis difficult, even if there are no technical problems. This chapter focuses on experimental design choices in terms of controls such as replicates and comparisons for microarray and proteomics. It also covers data validation and provides examples of studies using diverse experimental designs. The overall emphasis is on design efficiency. Though perhaps obvious, we also emphasize that design choices should be made so that biological questions are answered by clear data analysis.

Key Words: experimental design; gene expression; microarray, proteomics.

1. cDNA and DNA Microarrays

1.1. Background

How are complex organisms such as humans formed from a single cell? How are tissues differentiated? How do cells function in different environments? What changes occur in diseases? Such questions can be assessed using good biological models in combination with a comprehensive assessment of gene expression patterns. Whether you want to view the entire genome on a single array or focus on a target set of biologically relevant genes, microarrays allow the analysis of gene expression levels and can be applied in a broad spectrum of questions, including uncovering new regulatory pathways, validating drug targets, clarifying diseases, analyzing toxicological responses, or building robust databases *(1–9)*. Although the full potential of arrays is yet to be realized, these tools have shown great promise in deciphering complex diseases such

From: *Bioarrays: From Basics to Diagnostics*
Edited by: K. Appasani © Humana Press Inc., Totowa, NJ

as cancer *(10–13)*. Array technologies have varying limitations, which need to be kept in mind when choosing among them. This chapter covers two types of microarray platforms—cDNA and oligonucleotides—that are currently used, and both are effective for assessing gene expression patterns *(14,15)*.

Oligonucleotide microarrays use direct synthesis or deposition of oligonucleotides onto a solid surface and single color readout of gene expression from a test sample. Oligonucleotides offer greater specificity than cDNAs, because they can be tailored to minimize chances of cross-hybridization. Major advantages of these arrays include the uniformity of probe length and the ability to distinguish among splice variants. Another advantage particular to the commonly used Affymetrix GeneChip system (Affymetrix, Santa Clara, CA) *(16)* is the ability to perform multiple independent measurements of each transcript of interest, providing reliable assessments of each data point. In addition, this system allows the recovery of samples after hybridization to a chip and its sequential hybridization to multiple arrays, a considerable advantage when dealing with limited resources.

cDNA microarrays are typically limited by density and can analyze each transcript with only a single probe, compromising the robustness of the array. The primary benefit of cDNA arrays is that they can be made by individual investigators, are easily customized, and do not require *a priori* knowledge of a cDNA sequence, because clones can be used and then sequenced later if they are of interest.

1.2. One-Color Vs Two-Color Arrays

In one-color arrays, the experimental RNA sample is amplified enzymatically, biotin-labeled for detection, hybridized to the microarray, and detected through the binding of a fluorescent compound (streptavidin-phycoerythrin). Two-color arrays use the competitive hybridization of two messenger RNA (mRNA) samples labeled with dyes—cyanine (Cy)3 and Cy5—to measure the relative gene expression levels of the samples. Quantified signal intensities for Cy3 (R) and Cy5 (G) are intended to be proportionally consistent with the mRNA levels for the two samples across all spotted genes and slides. Inconsistencies in channel intensity result from various steps of microarray fabrication, RNA preparation, hybridization, scanning, or image processing. The ability to make a direct comparison between two RNA samples on the same microarray slide is a unique and powerful feature of the two-color arrays.

1.3. Experimental Control and Bias Issues

Experimental controls provide the standard means to implement quality assurance procedures in biological experiments. In addition, the experimental design should incorporate controls to minimize sources of bias in an experiment. In statistical terms, bias is defined as the difference in value between a sample and a population measurement. More generally, bias is any partiality that prevents the objective consideration of an issue or situation. Experimental controls provide a means to measure and normalize biases associated with sample conditions and technical aberrations. To eliminate unwanted bias, it is necessary to include controls for experimental conditions in the analysis *(17)*. Microarray experiments typically involve small numbers of replicates where the assumptions of classical statistics (normally distributed values) do not apply. Typically, researchers calculate *p* values without testing the normality of the data. A more important concern is to control for technical biases and erroneous data through the implementation of experimental controls.

1.3.1. Nonbiological Bias and Data Normalization:

In microarray experiments, systematic variation from a variety of nonbiological sources can affect measured gene expression levels. The process of normalization seeks to eliminate such variation and thus enhance the reliability of results obtained from subsequent higher-order statistical analysis of the data *(18,19)*. Important low-level sources of technical variation are fluorescent intensities between two channels and the physical locations of spots on a microarray slide. Technical bias is introduced during array printing, extraction, labeling, and hybridization. Sophisticated normalization methods adjust such spatial and intensity bias. The diagnosis of bias and the assessment of normalization methods are currently accomplished using various plots *(20)*. Using plots to assess normalization, however, does not specifically address how to order methods to best remove inconsistent bias patterns, how to compare methods statistically, or how to verify the quality of single arrays. In addition, postdata collection error may occur when bias is introduced by erroneous data normalization *(21)*. Therefore, it is desirable to avoid nonbiological bias by implementing a rigorous quality control during all steps of the experiments, including sample preparation, probe amplification, array spotting, hybridization, and signal detection. Thus, if using custom- or home-made arrays, it is important to perform a range of optimization experiments to ensure quality control before performing actual experiments. Ensuring quality control will allow the use of simple normalization methods such as normalization of log values to the median of each array.

1.3.2. Biological Variation

Variability is intrinsic to all organisms and is influenced by genetic or environmental factors. Thus, measurements taken from a particular cell culture are biased by that culture and may not represent a broader gene expression. Pooling samples will conceal biological variation in that expression levels for a gene may vary in each sample. But, once pooled, all the replicates should indicate the same level. In this case, you can assume that any remaining variation is nonbiological. Data from pooled RNA replicates (technical replicates) would be useful for assessing the quality of the original arrays and the hybridization conditions. From a biological perspective, pooling RNA eliminates an important experimental control as well as potentially useful gene expression information. For considering sample controls, it is better to use several pools and fewer replications than one pool of all the available samples and multiple replications *(22)*.

The main goal of replication is to generate independent measurements for the purpose of reducing a type of bias. For most questions, biological bias and technical bias should be considered before constructing statistical tests. Hence, good designs should incorporate replication at these levels *(22–25)*. For example, performing dye swaps in two-color array experiments provides technical replicates and minimizes technical bias caused by any difference in dye intensity.

Identifying the independent measurements in an experiment is a prerequisite for a proper statistical analysis. Details on how individual animals, cultures, or samples were handled through the course of an experiment can be important for identifying which biological samples and technical replicates are independent. Although it is tempting to avoid biological replicates in an experiment because the results seem to be more reproducible (23), it is useful to use them because they ensure that the results are biologically significant. In contrast, by eliminating biological replicates (e.g., pooling RNA), you can be somewhat certain that the remaining variation is technical (nonbiological). Replicate spots on the same slide provide an effective means to measure technical variation, which could represent array printing quality, hybridization conditions, or spot-reading software. When initially calibrating laboratory conditions, you may focus on reducing technical error by using pooled RNA. After the conditions are optimized, the focus of measure may turn to controlling the biological variance.

1.3.3. Reference Samples

The ability to make a direct comparison between two samples on the same microarray slide is a unique and powerful feature of the two-color microarray. Thus, the main consideration for cDNA arrays is which samples should be cohybridized (22). For example, if an investigator has unlimited amounts of reference material, but a limited amount of test RNA, then a comparison of the test sample with a reference sample should be the choice. Reference or control samples can be biological controls (untreated cell or animal) or universal reference RNA.

A useful tool for monitoring and controlling intra- and interexperimental variation in experiments applying a two-color microarray is universal reference RNA (URR) by providing a hybridization signal at each microarray probe location (spot). In this case, all the experimental RNAs, including experimental controls, are hybridized to a URR rather than to each other. Measuring the signal at each spot as the ratio of experimental RNA to reference RNA targets, rather than relying on absolute signal intensity, decreases the variability by normalizing the signal output in any two-color hybridization experiment (26). URR is prepared from pools of RNA derived from individual cell lines representing different tissues. Moreover, experiments using URR can be compared over time and across operators, because the reference RNA is the same (27).

1.4. Types of Design

Experiments should be designed to maximize the efficiency and reliability of the data obtained (28). Careful attention to the experimental design will ensure that the use of available resources is efficient, obvious biases are avoided, and the primary question is answered (22,29). For example, primary questions may include identifying differentially expressed genes, defining groups of genes with similar patterns of gene expression, and identifying tumor subclasses.

Consideration for a good experimental design should take into account the aims of the study, the choices of the biological model or clinical setting, sources of variability, replicates, and the optimal sample size (21,29–31). Biological resources and cost considerations will usually dictate the amount of RNA available and the number of replicates to be used respectively.

Sources of RNA can be either tissue samples or cell lines. How much RNA is available will affect the number of times the experiment can be repeated and its validation. Sample isolation, RNA extraction, and labeling also affect the number of replicates required. Data validation is discussed in greater detail (see Subheading 1.5.).

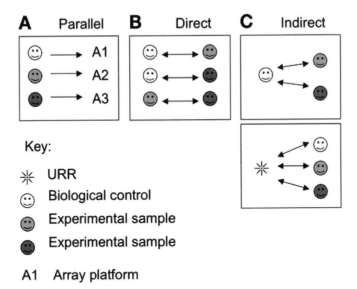

Fig. 1. Parallel, direct, and indirect designs. (**A**) In one-color arrays, control and experimental samples are hybridized to independent arrays, and the resulting data is analyzed in parallel. The main consideration for two-color arrays is which samples should be cohybridized. Two possible designs that compare gene expression can be applied using two-color arrays. (**B**) Direct comparison measures differential gene expression directly in the same slide. (**C**) In an indirect comparison, differential gene expression is measured in relation to a reference.

A single-color microarray compares samples in parallel manner, where each sample is probed in an independent array (**Fig. 1A**). Comparisons can be either direct or indirect when using a two-color microarray (*22*). A direct comparison occurs when two samples are compared in the same array (**Fig. 1B**). In an indirect comparison, the expression levels of the experimental samples are measured separately on different slides by using a reference sample (**Fig. 1C**). The type of experiment usually dictates the type of design, although there are experiments where diverse types of designs are suitable. Principles are needed for choosing one design from among several possible designs. A focus on the primary question is necessary in such experiments. You should think about which design is suitable to answer the most interesting question. One approach is to ask which comparisons are of greater and which are of lesser interest and then seek a design that gives greater precision to the former and less precision to the latter.

In general, if an experiment needs a small sample size, using two-color arrays and a direct comparison can be advantageous because of the lower statistical variance (*22,32*). A parallel comparison should be the option for a large number of samples, because of the higher specificity and reproducibility between replicates. Moreover, parallel comparisons facilitate data analysis by using unsophisticated statistical tools, which provide a considerable advantage for a large data set.

How do you make design choices based on the type of experiment? Broadly, experiments can be classified into one of the following categories: treatment regimen, disease classification, time courses, and factorial studies.

1.4.1. Treatment Regimen Studies

The use of an untreated control is obvious in experiments involving treatment regimens. In two-color microarrays, all treated samples should be hybridized with the untreated control. There are two fundamentally distinct designs in human studies to assess treatment effects *(32)*. In one design, the investigator assigns the exposure and then measures the outcome (experimental). In the other design, the exposure is measured (observational). Most retrospective studies fall under the latter category. Case-control design is a good option in studies of gene expression patterns where the treatment and collection of tissue samples are not designed in the laboratory. Cases are patients who died or had recurrence (high aggressive), and controls are those who lived beyond a time line after diagnosis and treatment (low aggressive). For example, in a recent study, 60 patients were selected from a total of 103 ER-positive, early stage cases presented to Massachusetts General Hospital between 1987 and 1997, from whom tumor specimens were snap-frozen and for whom minimal 5-yr follow-up was available. In this study, breast cancer patients treated with tamoxifen were compared for differential gene expression between tamoxifen responders and nonresponders *(33)*. Additional examples on using microarrays for the development and assessment of therapies are covered in previous reviews *(4,34–37)*.

1.4.2. Disease Classification Studies

Microarray studies allow class comparison, class discovery, or class prediction *(10)*. In class comparison, the study aims to establish whether gene expression profiles differ between classes. In class discovery, the goal is to elucidate unrecognized subclasses—such as new tumor subclasses—based on gene expression profiles. In class prediction studies, information from gene expression profiles is used to predict a phenotype *(10,27)*. Combinations between class discovery and class prediction have been used in several clinical studies to assess gene expression patterns identifying tumor subclasses *(38–44)*.

Human tumor samples are heterogeneous mixtures of diverse cell types, including malignant cells, stromal elements, blood vessels, and inflammatory cells. Because of this heterogeneity, the interpretation of gene expression studies is not always simple. Groups attempting to focus on differences between malignant and nonmalignant components of a tumor may use laser capture microdissection of individual cells from a tumor section to isolate cancer cell RNA for microarray experiments. Sgroi et al. *(45)* combined laser capture microdissection and cDNA microarray analysis to compare global gene expression in normal and tumor cells from the same tissue specimen. In other study, cell type-specific surface markers and magnetic beads have been used to isolate various cell types in a tissue sample for gene expression studies *(46,47)*. Designs using magnetic beads for cell type isolation allowed analysis of tumor cells and the surrounding microenvironment.

1.4.3. Time-Course Studies

A time-course experiment is a case of a multiple-slide experiment in which transcript abundance is monitored over time. Recently, several methods have been suggested to identify differentially expressed genes in multiple-slide microarray experiments based on statistical models such as the analysis of variance model and the mixed effect model *(48,49)*. In time-course experiments, the comparisons to be made might not be obvious. The comparisons of greatest interest should determine the best design (**Table 1**). For example, in a time-course experiment where the main question is about gene expression changes occurring in relation to the starting point, the use of a starting point (time zero) as a reference is an appropriate choice. Assessing the activity of indi-

Table 1
Time-Course Experiments

One-color array using oligonucleotide platform
• Parallel comparison

Sample	Array unit	Type of replicate	Strongness
T1	A1		One or more samples can be used as a reference
T2	A2		Easy data analysis
T3	A3		Flexibility
T1'	A4	Biological replicate	Highly reproducible
T2'	A5	Biological replicate	Multiple representation of genes in the arrays
T3'	A6	Biological replicate	

Two-color array
• Design I: using direct comparison of time points (Cy3-Cy5)

Sample	Array unit	Type of replicate	Strongness
T1-T2	A1		Comparison of time points in the same array
T1-T3	A2		
T2-T3	A3		
T2-T1 (dye swap)	A4	Technical replicate	
T3-T1 (dye swap)	A5	Technical replicate	
T3-T2 (dye swap)	A6	Technical replicate	

• Design II: indirect comparison using a reference sample (Cy3-Cy5)

Sample	Array unit	Type of replicate	Strongness
T1-R	A1		Apply universal reference sample
T2-R	A2		Easy analysis
T3-R	A3		
Plus dye swaps	A4-A6	Technical replicate	

• Design III: linear comparison

Sample	Array unit	Type of replicate	Strongness
T1-T2	A1		
T2-T3	A2		Uses less microarray slides
Plus dye swaps	A3-A4	Technical replicate	

• Design IV: using T1 as reference

Sample	Array unit	Type of replicate	Strongness
T1-T2	A1		Uses less microarray slides
T1-T3	A2		Simple analysis
Plus dye swaps	A3-A4	Technical replicate	

In an experiment including three time points (T1, T2, and T3) at least five possible experimental designs can be considered. The design choice should be made based on the time-point comparisons most important for answering the question. T1, time zero; T2 and T3 experimental time points.

vidual genes as a function of time is an example of time-course experiments. The oncogenic potential of cyclin D1 overexpression in a time dependent manner has been dissected by using gene expression profiling as a global assay *(2)*. Comparing temporal gene expression patterns, following the expression of wild-type and mutant cyclin D1 proteins, revealed an expression profile of cyclin D1 target genes. In this case, the gene expression was analyzed from time 0 to 24 h postinfection with adenovirus. The genes were selected by expression levels in comparison with time 0 and 4 h and showed maximum expression at 24 h. This empirically determined expression profile was used to further study the mechanistic basis of the oncogenic consequences of cyclin D1 overexpression in human tissues.

1.4.4. Factorial Studies

Gene expression profiling can capture the activity of individual genes by using over-expression or depletion systems. For example, Hughes et al. *(50)* investigated in yeast the functions of previously uncharacterized genes by matching gene expression changes induced by their deletion against a collection of yeast reference profiles. Multifactorial designs consider differences that not only are caused by single factors but also result from the interaction of two or more factors. In such experiments, there are three major questions to be addressed: 1) How do individual factors affect cells (unique pathways affected), 2) What are the common pathways affected, and 3) What are the combinatorial effects (**Table 2**). One possible design would be the comparison of single treated or combination treated with a reference biological control (untreated). Parallel designs have the advantage of allowing comparisons between any of the experimental samples, easy analysis, and interpretation without the need for statistical tools. Direct designs can generate results where a lower variance of estimated effects is obtained compared with indirect designs *(22)*. A balance between direct and indirect designs can be used as well. In this case, the main comparisons should be direct, whereas the secondary comparisons can be indirect. Examples of multifactorial studies include treatment combination *(51–54)* and double knockouts *(55,56)*.

1.5. Verification

The amount of verification can influence the choice of statistical method and sample size. The main goal for verifying the array data is to identify bona fide gene expression changes. Verification can be broad or confined to the original samples used to acquire the data. When array verification is confined to the samples used for the generation of the data, it is usually focused on one or a small group of genes. A group of genes can be assessed using reverse transcription-PCR, Northern blots, Westerns blots, or tissue immunostaining, providing a validation of patterns of gene expression. Verification also can be extended to an independent data set to validate the ability of selected sets of genes to make a certain distinction, such as a treatment response or a clinical outcome. This strategy has been used to unveil gene expression profiles that predict clinical outcomes of breast cancer *(43)*. In this study, a group of 70 genes was identified as a molecular signature for a poor prognosis in early stage breast cancer patients. This molecular signature was validated in a second study by using an independent data set *(57)*. Confirmation using an independent data set enhances confidence in the obtained results.

Data verification and validation also can use gene expression manipulation (e.g., expression systems, knockout, and small-interfering RNA). Studies that aim to discover relevant gene expression associated with a phenotype or to elucidate a gene function should use this approach.

An experimentally generated data set also can be compared with published data available online in public databases. One of the advantages of validation using gene expression databases is the fast assessment of data. However, the investigator should keep in mind that the variability in the quality and reproducibility between array platforms is a limiting factor. The choice of the verification method will depend on the biological resources and budget available. A combination of verification levels is usually a good choice.

Table 2
Multifactorial Experiments

One-color array by using oligonucleotide platform
- Parallel comparison

Sample	Array unit	Type of replicate	Strongness
R	A1		One or more samples can be used as a reference
A	A2		Easy data analysis
B	A3		Flexibility
AB	A4		Highly reproducible
R'	A5	Biological replicate	Multiple representation of genes in the arrays
A'	A6	Biological replicate	
B'	A7	Biological replicate	
AB'	A8	Biological replicate	

Two-color array
- Design I: indirect comparison using a reference sample

Sample	Array unit	Type of replicate	Strongness
A-R	A1		
B-R	A2		Uses less microarray slides
AB-R	A3		
dye swaps	A4	Technical replicate	
dye swaps	A5	Technical replicate	
dye swaps	A6	Technical replicate	

- Design II: loop

Sample	Array unit	Type of replicate	Strongness
R-A	A1		Mix direct and indirect comparisons
A-B	A2		
B-AB	A3		
AB-R	A4		
dye swaps	A5	Technical replicate	
dye swaps	A6	Technical replicate	
dye swaps	A7	Technical replicate	
dye swaps	A8	Technical replicate	

In multifactorial experiments there are 3 major questions to be addressed: how individual factors affect cells (the unique pathways affected), the common pathways affected, and combinatorial effects.

R, not treated; A, treatment 1; B, treatment 2; AB, combinatorial treatment.

2. Proteomics

2.1. Background

Proteins are the functional units of cells and represent the end products of gene expression. Proteomics is a set of methods and tools for studying a cell's proteome, the proteins expressed by a cell's genome at a given time. This approach permits the study of posttranslational modifications and quantifies levels of protein expression over time in response to changes in environment or disease. In proteomics methodology, proteins are identified mostly by the use of two-dimensional electrophoresis (2DE) coupled with mass spectrometry (MS) and computer search algorithms. These combined tools are able to identify a purified, digested protein by comparing the mass spectra of the peptide with *in silico* digestion of proteins contained in a database (e.g., Swiss–Prot).

New advances through proteomics include cancer biomarkers, protein signatures, interactions between protein networks, and their relation to multicellular functions. A specific biomolecule in the body that can be used for detecting a disease and measuring its progress or response to a treatment constitutes a biomarker. Even though many advances have been made in mass spectrometry-based proteomics, it still lags behind microarray technology when dealing with experimental design and reproducibility. Integrating proteomic data in global databases and standardizing protocols is still a subject of great discussion, because results vary with different mass spectrometers.

2.2. Two-Dimensional Electrophoresis

2DE is a common and efficient technique for isolating proteins for subsequent MS identification. Numerous studies involving 2DE and MS have already identified various disease-related changes in the levels of protein expression. 2DE gels are prepared with the biological samples of healthy and sick subjects, so the identification of differentially expressed proteins can be done visually or by computer gel comparison algorithms. This powerful method is capable of separating several thousand proteins per experiment, according to two independent physicochemical properties of the protein: isoelectric point in the first dimension and molecular weight in the second dimension *(58)*. Reproducibility is of the highest importance so that variations in the level of protein expression between samples can be studied. The use of immobilized pH gradients for the first dimension *(59)*, as an alternative to carrier ampholytes, has made 2DE reproducible enough for proteome analysis. Although it is common to separate proteins through the linear pI range of 3.5–10 (isoeletric focusing), there are commercially available immobilized pH gradients with many different ranges so that the separation of proteins can be optimized to a desired range. Note that highly hydrophobic proteins are hard to keep in solution; thus, they can be lost during sample preparation and isoeletric focusing.

The molecular weight separation can range from 6 to 300 kDa, and gel staining can be done through various techniques, with silver *(60)* and Coomassie blue being the most popular. To find a protein of interest, different gels are created (i.e., healthy and patient subjects). Once a differentially expressed spot is found, it can be further analyzed. MS should then be used to identify the protein and its possible function.

2.3. Protein Identification Through Matrix-Assisted
Laser Desorption Ionization-Time of Flight (MALDI-TOF)

MALDI-TOF MS is the method of choice for analyzing peptide mixtures because of its fast data acquisition (**Fig. 2**). After the spectra are obtained from the mixture, program filters such as smooth and background cutoffs can be applied to the spectra data. Other options, such as the selection of spectra mono-isotopic peaks, also should be used when exporting the data to Web interfaces for protein identification. Among such programs, Protein Prospector *(61)* and Mascot (www.matrixscience.com) have become popular, because they are freely available through the World Wide Web. After receiving a peptide mass list through a Web form, these programs perform an *in silico* digestion of the proteins available in their database and calculate the mass of each theoretical peptide. Possible identifications are presented as a report. If an ambiguous identification occurs, tandem mass spectrometry can be done to obtain the amino acid sequence or the tag of a small peptide. By inputting the sequence tag, together with the mass spectra data, the protein should be properly identified if it is contained in the database.

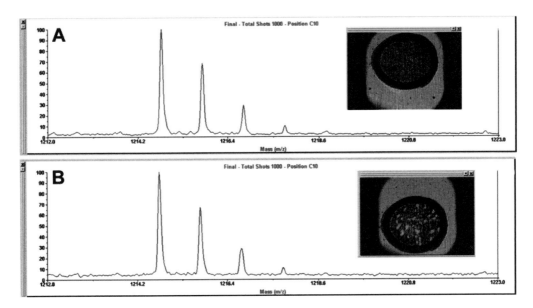

Fig. 2. Figure-laser intensity. This image shows two spectra obtained with the Applied Biosystems 4700 proteomics analyzer with TOF/TOF optics. The magnified view of the sample spotted on the MALDI plate is shown in the top right corner of each spectrum after its acquisition. Spectrum A was obtained with a good laser intensity. The peaks are sharp and there is low background noise. Spectrum B was obtained with a higher laser intensity. Sample damage may be noticed. If the spot is used for further studies—such as the MS/MS of intense peaks—the results could suffer from laser damage. For even higher intensities, more background noise and lower peak sharpness will occur.

2.4. Differential Spectra Analysis for Cancer Biomarker Characterization

Generally, cancer is diagnosed and treated too late, after cancer cells have already invaded and formed metastasis *(62)*. Serum tests based on single biomarkers, such as the prostate-specific antigen for the prostate, can sometimes give misleading results *(63)*. Protein expression analysis in body fluids by using MS offers great hope for early cancer diagnosis screening tests by characterizing pathological protein patterns. These protein expression signatures can also be used in therapy response predictions, allowing appropriate medication selection *(63)*. The quest for new biomarkers, so protein arrays can be used instead of a single target, is the hope for improving specificity and sensitivity in early cancer detection.

One experimental approach for early diagnosis by using differential protein patterns is based on the construction of antibody chips that assay the levels of several biomarkers simultaneously *(64)*. In this case, the diagnosis is based on a "larger picture," rather than on the expression of a single protein that could mislead a test. This methodology is straightforward; and the chip is simply a set of antigen tests that run in parallel.

Another experimental design is based on differential mass spectra analysis and consists of monitoring specific spectrum peaks, where peak intensity variation could suggest a pathology diagnosis *(65)*. Such an approach could be limited, because it does not detect variations in other regions of the mass spectrum. A multidisciplinary approach to performing pathology classification based on protein profiling is attained by using artificial intelligence to scan across the entire spectrum. Machine-learning techniques—such as neural networks, and recently, supporting vector machines—can correctly classify an unknown spectrum among known groups.

Tumor biomarkers are usually secreted in extremely low quantities into human blood. To better identify the differentially expressed proteins, experimental approaches are targeted at collecting and studying directly what is being secreted by tumor tissues (*66*). This approach provides key insight into "what to look for" when trying to spot diluted tumor biomarkers in the serum. Detection in a serum is important because most medical diagnosis strategies use easily accessed body fluids.

2.5. Precautions During Experimental Design

Experimental design is a key issue when dealing with proteomics. The critical issues are eliminating bias, providing a careful analysis of the data without overfitting, and recognizing its limitations. Proteomic experiments deal with many experimental design challenges, such as limited and variable sample amounts, sample degradation, posttranslational modifications, and disease or drug alterations. When performing differential spectra analysis or protein identification, a few key points should be considered for experimental design:

1. It has been shown that both sensitivity and specificity decline significantly when samples are processed in the same laboratory after a delay of several months. Always keep samples at –80°C.

2. When using liquid chromatography coupled with an electron spray, randomize the sample order between control subject and patients, and, if possible, use several sample pools.

3. Always calibrate the mass spectrometer before an experiment. In MALDI-TOF, the use of commercially available controls that contain known masses can be mixed with the analyte and used as internal controls, or they can be spotted next to the analyte for MS/MALDI calibration.

4. In differential spectra analysis, biomarkers are expressed in very low amounts, so whether or not to deplete high abundant proteins (i.e., Albumin) is a key issue in experimental design. If such proteins are depleted, detection sensitivity for other proteins could be improved. However, many other proteins that stick to albumin also are removed, resulting in the loss of information and possible biomarkers.

5. The removal of contaminant peaks is an important step in protein identification. Trypsin autolysis and human keratin are among the most common contaminants having well-known and characterized peaks. In the majority of spot identification cases, the removal of contaminant peaks from the mass spectrum will result in better spot identification with higher scores (*67*). It is important to note that, although rare, the protein of interest may have masses identical to one or more of these peaks, and its removal may reduce the match score.

6. In differential spectra analysis, choose the best method of data normalization (e.g., total ion current, maximum peak, intensity, etc.).

7. Optimize the MALDI-TOF laser frequency and intensity for a given experiment. Lower frequencies allow the detector to better recover; high laser intensity can result in a "noisy" spectrum.

8. Validate the experiment by using antibodies. Keep in mind that differentially expressed protein peaks could be related to well-known acute phase proteins and thus might not represent putative biomarkers.

9. When performing protein identification by using Web programs, always consider possible posttranslational modifications and tryptic missed cleavages.

3. Conclusions

As informative as DNA microarray expression studies are, it has been shown that changes in mRNA expression often correlate poorly with changes in protein expression. However, both technologies give insight into pathology prediction and treatment *(63)*. The future holds great challenges in proteomics and transcriptomics, especially when integrating large and disparate data types. Proteomics promises a major role in creating a predictive, preventative, and personalized approach to medicine. Great efforts are being made to standardize protocols so that a global proteome database can be constructed. This standardization is not currently possible, because spectra reproducibility, especially of complex mixtures among different mass spectrometers, is still a challenge. Although great advances in MS methods and bioinformatics have been made recently, much more effort is necessary to decipher the secrets encrypted within the large volumes of data generated by modern proteomic techniques, to unravel the true power of proteomics, and to find the hidden links with transcriptomics.

Microarray applications towards the diagnostics or the pathogen-derived DNA/proteome chips will have tremendous clinical value and they are not far away from the clinic. As of today, the lab-made Oligo or cDNA arrays are currently useful in research labs either for the classification of cancers or for infectious disease diagnosis, but their real applications did not reach the clinic. We hope that it is not too far from reality since US Food and Administration has approved Cytochrome P450 Affymetrix gene chip for the xenobiotics (drug resistance in liver disease) studies, which are commercially available through Roche. On the other hand, protein chips are available from Invitrogen (Carlsbad, CA) for the study of protein–protein interactions; however their clinical applications also have not reached the pathology laboratories yet. Several groups, including Michael Snyder's (Yale), Joshua LaBaer's (Harvard), DeResi's (UCSF), Liotta and Petricoin's (George Mason University,) and companies such as Affymetrix, Illumina, JPT Peptide Technologies, GE Healthcare, Pierce Biotechnology, CombiMatrix, Agilent and others, are developing both DNA, protein, and glycoarrays for the diagnosis of infectious diseases (like coronavirus, *Francisella tularensis*, *Vibrio cholerae*). In recent years, contributions from the laboratory of Patrick Brown's group at Stanford are well appreciated in the field of microarrays, since he has looked at the global analysis of gene expression patterns in normal human tissues/cell lines, including several human disease tissues such as various forms of cancers, fibroblasts, peripheral blood, placenta and various compartments of eye, and developed a comprehensive "global atlas of gene expression patterns" *(68)*, which will definitely be useful in the clinic in the coming decade.

Acknowledgments

We thank James Garbe for discussions and Robert Cowles for critical reading of this chapter. We apologize for inadvertently omitting the work of many investigators who could not be cited because of space limitations.

References

1. Debouck, C., Goodfellow, P. N. (1999) DNA microarrays in drug discovery and development. *Nat. Genet.* **21,** 48–50.
2. Lamb, J., Ramaswamy, S., Ford, H. L., et al. (2003) A mechanism of cyclin D1 action encoded in the patterns of gene expression in human cancer. *Cell* **114,** 323–334.

3. Bild, A. H., Yao, G., Chang, J. T., et al. (2006) Oncogenic pathway signatures in human cancers as a guide to targeted therapies. *Nature* **439,** 353–357.

4. Bild, A. H., Potti, A., Nevins, J. R. (2006) Linking oncogenic pathways with therapeutic opportunities. *Nat. Rev. Cancer* **6,** 735–741.

5. Bejjani, B. A., Shaffer, L. G. (2006) Application of array-based comparative genomic hybridization to clinical diagnostics. *J. Mol. Diagn.* **8,** 528–533.

6. Jayapal, M., Melendez, A. J. (2006) DNA microarray technology for target identification and validation. *Clin. Exp. Pharmacol. Physiol.* **33,** 496–503.

7. Wulfkuhle, J. D., Edmiston, K. H., Liotta, L. A., Petricoin, E. F., 3rd (2006) Technology insight: pharmacoproteomics for cancer–promises of patient-tailored medicine using protein microarrays. *Nat. Clin. Pract. Oncol.* **3,** 256–268.

8. Liu, E. T., Kuznetsov, V. A., Miller, L. D. (2006) In the pursuit of complexity: systems medicine in cancer biology. *Cancer Cell* **9,** 245–247.

9. Fournier, M. V., Martin, K. J. (2006) Transcriptome profiling in clinical breast cancer: From 3D culture models to prognostic signatures. *J. Cell. Physiol.* **209,** 625–630.

10. Ramaswamy, S., Golub, T. R. (2002) DNA microarrays in clinical oncology. *J. Clin. Oncol.* **20,** 1932–1941.

11. Quackenbush, J. (2006) Microarray analysis and tumor classification. *N. Engl. J. Med.* **354,** 2463–2472.

12. Hayes, D. N., Monti, S., Parmigiani, G., et al. (2006) Gene expression profiling reveals reproducible human lung adenocarcinoma subtypes in multiple independent patient cohorts. *J. Clin. Oncol.* **24,** 5079–5090.

13. Fan, C., Oh, D. S., Wessels, L., et al. (2006) Concordance among gene-expression-based predictors for breast cancer. *N. Engl. J. Med.* **355,** 560–569.

14. Bowtell, D. D. (1999) Options available–from start to finish–for obtaining expression data by microarray. *Nat. Genet.* **21,** 25–32.

15. Holloway, A. J., van Laar, R. K., Tothill, R. W., Bowtell, D. D. (2002) Options available–from start to finish–for obtaining data from DNA microarrays II. *Nat. Genet.* **32 Suppl,** 481–489.

16. Dalma-Weiszhausz, D. D., Warrington, J., Tanimoto, E. Y., Miyada, C. G. (2006) The affymetrix GeneChip platform: an overview. *Methods Enzymol.* **410,** 3–28.

17. Novak, J. P., Sladek, R., Hudson, T. J. (2002) Characterization of variability in large-scale gene expression data: implications for study design. *Genomics* **79,** 104–113.

18. Quackenbush, J. (2002) Microarray data normalization and transformation. *Nat. Genet.* **32 Suppl,** 496–501.

19. Yang, Y. H., Dudoit, S., Luu, P., et al. (2002) Normalization for cDNA microarray data: a robust composite method addressing single and multiple slide systematic variation. *Nucleic Acids Res.* **30,** e15.

20. Park, T., Yi, S. G., Kang, S. H., et al. (2003) Evaluation of normalization methods for microarray data. *BMC Bioinformatics* **4,** 33.

21. Morrison, D. A., Ellis, J. T. (2003) The design and analysis of microarray experiments: applications in parasitology. *DNA Cell. Biol.* **22,** 357–394.

22. Yang, Y. H., Speed, T. (2002) Design issues for cDNA microarray experiments. *Nat. Rev. Genet.* **3,** 579–588.

23. Churchill, G. A. (2002) Fundamentals of experimental design for cDNA microarrays. *Nat Genet* **32 Suppl,** 490–495.

24. Pan, W., Lin, J., Le, C. T. (2002) How many replicates of arrays are required to detect gene expression changes in microarray experiments? A mixture model approach. *Genome Biol.* **3,** research0022.

25. Lee, M. L., Kuo, F. C., Whitmore, G. A., Sklar, J. (2000) Importance of replication in microarray gene expression studies: statistical methods and evidence from repetitive cDNA hybridizations. *Proc. Natl. Acad. Sci. USA* **97,** 9834–9839.

26. Novoradovskaya, N., Whitfield, M. L., Basehore, L. S., et al. (2004) Universal Reference RNA as a standard for microarray experiments. *BMC Genomics* **5**, 20.
27. Miller, L. D., Long, P. M., Wong, L., et al. (2002) Optimal gene expression analysis by microarrays. *Cancer Cell* **2**, 353–361.
28. Gaasterland, T., Bekiranov, S. (2000) Making the most of microarray data. *Nat. Genet.* **24**, 204–206.
29. Kerr, M. K., Churchill, G. A. (2001) Experimental design for gene expression microarrays. *Biostatistics* **2**, 183–201.
30. Kerr, M. K., Churchill, G. A. (2001) Statistical design and the analysis of gene expression microarray data. *Genet. Res.* **77**, 123–128.
31. Simon, R., Radmacher, M. D., Dobbin, K. (2002) Design of studies using DNA microarrays. *Genet. Epidemiol.* **23**, 21–36.
32. Yang, P., Sun, Z., Aubry, M. C., et al. (2004) Study design considerations in clinical outcome research of lung cancer using microarray analysis. *Lung Cancer* **46**, 215–226.
33. Ma, X. J., Wang, Z., Ryan, P. D., et al. (2004) A two-gene expression ratio predicts clinical outcome in breast cancer patients treated with tamoxifen. *Cancer Cell* **5**, 607–616.
34. Clarke, P. A., te Poele, R., Workman, P. (2004) Gene expression microarray technologies in the development of new therapeutic agents. *Eur. J. Cancer* **40**, 2560–2591.
35. Nelson, P. S. (2004) Predicting prostate cancer behavior using transcript profiles. *J. Urol.* **172**, S28–32; discussion S33.
36. Mischel, P. S., Cloughesy, T. F., Nelson, S. F. (2004) DNA-microarray analysis of brain cancer: molecular classification for therapy. *Nat. Rev. Neurosci.* **5**, 782–792.
37. Lee, C. H., Macgregor, P. F. (2004) Using microarrays to predict resistance to chemotherapy in cancer patients. *Pharmacogenomics* **5**, 611–625.
38. Perou, C. M., Sorlie, T., Eisen, M. B., et al. (2000) Molecular portraits of human breast tumours. *Nature* **406**, 747–752.
39. Ramaswamy, S., Tamayo, P., Rifkin, R., et al. (2001) Multiclass cancer diagnosis using tumor gene expression signatures. *Proc. Natl. Acad. Sci. USA* **98**, 15,149–15,154.
40. Sorlie, T., Perou, C. M., Tibshirani, R., et al. (2001) Gene expression patterns of breast carcinomas distinguish tumor subclasses with clinical implications. *Proc. Natl. Acad. Sci. USA* **98**, 10,869–10,874.
41. Wang, Y., Klijn, J. G., Zhang, Y., et al. (2005) Gene-expression profiles to predict distant metastasis of lymph-node-negative primary breast cancer. *Lancet* **365**, 671–679.
42. Dai, H., van't Veer, L., Lamb, J., et al. (2005) A cell proliferation signature is a marker of extremely poor outcome in a subpopulation of breast cancer patients. *Cancer Res.* **65**, 4059–4066.
43. van 't Veer, L. J., Dai, H., van de Vijver, M. J., et al. (2002) Gene expression profiling predicts clinical outcome of breast cancer. *Nature* **415**, 530–536.
44. Fournier, M. V., Martin, K. J., Kenny, P. A., et al. (2006) Gene expression signature in organized and growth-arrested mammary acini predicts good outcome in breast cancer. *Cancer Res.* **66**, 7095–7102.
45. Sgroi, D. C., Teng, S., Robinson, G., et al. (1999) In vivo gene expression profile analysis of human breast cancer progression. *Cancer Res.* **59**, 5656–5661.
46. Allinen, M., Beroukhim, R., Cai, L., et al. (2004) Molecular characterization of the tumor microenvironment in breast cancer. *Cancer Cell* **6**, 17–32.
47. Klein, C. A., Seidl, S., Petat-Dutter, K., et al. (2002) Combined transcriptome and genome analysis of single micrometastatic cells. *Nat. Biotechnol.* **20**, 387–392.
48. Park, T., Yi, S. G., Lee, S., et al. (2003) Statistical tests for identifying differentially expressed genes in time-course microarray experiments. *Bioinformatics* **19**, 694–703.
49. Churchill, G. A. (2004) Using ANOVA to analyze microarray data. *Biotechniques* **37**, 173–175, 177.

50. Hughes, T. R., Marton, M. J., Jones, A. R., et al. (2000) Functional discovery via a compendium of expression profiles. *Cell* **102**, 109–126.
51. Taxman, D. J., MacKeigan, J. P., Clements, C., Bergstralh, D. T., Ting, J. P. (2003) Transcriptional profiling of targets for combination therapy of lung carcinoma with paclitaxel and mitogen-activated protein/extracellular signal-regulated kinase kinase inhibitor. *Cancer Res.* **63**, 5095–5104.
52. Nasr, R., Rosenwald, A., El-Sabban, M. E., et al. (2003) Arsenic/interferon specifically reverses 2 distinct gene networks critical for the survival of HTLV–1-infected leukemic cells. *Blood* **101**, 4576–4582.
53. Zhelev, Z., Bakalova, R., Ohba, H., et al. (2004) Suppression of bcr-abl synthesis by siRNAs or tyrosine kinase activity by Glivec alters different oncogenes, apoptotic/antiapoptotic genes and cell proliferation factors (microarray study). *FEBS Lett.* **570**, 195–204.
54. Ruddy, M. J., Wong, G. C., Liu, X. K., et al. (2004) Functional cooperation between interleukin-17 and tumor necrosis factor-alpha is mediated by CCAAT/enhancer-binding protein family members. *J. Biol. Chem.* **279**, 2559–2567.
55. Oishi, K., Miyazaki, K., Kadota, K., et al. (2003) Genome-wide expression analysis of mouse liver reveals CLOCK-regulated circadian output genes. *J. Biol. Chem.* **278**, 41,519–41,527.
56. Kwak, M. K., Wakabayashi, N., Itoh, K., et al. (2003) Modulation of gene expression by cancer chemopreventive dithiolethiones through the Keap1-Nrf2 pathway. Identification of novel gene clusters for cell survival. *J. Biol. Chem.* **278**, 8135–8145.
57. van de Vijver, M. J., He, Y. D., van't Veer, L. J., et al. (2002) A gene-expression signature as a predictor of survival in breast cancer. *N. Engl. J. Med.* **347**, 1999–2009.
58. Righetti, P. G. (1990) Recent developments in electrophoretic methods. *J. Chromatogr.* **516**, 3–22.
59. Bjellqvist, B., Ek, K., Righetti, P. G., et al. (1982) Isoelectric focusing in immobilized pH gradients: principle, methodology and some applications. *J. Biochem. Biophys. Methods* **6**, 317–339.
60. Rabilloud, T., Vuillard, L., Gilly, C., Lawrence, J. J. (1994) Silver-staining of proteins in polyacrylamide gels: a general overview. *Cell Mol. Biol. (Noisy-le-grand)* **40**, 57–75.
61. Clauser, K. R., Baker, P., Burlingame, A. L. (1999) Role of accurate mass measurement (+/- 10 ppm) in protein identification strategies employing MS or MS/MS and database searching. *Anal. Chem.* **71**, 2871–2882.
62. Petricoin, E. F., Zoon, K. C., Kohn, E. C., Barrett, J. C., Liotta, L. A. (2002) Clinical proteomics: translating benchside promise into bedside reality. *Nat. Rev. Drug Discov.* **1**, 683–695.
63. Weston, A. D., Hood, L. (2004) Systems biology, proteomics, and the future of health care: toward predictive, preventative, and personalized medicine. *J. Proteome Res.* **3**, 179–196.
64. Sun, Z., Fu, X., Zhang, L., et al. (2004) A protein chip system for parallel analysis of multitumor markers and its application in cancer detection. *Anticancer Res.* **24**, 1159–1165.
65. Nomura, F., Tomonaga, T., Sogawa, K., et al. (2004) Identification of novel and downregulated biomarkers for alcoholism by surface enhanced laser desorption/ionization-mass spectrometry. *Proteomics* **4**, 1187–1194.
66. Volmer, M. W., Radacz, Y., Hahn, S. A., et al. (2004) Tumor suppressor Smad4 mediates downregulation of the anti-adhesive invasion-promoting matricellular protein SPARC: Landscaping activity of Smad4 as revealed by a "secretome" analysis. *Proteomics* **4**, 1324–1334.
67. Krah, A., Schmidt, F., Becher, D., et al. (2003) Analysis of automatically generated peptide mass fingerprints of cellular proteins and antigens from *Helicobacter pylori* 26695 separated by two-dimensional electrophoresis. *Mol. Cell Proteomics*, 1271–1283.
68. Brown, P.O. (2006). Exploring along a crooked path. *Am. J. Human Genet.* **79**, 429–433.

4

From Microarrays to Gene Networks

Hasan H. Otu and Towia A. Libermann

Summary

Understanding the roles and functions of genes and proteins through their interactions with each other and the environment has been reshaped with technological advancements such as gene chips and protein arrays. These techniques simultaneously probe thousands of molecules at any given time. Interrogating the network as opposed to a single entity as in traditional methods necessitates a departure from reductionism and requires developing biological insight in a networks setting. A fundamental challenge is to develop computational methods to analyze this vast amount of data and transform it into meaningful biological knowledge. Because of the nature of the data and the system under investigation, this goal can be accomplished by considering high-throughput data analysis in the context of biological networks. In this chapter, we describe the foundations of this methodology through an overview of the problems encountered along the way and a summary of basic biological and mathematical concepts.

Key Words: Functional genomics; high-throughput; microarray; network; time series.

1. Introduction

In the early years of DNA sequencing, the objective was to sequence only a fragment of the chromosome. With the discovery of new techniques in automated sequencing (such as shotgun sequencing), biologists were able to obtain longer DNA sequences, even the whole chromosome. In 1995, the complete genome of *Haemophilus influenza* was sequenced, which marked the first genome sequence of any free-living organism. It was soon followed by the whole genomes of other bacteria and eukaryotes such as *Saccharomyces cerevisiae* (yeast), *Caenorhabtidis elegans* (nematode), and *Drosophila melanogaster* (fruit fly). In 2001, the long-awaited working draft of the human genome was sequenced separately by a private genomics company and the public human genome project.

Following the advances in sequencing of genetic information, biology has become an increasingly information-intensive discipline. One of the most important consequences of having whole genome sequences of various organisms is the development of high-throughput systems such as microarrays that can measure the expression of thousands of genes in a given sample. This new technology has been placed at the heart of functional genomics, which tries to understand the functionality and interrelationships of genes identified through sequencing projects. Microarray technology has been applied to finding differentially expressed genes in two different biological states (e.g., normal cells vs tumor cells, treated cells vs untreated cells, or cells with a certain mutated gene vs cells

From: *Bioarrays: From Basics to Diagnostics*
Edited by: K. Appasani © Humana Press Inc., Totowa, NJ

with no mutation), characterizing groups of genes that define a certain type of cells or a clinical phenotype, and identifying gene regulatory mechanisms.

From a biological perspective, microarray technology provides a peek at the functional organization of the cell. It gives a snapshot of the enormous interaction network that defines the working mechanisms in the cell by measuring thousands of data points at a given time. This technology creates the possibility (and challenge) to reverse engineer biological networks by using high-throughput systems such as by using gene expression data to infer genetic regulatory networks *(1)*. The tools for this new quest come from graph theoretic analysis of networks, which was initiated by Euler in the 18th century and further developed by Erdös and Rènyi in the second part of 20th century.

In the 1990s, "the science of a connected age" has rekindled the attention in graph theory *(2)*, and the effort of trying to reassemble the pieces of nature divided and conquered through reductionism has found new pastures in business, the World Wide Web, biology, intelligence, and so on in the context of "networks." As new theoretical methods became available, properties of real networks have been investigated intensively *(3)*. In these analyses, networks are parameterized using measures that characterize the topology, interconnectivity, and density of the network. This new and exciting research provides solid means to understand the functional organization of living cells.

Analysis of biological networks has provided insight to the scientific community. The emerging theme in these networks is the existence of hubs (highly connected members in a graph); tight local clustering of entities, suggesting modularity; and relatively short distances between any two elements, suggesting robustness against errors in the network. The next step is trying to analyze microarray data in a networks setting, which is a natural extension of the properties of the technology. The focus in analyzing microarray data in a networks setting has been on extracting possible causal correlative relationships between gene expression levels to be used in dynamic models for time series data sets *(4-6)*. However, these algorithmic approaches are at early stages, and there is still a need to integrate their results with other data types and with existing biological networks.

In this chapter, we discuss the framework through which functional genomics problems can be studied using a graph theoretic approach. We give an overview of the technology and the computational challenges faced in analyzing microarray data. We also establish the basic foundation for the mathematical concepts used in studying large-scale characteristics of cellular networks and provide results of such research efforts. Finally, the application of microarray data analysis in a networks setting is discussed, and pointers for current challenges and future research are given.

2. Background

2.1. Biological Background

Proteins are building blocks of all living cells and include many substances necessary for the proper functioning of the cell. The main step that precedes protein synthesis is called transcription, during which gene sequences on the DNA are copied to the mRNA molecules. Transcription is followed by translation, which consists of using the sequence on messenger RNA (mRNA) as a template to link the amino acids carried by tRNA molecules to make the protein. Microarrays measure gene expression levels by quantifying the abundance of mRNA molecules in a given sample, a measure that correlates with protein expression in all but few cases. This process is done in a parallel manner by simultaneously measuring thousands of mRNA levels on a small surface called a chip.

The chip surface is generally made of glass and contains spots, each of which holds thousands of probes representing a gene. Probes are cDNA copies of the corresponding sequences of genes. Fluorescent dye-labeled cDNA or cRNA representations of cellular mRNA are hybridized to the probes on the chip, and each mRNA molecule binds to its corresponding spot. A laser excitement of the chip generates an illumination at each spot, which is then scanned to generate a digital image. The premise is that the more mRNA molecules in the sample, the more hybridization that occurs at the corresponding spot, which yields a brighter region on the scanned image. The final digital image is partitioned using a grid to determine the intensity value for each probe.

Two types of chips have emerged as popular methods to measure gene expression: cDNA or oligonucleotide (oligo) microarrays and synthetic oligo microarrays (primarily used by Affymetrix, Santa Clara, CA). The cDNA microarray is built using a glass slide where spotting of cDNAs or oligos is done by high-speed robots *(7)*. The oligo microarray uses a silicon chip where the synthesis and immobilization of the oligonucleotide sequences on the chip are done using photolithography, similar to the process used in producing microchips in the electronics industry *(8)*.

The first step in data analysis of microarray experiments is extracting probe intensities from digital images that represent hybridization strength. There have been different strategies for cDNA *(9,10)* and oligo microarrays *(11,12)*, accounting for fitting a grid to the image and calculating the intensity for each spot. Going from raw probe intensities to a representative signal value for the expression of each gene requires normalization. The goal of normalization is to minimize the effects of experimental errors and to make the signal values extracted from two different chips (or dyes in cDNA arrays) comparable. These experimental errors can be because of chemical properties (e.g., cystine [Cy]3 is consistently less intense than Cy5); human interventions; or differences in time, place, or batch involved in running the chips *(13)*.

The simplest normalization method assumes equal amounts of mRNA hybridization on the chips and therefore brings the average signal values obtained from the chip images to the same level. An improvement to this idea is to fit a linear regression on the scatter plot of the gene expression levels and to use these estimated parameters to normalize the two chips. The premise behind this approach is that most of the gene expression does not change between the two conditions. These ideas have been extended to include nonlinear transformations and nonparametric regression techniques *(9,13)*. Normalization can be applied to either a subset (e.g., housekeeping genes or genes in a certain interval of the expression range) or to the ensemble of the genes on the chip. In experiments involving more than two chips, duplicate measurements, or more than two experimental conditions, more complicated and specific methods are required *(9,14)*.

The aforementioned techniques are applied to the signal values obtained after the image analysis. In cDNA arrays, this step is more straightforward, because there is only one spot per gene. For Affymetrix oligo arrays, more complicated techniques exist for signal value calculation and normalization, because there are several probe pairs that measure a gene's expression level. These methods have been justified, particularly because different probes for a given gene show different affinities to their target mRNA molecule.

In addition to the Affymetrix Microarray Suite 5.0 *(15)* algorithm, the two prevailing techniques are dChip *(16,17)* and robust multichip average (RMA) *(18)*, which are applied to the probe level data. Although comparable, performance differences between the three approaches have been demonstrated, especially for detection and comparison of genes with low expression values *(19)*. Affymetrix Microarray Suite 5.0 and dChip provide comparison of two conditions wherein the former technique can compare the expression levels of the genes on two chips at a time, and the latter technique can use any number of chips in the two conditions. The RMA method has not been designed to make direct comparisons between the chips, but you can use the signal values obtained with RMA in applying other algorithms for comparing gene expression levels.

Although often overlooked, one of the most critical steps in using microarrays is the right experimental design to exploit the technique's potential in answering a scientific question. Because of differences in probe design and target preparation, cDNA and oligo arrays can require peculiar attention based on the experimental conditions. For example, in cDNA arrays you can compare two samples directly on a chip (competitive hybridization), or you can do an indirect comparison, where the two samples are hybridized to two different chips by using the same reference sample. Other issues to be considered are number of replicates required for each condition and instances where pooling of samples is required, either owing to insufficient amounts of mRNA or financial restrictions *(20)*. Design of experiments depends on the method of analysis to be applied to the chip results, which makes the problem more difficult because there is no agreed upon workflow on the steps involving choice of platform, method of signal extraction, normalization, data preprocessing, filtering, comparative analysis, and so on.

When two well-defined conditions exist, the main problem in microarray analysis is finding genes whose expression levels have changed significantly. There are numerous methods used for this purpose, ranging from simple -fold change analysis to more complicated statistics techniques such as permutation tests or Bayesian methods *(21,22)*. When there are multiple conditions, the problem can be grouped in two subtopics: class prediction and class discovery. Here, we refer to the word "class" in the sense of a phenotype that defines a condition in the experiment, for example, tumor samples in a cancer study.

In class prediction, the goal is to find a set of genes that can successfully predict samples from a certain condition in the experiment. The preferred approach is to have independent training and test data sets. The training data set is used for identifying the predictive genes, and these genes are applied blindly on the test data set to validate the accuracy of the predictor. However, because of financial and biological restrictions, experiments involving large amounts of chips are not always possible. In such cases, you can resort to the idea of leave-one-out cross-validation. This procedure starts with a number of classes and a set of genes. A sample is left out, and a predictor set of genes that differentiate between the groups is built. The sample that is left out is then classified as one of the groups by using the predictive genes. This predictor is cycled through all samples individually. The accuracy of the predictor is assessed by the total number of correct predictions; however, this approach has the danger of overfitting the model to data and enhances the results artificially.

Class discovery rests on the assumption that molecular profiles, which are gene expression patterns in microarray analysis, can help discover new classes that are otherwise unknown. The main difference between these two approaches is the labeling of the samples. In class prediction, the assignment of samples to categories is known, and the goal is to find classification rules by using the available data. In class discovery, the samples are not allocated into groups, and the problem requires finding clustering methods to organize the sample set based on the molecular profiles. The approaches used to handle class prediction and discovery are supervised and unsupervised learning techniques, respectively.

Prominent supervised learning methods used in microarray analysis are Bayesian classifiers, linear discriminant analysis, nearest neighbor classification trees, and support vector machines *(23)*. In the unsupervised learning counterpart, the prevailing methods have been hierarchical clustering, k-means clustering, self-organizing maps, principal components analysis, and variations of these techniques *(22,24)*. These algorithms can be used to cluster the samples, genes, or both used in the experiment. Having identified gene profiles that correlate with differential gene expression, class prediction, or class discovery leads to the most challenging questions of biological pathways and gene network relationships that have only recently started to be unraveled.

One important type of microarray data that has posed a challenge to the scientific community is the analysis of time-series data, which consists of transcription profiling of one or more experimental conditions at different time-points. Such temporal profiling can be used to identify the regulatory relations between genes and conditions. Most supervised and unsupervised learning methods are not particularly suited for time-series data, because these methods use distance metrics that are not sensitive to the order of inputs, such as Euclidean distance or Pearson's correlation. Moreover, these methods assume some sort of independence between the classes (and samples within classes) that may not hold true for time-series microarray data. Finally, these generic algorithms lack inherent dynamic modeling, which constitutes the very nature of temporal profiling experiments.

Time-series experiments provide an opportunity to consider high-throughput data analysis in a networks model. Under **Subheading 2.2.**, we provide basic mathematical concepts to form the computational foundation required for such approaches.

2.2. Mathematical Background

The computational tools at our disposal to study networks use ideas from graph theory. Graph theory is a branch of mathematics concerned with how networks can be encoded and their properties measured. Here, we briefly summarize some of the basic and key definitions and results.

- Graph: A graph G is a set of vertices (nodes) $V = \{v_1, \ldots, v_N\}$ connected by edges (links) $E = \{e_1, \ldots, e_M\}$. Hence, $G = (V,E)$.

- Vertex (node): A node v is a terminal point of a graph. It is the abstraction of an entity such as a gene or protein.

- Edge (link): An edge e is a link between two nodes. The link $e_k = (i, j)$ represents an edge from node i *to* node j.

  An edge can be either directed or undirected. An undirected edge does not convey any information regarding the direction of the interaction. For example "binding" relation between two proteins is irrelevant of the direction of the interaction, whereas in expression regulation networks, the direction between two nodes (genes) is relevant, such as a transcription factor regulating the expression of a gene.

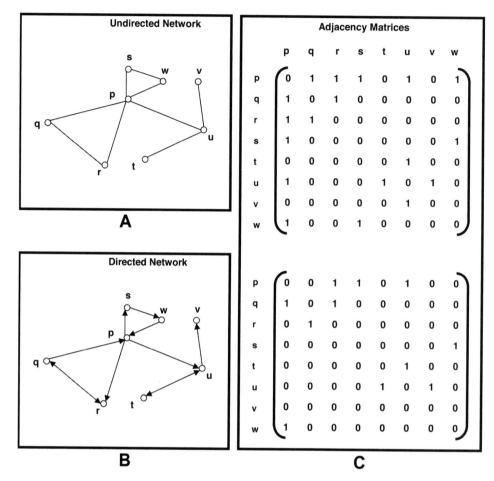

Fig. 1. Examples of undirected (**A**) and directed networks (**B**). The adjacency matrix for the undirected network (**C, top**) is symmetrical, whereas the adjacency matrix for the directed network (**C, bottom**) is not.

- (Un)directed network: A(n) (un)directed network consists of (un)directed edges only. In **Fig. 1A,B**, we show examples of undirected and directed networks, respectively. In directed networks, the direction of the interaction is indicated with an arrow. The interactions are either uni- or bidirectional.

- Adjacency matrix: An adjacency matrix is a binary $N \times N$ matrix, where N is the number of nodes in the graph. If the (i, j)th entry in the matrix is 1, then there is a link from node i to node j; otherwise, there is no link. The adjacency matrix of an undirected graph is always symmetrical.

    The adjacency matrices for the networks shown in **Fig. 1A, A** are represented in **Fig. 1C**. For example, consider the link between node p and node u in the undirected network. Because there is no direction in the edge, it is viewed as a link from p to u as well as a link from u to p. This link is represented in the corresponding adjacency matrix by placing a 1 at two positions: the cell represented by row p, column u and the cell represented by row u, column p. Repeating this process for each link results in a symmetric matrix. The same link between node p and u is unidirectional for the directed network shown in **Fig. 1B**.

Because the link goes from node p to node u only, in the corresponding adjacency matrix, we place a 1 at the cell represented by row p, column u. The cell represented by row u, column p is given a value of 0, resulting in an asymmetric matrix.

- Degree: The degree, k, of a vertex v is the number of edges touching v. If M is the total number of edges in a graph, then

$$\sum_{i=1}^{N} k_i = 2M \tag{1}$$

where k_i is the degree of the ith node v_i, and N is the total number of nodes in the graph. The average degree of a network $\bar{k}$ is $2M = N$.

For example, node p in the network shown in **Fig. 1A**, has a degree of 5, and node t has a degree of 1, i.e., $k_p = 5$ and $k_t = 1$. There are nine links and eight nodes in this network, so the average degree of the network, $\bar{k}$, is 2:25.

Degree distribution: The degree distribution, $P(k)$, is the probability that a given node has degree k. The degree distribution of a random network follows a Poisson distribution

$$P(k) \approx \frac{z^k e^{-z}}{k!} \tag{2}$$

where $z = \bar{k}$. This asymmetrical bell-shaped curve peaks around the characteristic degree or "scale" $\bar{k}$ of the graph, meaning that there are no vertices with a large degree as opposed to a scale-free network, which has a power law degree distribution $P(k) \approx k^{-\lambda}$, implying a few "hubs" in the network that have very large degrees.

Examples of these networks are shown in **Fig. 2**. **Figure 2A** shows a random network where every node has similar degree, and there are no nodes that differ significantly in connectivity. Alternatively, the scale-free network shown in **Fig. 2B** has hubs indicated by black nodes. There are only a few of these nodes, and their connectivity is significantly larger than the remaining nodes in the network.

This phenomenon is apparent from the degree distributions of the two networks. The Poisson degree distribution for the random network shown in **Fig. 2C** implies that the probability of getting a node with the "typical" degree of the network is high, and the probability of finding nodes with higher degrees converges rapidly to zero. The power law degree distribution for the scale-free network shown in **Fig. 2D** represents existence of few hubs in the network that are highly connected.

- Path length: The path length d_{ij} between two nodes is the number of links required to travel to get to node j starting from node i. Note that there can be more than one path between two nodes. For the sake of simplicity, we call d_{ij} the minimum of the lengths of such paths. In an undirected network, $d_{ij} = d_{ji}$, whereas this relationship is not necessarily true for directed networks.

- Diameter: The diameter of a network d is the average of the shortest path lengths between all possible pairs in the graph.

$$d = \frac{\sum_i \sum_j d_{ij}}{N^2} \tag{3}$$

where N is the total number of nodes in the graph. The diameter of a network characterizes its overall navigability.

- Clustering coefficient: Let n_i be the set of neighbors of the node i, i.e., the set of nodes that i is linked to. The maximum number of links between the elements of n_i is

$$M_i = \frac{|n_i|(|n_i|-1)}{2} \tag{4}$$

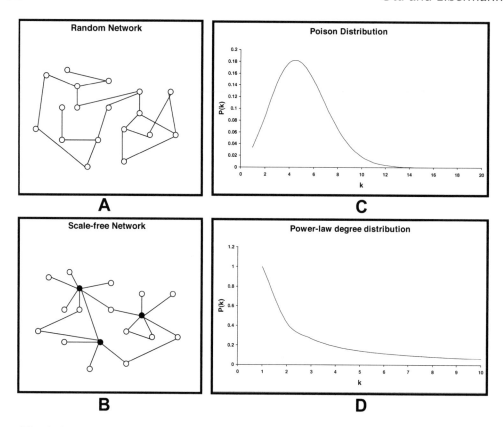

Fig. 2. Sample random network (**A**) and scale-free network (**B**). The hubs in the scale-free network are shown in black. The degree distribution for random networks is Poisson (**C**), whereas it follows a power law (**D**) for scale-free networks.

which is realized when every neighbor of i is linked to each other. Let L_i be the actual number of links between the neighbors of node i. The clustering coefficient for I is $C_i = L_i = M_i$. The average clustering coefficient for a network is defined as

$$\bar{C} = \frac{1}{N} \sum_i C_i \qquad (5)$$

For example, consider node p of the undirected network shown in **Fig. 1A**. There are five neighbors of p: nodes q, r, s, u, and w. The number of links between these nodes is 2; q is connected to r, and s is connected to w. The maximum number of possible links among these neighbors is $5 \times 4/2 = 10$. Therefore the clustering coefficient for node p, C_p, is 0.2. The clustering coefficient for node u is 0, because none of its neighbors are connected to each other, where C_q is 1 because all of q's neighbors are connected to each other.

The average clustering coefficient is a measure of the likelihood of a network to form clusters. For example, social networks have high clustering coefficients because people we know usually know each other. The clustering coefficient distribution C_k, which is defined as the average clustering coefficient of nodes with k links, is approximately k^{-1} for most real networks, indicating a hierarchical organization of the network.

Note that, $\bar{k}$, d, and $\bar{C}$ depend on the number of nodes and links of the network, whereas $P(k)$ and $C(k)$ are independent of network's size.

3. Biological Network Models

As new theoretical methods became available, characteristics of real networks have been investigated intensively *(3)*. Real networks were found to vary drastically from random networks. In these analyses, the networks are parameterized using the measures described under **Subheading 2.2.**, such as $P(k)$, $C(k)$, d, $\bar{C}$, and so on. These parameters characterize the topology, interconnectivity, and density of the network.

When metabolic networks of 43 distinct organisms were analyzed, these networks were shown to be scale-free with few highly connected nodes (hubs), yet they exhibited small-world properties *(25)*. The 43 analyzed organisms represented eukaryotes, bacteria, and archaea. The degree distribution of all the networks followed a power law, $P(k) \approx k^{-\gamma}$, with $\gamma \approx 2{:}2$, suggesting a scale-free topology. In contrast, the average clustering coefficient $\bar{C}$ for these metabolic networks was an order of magnitude larger than the average clustering coefficient of a typical scale-free network with similar size. Moreover, the clustering coefficient remained unchanged for the metabolic networks when the sizes of the networks changed. This finding was in contrast to the exponential decay in $\bar{C}$ expected in scale-free networks with increasing size. Another common property of the analyzed networks was relatively small diameters, changing between 2.5 and 4. Like $\bar{C}$, d remained virtually unchanged as the network size changed in different organisms.

These findings posed conflicting results at first, because small diameters imply small-world networks and high clustering coefficients are characteristics of tightly connected networks with high modularity, whereas power law degree distributions are specific to scale-free networks, which do not typically possess these qualities. The conservation of small diameters with increasing size of the organism's metabolic network (i.e., the organisms get more complex) can be explained by the average number of reactions a given substrate participates in increasing with the total number of substrates found in the organism. The modularity is also understandable because biological entities work in groups to perform relatively distinct biological functions. That there are indeed certain substrates that participate in a large number of reactions, such as pyruvate or coenzyme A, suggests hubs in the network, implying power law degree distributions and hence scale-free network topology.

These arguments led to the idea that metabolic networks exhibit a "hierarchical modularity" and motivated models proposed to generate such a network *(25,26)*. In this model, you can start with a tightly clustered small network having a central node and add replicates of this small network by connecting the very first central node to every peripheral node added. The proposed network thereby resembles the parametric characterizations of the analyzed metabolic networks.

The behavior seen in metabolic networks holds true for most of the biological networks analyzed *(27)*. These networks include but are not limited to protein–protein interaction networks, gene regulatory networks, and protein domain networks. One prevailing property is that these networks show topological resilience to errors. This resilience can be explained by the small diameters and hub-like structures. The deletion of a subset of the nodes does not disintegrate the whole network topology. Evolutionary development has placed alternative paths between any pair of nodes as seen in short path lengths, and the scale-free structure statistically makes it almost impossible for potential deletions to target the hubs.

It is also important to understand the processes that would result in the observed biological networks. In general, the dominating model for existence of such networks can be summarized as growth with preferential attachment, a dynamic growth model in which nodes that have high degrees are more likely to attract links as new nodes are added to the network. For example, a careful inspection of metabolic networks reveals that the remnants of the RNA world are among the most connected nodes.

4. Analysis of Gene Expression Data in a Networks Setting

There is a growing interest in using gene expression data to construct gene interaction networks. These networks have direct implications on protein networks, because proteins are end results of gene expression. Microarray data from several different publicly available data sets have been used to form networks by using coregulated genes *(28)*. In this setting, the expression profiles of genes that are close in the Euclidean sense are linked to each other. The resulting networks in all analyzed data sets have revealed small-world networks with scale-free properties. In disease-related studies, such as identifying mechanisms related to cancer, these results have suggested that the cause for malfunctioning is tightly linked to disruption in the functionality of genes that act as hubs in the gene expression networks. These genes either regulate the expression of a large number of genes or connect several large clusters in the network, each of which represents a gene regulatory pathway. Alternatively, the robustness because of the scale-free character and the ability to have fast response and quick adaptability because of small-world effects fit in the evolutionary development of biological systems.

When yeast gene expression system was analyzed in 273 gene deletion mutants *(29)*, the resulting networks were shown to exhibit very similar behaviors. In this data set, the deletion of a gene represents "poking" a network component, and the set of genes that change because of this perturbation implies a network connection. An expression network was constructed based on this premise by using the transcriptional profiling data obtained for each deletion and by comparing them to a control sample. In addition to the concordant results of the topology of this network to other previously analyzed biological networks, the yeast expression network was shown to be more tightly connected than other biological networks.

Networks analyses also have been applied to time-series data from gene expression experiments. These analyses provides a rich framework because time-series data are difficult to analyze using classical methods, and network analyses enable the use of dynamic models to infer causal relationships. Herein, we present details of such an analysis.

Let the column vector $\overrightarrow{a(t)}$ of size m represent the expression levels of the genes at time t. We assume that the expression level of gene i at time t, $a_i(t)$, is a linear combination of expression levels of all genes at time $t-1$. This assumption is equivalent to modeling the data by a first-order Markov process and can be represented as

$$a_i(t) = \sum_{j=1}^{m} \lambda_{ij} a_j(t-1) \tag{6}$$

This defines an $m \times m$ transition matrix Λ, whose elements λ_{ij} are assumed to be time independent. Note that λ_{ij} represents the weight of the effect of gene j at time $t-1$ on gene i at time t. If we have n samples in our time series data set, the aforementioned equation can be written as

$$A(t) = \Lambda A(t\text{-}1) \tag{7}$$

where

$$A(t) = \left[\overrightarrow{a(2)}, \overrightarrow{a(3)}, ..., \overrightarrow{a(n)} \right] \text{ and } A(t\text{-}1) = \left[\overrightarrow{a(1)}, \overrightarrow{a(2)}, ..., \overrightarrow{a(n-1)} \right] \tag{8}$$

To calculate for Λ, we need to multiply—on the right—both sides of the aforementioned equation by the inverse of the matrix $A(t-1)$. However, $A(t-1)$, which is of size $m \times n$, is almost never a square matrix, because we usually have tens of thousands of genes (m) as opposed to tens of samples (n).

Using singular value decomposition, we can represent A in terms of its eigenvectors. Let U and V be the matrices of left and right eigenvectors of A, respectively, and let E be the diagonal matrix of eigenvalues of A. Then, A = UEVT, where VT is the transpose of the matrix V. Substituting this representation in the aforementioned equation (after applying singular value decomposition on $A(t-1)$ and using an approximate least squares solution, we find that

$$\Lambda = A(t) V E^{-1} U^T \tag{9}$$

The model can be further developed to account for nonlinear kinetic terms. Consider the covariance matrix between samples (time-points) $Ct = A^T(t-1)A(t-1)$ and between genes $Cg = A(t-1)A^T(t-1)$. We can now rewrite our model as

$$A(t) = \Lambda_1 A(t\text{-}1) + \Lambda_2 C_t + C_g \Lambda_3 \tag{10}$$

In this model, Λ_1 also accounts for the first-order Markov model, and Λ_2 and Λ_3 account for nonlinear terms contained in the time and gene correlations, respectively.

Any of the aforementioned transition matrices, which define a corresponding relationship between genes measured on the array, can be thought of as an adjacency matrix. This process is done by using a cutoff, where entries below it are set to 0 and entries above it are set to 1. In this way, we are able to view the interaction between gene expression in a networks setting. Methods along the lines of this approach have been applied to diauxic shift and cell cycle of yeast expression data (*30,31*) as well as a model of T-cell activation data (*32*). These applications (*4–6*) use state–space models to reverse engineer the transcription regulation networks.

The conclusions regarding the network properties in these studies were similar to earlier analysis of other biological networks. The resulting networks showed high clustering coefficients, implying a tightly clustered modularity; low diameters, suggesting small-world effects; and power law degree distributions, showing a scale-free behavior in the underlying network. The methods suggested to simulate the formation of networks with the aforementioned properties were also in concordance with earlier work. These methods, in short, follow a dynamic model in which the network evolution is governed by the "growth with preferential attachment" rule.

5. Conclusions

Network theory provides a very promising framework to study biological systems and will have broad applications for microarray analysis. This framework is especially important because biology is constantly undergoing paradigm shifts, particularly with the fast advancements in technology, which makes it possible to interrogate the biological entity in question from different angles with tens of thousands of data points. Moreover, graphs are the natural choice of method to approach biological systems because a particular function in a cell results from interactions between a number of molecules. Microarrays provide a promising tool in analyzing biological systems in a networks setting.

Although there are still problems to overcome, such as standardization of the experimental and analysis methods, the ability to take a snapshot of the gene expression network holds great potential to understand the underlying network's dynamics. In this chapter, we presented such a framework in the context of analyzing time-series data. This method answers a significant challenge in microarray data analysis and brings together two quickly developing fields. From the results of different approaches mentioned in this chapter, there is an undeniable emerging theme in the cell's functional organization. Whether it be metabolic networks, protein–protein interactions, or gene regulation networks, similar properties surface. These networks have power law degree distributions, suggesting highly connected hubs. This distribution in turn brings robustness to the system, because if a node is removed randomly, it is unlikely to be a hub. Unless this happens, the system can still hold its integrity and functionality. Other common characteristics are high clustering coefficients, implying a modular structure, and small diameters, resulting in fast reaction and adaptation.

Despite promising results, the field is still at its infancy, and there are big challenges lying ahead. The most important of these challenges is development of new nomenclature and algorithms to define the finer structures of the networks. For example, the disassortativity in cellular networks still remains unexplained. Another open challenge is finding ways to partition the different modules in the network. A potential lead in this direction is identifying motifs occurring in a given network topology, but it is unclear how these often overlapping modules function and interact.

Network biology feeds itself from accurate and precise data acquisition. Another obstacle along the way to understanding biological networks is development of sensitive tools for data quantification at high resolution with reduced noise. Finally, data integration becomes an important area to provide a complete story. Almost all the approaches discussed in this chapter consider a certain kind of biological network in a given condition. However, a cell's state, context of physical existence, and factors as such do affect the different interactions being observed. Thus, we need a semantic integration at the data level to simultaneously analyze the large amount of knowledge obtained to represent these interactions.

References

1. van Someren, E. Wessels, L., Backer, E., and Reinders, M. (2002) Genetic network modeling. *Pharmacogenomics* **3**, 507–525.
2. Watts, D. J. (2004) *Six Degrees*, Norton, New York, NY.
3. Barabasi, A.-L. (2003) *Linked*, Penguin Group, New York, NY.
4. Dewey, T. G., and Galas, D. J. (2001) Dynamic models of gene expression and classifcation. *Funct. Integr. Genomics* **1**, 269–278.
5. Bhan, A., Galas, D. J., and Dewey, T. G. (2002) A duplication growth model of gene expression networks. *Funct. Integr. Genomics* **18**, 1486–1493.

6. Rangel, C., Angus, J., Ghahramani, Z., et al. (2004) Modeling T-cell activation using gene expression profiling and state-space models. *Bioinformatics* **20,** 1361–1372.

7. Shena, M., Shalon, D., Davis, R. W., and Brown, P. O. (1995) Quantitative monitoring of gene expression patterns with a complementary DNA microarray. *Nucleic Acids Res.* **270,** 467–470.

8. Ekins, R., and Chu, F. W. (1999) Microarrays: their origins and applications. *Trends Biotechnol.* **17,** 217–218.

9. Yang, Y. H., Dudoit, S., Luu, P., and Speed, T. P. (2001) Normalization for cDNA microarray data. In: *Microarrays: Optical Technologies and Informatics* (M. L. Bittner, Y. Chen, A. N. Dorsel, and E .R. Dougherty, eds.), vol. 4266, Proceedings of Society of Photo-Optical Instrumentation Engineers, Bellingham, WA.

10. Smyth, G. K., Yang, Y. H., and Speed, T. (2003) Statistical issues in cDNA microarray data analysis. In: *Functional Genomics: Methods and Protocols* (M. J. Brownstein and A. B. Khodursky, eds.), Humana, Totowa, NJ.

11. Schadt, E. E. Li, C., Su, C., and Wong, W. H. (2000) Analyzing high-density oligonucleotide gene expression array data. *J. Cell. Biochem.* **80,** 192–202.

12. Zuzan, H., Blanchette, C., Dressman, H., et al. (2001) Estimation of probe cell locations in high-density synthetic-oligonucleotide DNA microarrays. Technical report, Institute of Statistics and Decision Sciences, Duke Univerity, Durham, NC.

13. Quackenbush, J. (2001) Computational analysis of microarray data. *Nat. Rev. Genet.* **2,** 418–427.

14. Golub, T. Slonim, D. Tamayo, P., et al. (1999) Molecular classification of cancer: class discovery and class prediction by gene expression monitoring. *Science* **286,** 531–537.

15. Affymetrix. (2002) Statistical algorithms description document. http://www.affymetrix. com/support/technical/whitepapers/sadd whitepaper.pdf.

16. Li, C., and Wong, W. H. (2001) Model-based analysis of oligonucleotide arrays: expression index computation and outlier detection. *Proc. Natl. Acad. Sci. USA* **98,** 31–36.

17. Li, C., and Wong, W. H. (2001) Model-based analysis of oligonucleotide arrays: model validation, design issues and standard error application. *Genome Biol.* **2,** research 0032.1– 0032.11.

18. Irizarry, R. A., Hobbs, B., Collin, F., et al. (2003). Exploration, normalization, and summaries of high density oligonucleotide array probe level data. *Biostatistics* **4,** 249–264.

19. Barash, Y., Dehan, E., Krupsky, M. W., et al. (2004) Comparative analysis of algorithms for signal quantitation from oligonu-cleotide microarrays. *Bioinformatics* **20,** 839–846.

20. Yang, Y. H., and Speed, T. P. (2002) Design issues for cDNA microarray experiments. *Nat. Rev. Genet.* 3, 579–588.

21. Dudoit, S., Yang, Y., Callow, M., and Speed, T. (2002) Statistical methods for identifying differentially expressed genes in replicated cDNA microarray experiments. *Stat. Sin.* **12,** 111–139.

22. Sebastiani, P., Gussoni, E., Kohane, I. S., and Ramoni, M. (2003) Statistical challenges in functional genomics. *Stat. Sci.* **18,** 33–70.

23. Dudoit, S. Fridlyand, J., and Speed, T. (2002) Comparison of discrimination methods for the classifcation of tumors using gene expression data. *J. Am. Stat. Assoc.* **97,** 77–87.

24. Slonim, D. K. (2002) From patterns to pathways: gene expression data analysis comes of age. *Nat. Genet.* **32,** 502–508.

25. Jeong, H., Tombor, B., Albert, R., Oltvai, Z. N., and Barabasi, A.-L. (2000) The large-scale organization of metabolic networks. *Nature* **407,** 651–654.

26. Ravasz, E., Somera, A. L., Mongru, D. A., Oltvai, Z. N., and Barabasi, A.-L. (2002) Hierarchical organization of modularity in metabolic networks. *Science* **297,** 1551–1555.

27. Barabasi, A. L., and Oltvai, Z. N. (2004) Network biology: understanding the cell's functional organization. *Nat. Rev. Genet.* **5,** 101–113.

28. Agrawal, H. (2002) Extreme self-organization in networks constructed from gene expression data. *Phys. Rev. Lett.* **89,** 268702.
29. Featherstone, D. E., and Broadie, K. (2002) Wrestling with pleiotropy: genomic and topological analysis of the yeast gene expression network. *Bioessays* **24,** 267–274.
30. DeRisi, J. L., Iyer, V. R., and Brown, P. O. (1997) Exploring the metabolic and genetic control of gene expression on a genomic scale. *Science* **278,** 680–686.
31. Spellman, P. T., Sherlock, G., Zhang, M. Q., et al. (1999) Comprehensive identifcation of cell cycle-regulated genes of the yeast *Saccharomyces cerevisiae* by microarray hybridization. *Mol. Biol. Cell* **9,** 3273–3297.
32. Rangel, C., Wild, D. L., Falciani, F., Ghahramani, Z., and Gaiba, A. (2001) Modelling biological responses using gene expression profiling and linear dynamical systems. In: *Proceedings of the 2nd International Conference on Systems Biology*, pp. 248–256.

PART II

BIOMARKERS AND CLINICAL GENOMICS

Krishnarao Appasani

Gene expression studies in neuronal cells is the principal focus of Part II, which details approaches for isolating neuronal-specific and aging-specific biomarkers. In addition, Part II describes the isolation of biomarkers in various cancers, endothelial cells, and infectious diseases. Brain is the most complex organ in the body, possessing many anatomically distinct regions. Neighboring cells, especially neurons, can have divergent patterns of gene expression because of their neuronal connections and anatomical localizations. Generally, the enrichment procedures that apply to the various populations in blood tissue are not applicable to the isolation of neuronal cells, which are divergent in function. Therefore, laser capture microdissection has been used as an alternative approach to isolate cell subpopulation or pools of single cells from fixed brain tissue sections. In Chapter 5, Williams et al. describe results from microarray analysis performed on these isolated neurons, which allowed them to detect low-abundance neuronal-specific genes.

Many features of normal aging of the brain and central nervous system are common to diverse mammalian species, including atrophy of pyramidal neurons, synaptic atrophy, deterioration in memory functions, and personality disorders. Aging of the brain is a major risk factor for neurological and psychiatric disorders and is a process associated with impairments in cognitive and motor function. A better understanding of the molecular effects of aging in the brain may help to reveal processes that lead to age-related brain dysfunction. Gene expression profiling in such aging processed tissues may reveal components involved in aging that may be potential therapeutic targets for pathological aging. To understand the biology of aging, Erraji-Benchekroun et al. describe in Chapter 6 a systematic implementation of the microarrays approach in the normal brain, and they identify putative markers during the selective reorganization of glial and neuronal cells.

Tumor-specific molecular markers that identify circulating cancer cells in peripheral blood or tumor-associated proteins in serum are useful for early diagnosis and prognosis. Papillary thyroid carcinoma is the most common type of malignant thyroid tumor, representing 80 to 90% of all thyroid malignancies. In Chapter 7, Uchida describes microarray analyses of thyroid cancers, identifies several potential biomarkers, and validates them with immunohistochemical methods. Multidrug-resistant tuberculosis has become a global threat, and it is important to understand its molecular basis so that better treatment regimens can be developed. In Chapter 8, Bashyam and Hasnain compare the cDNA approach with array-based comparative genomic hybridization, especially in the characterization of mycobacterial strains.

Endothelial cells line the inside of all blood and lymphatic vessels, forming structurally and functionally heterogeneous populations of cells in a large network of vascular channels. To understand the disease biology of endothelial cells, Chi et al. in Chapter 9 undertake a gigantic study of gene expression patterns between macrovascular and microvascular and between arterial and venous endothelial cells. They describe how a higher number of vein-specific genes were identified compared with artery-specific genes.

5

Reduction in Sample Heterogeneity
Leads to Increased Microarray Sensitivity

Amanda J. Williams, Kevin W. Hagan, Steve G. Culp, Amy Medd,
Ladislav Mrzljak, Tom R. Defay, and Michael A. Mallamaci

Summary

DNA microarrays are most useful for pharmacogenomic discovery when a clear relationship can be made between gene expression in a targeted tissue and drug affect. Unfortunately, the true target of the drug affect is most often a subpopulation of cells within the tissue. Thus, when heterogeneous tissues containing many diverse cell types are profiled, expression changes, especially in low-abundance genes, are often obscured. In this chapter, two examples are presented where a cellular subpopulation is isolated from its complex background, with minimal cellular activation, resulting in increased microarray detection sensitivity. In the first example, erythrocytes (the most abundant cell population in blood) were removed or whole blood was immediately stabilized before RNA isolation. The removal of erythrocytes resulted in a twofold increase in the detectability of leukocyte-specific genes. During the study, protocols for RNA isolation from rat blood were validated. In addition, a list of 91 genes was generated whose expression correlated with the level of erythrocyte contamination in rat blood. In the second example, laser microbeam microdissection (LMM) was used to isolate a specific neuronal population. Our LMM amplification technique was first validated for reproducibility. After validation, data obtained from pooled neurons, cortical tissue slices, and whole brain were compared. Overall, 20% of the transcripts detected in whole brain and 13% of the transcripts detected in tissue slices were not detected in LMM neurons. Many of these transcripts were specific to neuroglial support cells or noncortical neurons, verifying that our LMM technique captured only the neurons of interest. Conversely, 10% of the transcripts detected in LMM neurons were not detected in cortical tissue slices, and 14% were not detected in whole brain. As expected, these transcripts were neuronal specific and were presumably still present in the broader tissue regions. However, in neurons isolated by LMM, the effective concentration of these previously undetectable transcripts was raised because of the elimination of competing signal noise from extraneous cell types, reinforcing the claim that microdissection can be used to increase microarray sensitivity.

Key Words: Brain; blood; detection sensitivity; DNA microarray; erythrocyte; laser capture microdissection; leukocyte; neuron.

1. Introduction

Gene expression profiling by using DNA microarrays has become an integral part of basic and applied research in both the academic and industrial scientific communities. This technology has been successfully used for many distinct applications ranging from disease classification and functional genomics to pharmacogenomics biomarker identification and single-nucleotide polymorphism analysis *(1)*. Because of the relatively

From: *Bioarrays: From Basics to Diagnostics*
Edited by: K. Appasani © Humana Press Inc., Totowa, NJ

large quantity of starting material necessary for microarray use, initial studies primarily focused on animal disease models or cultured cells. Profiling experiments on human tissue required macrodissected regions to generate sufficient starting material. Unfortunately, owing to the heterogeneous mixture of cells present in complex tissues, it is difficult for such studies to detect genes that are expressed at low levels or within rare subpopulations. When genes are detectable, it is difficult to compare the relative levels of gene expression between two or more samples. Part of the problem is the extraneous signal noise contributed by cell types that do not express the genes of interest. In addition, variability in the cellular composition of each sample can obscure changes that are occurring within one cell type. Recent technical advances in small-scale RNA isolation and amplification, laser microdissection, and RNA stabilization have now made it possible to stratify, with minimal cellular activation, specific cell populations within complex tissue samples. Thus, the expression profiles of cellular subpopulations previously lost in the transcriptional complexity of heterogeneous tissues can now be uncovered.

In this chapter, we provide two examples in which the reduction of biological heterogeneity within a sample is accompanied by an increase in gene expression detectability by using microarrays. In the first example, the advantages and disadvantages of reducing cellular heterogeneity in whole blood are explored. In the second example, the challenges and benefits of coupling capture of neurons by laser microdissection to microarray gene expression analysis are examined. In both examples, restricted subpopulations of cells are effectively isolated from heterogeneous tissues, revealing previously undetectable transcripts. The two examples detail different methods to reduce sample heterogeneity without a major disruption in the native environment of a cell.

The aim of the first study was to identify a biomarker from whole blood. Because blood is easy to access, it is an attractive tissue for the identification and subsequent use of biomarkers. In whole blood, leukocytes (specifically the lymphocyte subpopulation) were the cells of interest. Because whole blood is a complex mixture composed of many distinct cell types at various levels of differentiation, some form of enrichment or isolation was necessary. An evaluation of the effectiveness of that enrichment procedure is presented here.

The most abundant cell type in blood is the red blood cell (erythrocyte), constituting 5.0 to 5.5×10^6 cells/mm^3 of whole blood (2). In comparison, white blood cells (leukocytes) make up only 5000 to 9000 cells/mm^3 of blood (2). Numerous techniques have been developed to purify leukocyte fractions from whole blood. Two of the most common methods are density gradient separation and red blood cell-specific lysis, both of which require multiple manipulations of the cells outside of their natural environment. Unfortunately, studies have shown these manipulations can lead to gene expression changes. For example, peripheral blood mononuclear cell density separation by using Ficoll-Hypaque results in upregulation of cytokines, implying preactivation of immune cells (3). Alteration of leukocyte size and buoyant density also has been observed (4). Cell separation techniques are not the only variables that can affect blood phenotype. Differences in blood collection methods, such as withdrawal speed, presence of anticoagulants, and storage time and temperature, also have been linked to white blood cell activation, gene expression changes, and reduced cell viability (3–8).

To avoid unintended gene expression changes, whole blood can be stabilized after withdrawal by adding a phenol/guanidine isothiocyanate lysis solution (TRIzol LS; Invitrogen, Carlsbad, CA) or by collecting the blood into PAXgene (PreAnalytiX, distributed by QIAGEN, Valencia, CA) tubes containing an RNA stabilization reagent. Compared with prolonged blood storage in EDTA tubes, both stabilization methods have been shown to eliminate or reduce alterations in expression for several genes *(9,10)*. Although immediate whole blood stabilization avoids introducing gene expression artifacts because of blood handling and processing, enrichment for the white blood cell of interest is no longer possible.

Biomarker discovery for CNS begins with model organisms, such as the rat, in which peripheral biomarker expression can be linked to relevant mechanistic changes in the brain. Because most blood RNA isolation methods are validated only for human samples and animal blood often has an increased cell density compared with human blood, we first validated the use of PAXgene collection tubes by comparing the RNA quality and gene expression profiles obtained from rat samples with those from human samples. In the definitive experiment of the first study, profiles from rat whole blood prepared using the PAXgene system were compared with a purified subset of cells where the majority of erythrocytes had been selectively lysed using the QIAamp RNA blood mini kit (QIAGEN). Subsequently, the confounding of microarray results by variability in blood cell purification efficiency also was addressed.

The aim of the second study was to identify genes that are specific to pyramidal neurons in the prefrontal cortex. Although this goal was more straightforward than that of the lymphocyte experiment, the technical challenges were significantly greater. The brain is the most complex organ in the body, possessing many anatomically distinct regions *(11)*. Neighboring cells, especially neurons, can have divergent patterns of gene expression because of their neuronal connections and anatomical localizations *(12)*. Highlighting the problem with profiling complex tissues such as brain, one study estimated that when gross dissections are used, only 30% of the transcripts expressed in rat hippocampus are reliably detected (using Affymetrix oligonucleotide arrays, Affymetrix, Santa Clara, CA) *(13)*. The majority of the detected genes were of medium and high abundance. This study and others like it highlight the need for a method to selectively enrich for cell types of interest while maintaining the in vivo character of the gene expression originating from the targeted cells. The approach that was used for the erythrocytes is not practical here, because there is no way to selectively lyse or bulk separate different populations of neurons without profoundly disturbing gene expression. Laser capture microdissection (LCM), which was developed to isolate cell subpopulations or pools of single cells from fixed tissue sections *(14)*, offers a way forward. Commercial instruments are available from Arcturus Engineering (Molecular Devices, Inc., Mountain View, CA) (www.arctur.com), and a similar microdissection technology, laser microbeam microdissection (LMM), is available from P.A.L.M. Microlaser Technologies AG (Bernreid, Germany) (www.palm-microlaser.com). The earliest and by far the main application of LCM coupled with microarray expression profiling has been the identification of marker genes for the early detection and prevention of cancer *(15)*, but the list has grown to include a wider range of applications across numerous diseases *(16–19)*.

The secondary purpose of this study, and the focus of this chapter, was to provide evidence that isolation of a defined neuronal subtype from its complex tissue background will increase the sensitivity of the microarray platform. This improved sensitivity would be observed as an increase in the detection of low-abundance neuronal-specific genes that are normally undetectable using traditional sample preparation. To accomplish this goal, many validation experiments were required to ensure that the microdissection technique generated sufficient amounts of quality total RNA and that the amplification method was both reproducible and robust. Using LMM, individual human pyramidal neurons from the layer III prefrontal cortex (Brodmann area 9) were captured and pooled. As a control, RNA was purified from 10 μM prefrontal cortex (PFC) tissue slices. Samples of pooled single cells, PFC slices, and whole brain tissue were then processed and profiled using DNA oligonucleotide microarrays. The absence, presence, and relative levels of expression, for several verified neuronal and neuroglia cell markers in the different sample preparations, were determined.

2. Materials and Methods

2.1. Sample Staining, LMM, and RNA Preparation

Fresh frozen, unfixed brain tissue (postmortem interval of 15–20 h) was obtained from the Harvard Brain Tissue Resource Center (Belmont, MA). Sections (10 μM) were adhered to glass slides with a PEN membrane (P.A.L.M. Microlaser Technologies AG). Slides were stored at –70°C and warmed at room temperature for 1 h before use. Slides were then fixed with 70% ethanol, stained with 1% methylene blue, and rinsed briefly with water followed by 70% ethanol. The slides were dried in a 37°C oven for 15 min before laser microdissection. LMM followed by laser pressure catapulting (LPC) was performed using a P.A.L.M. microbeam laser dissection system. Pyramidal neurons of layer III BA9 human cortex were excised and catapulted into caps containing 20 µL of lysis buffer from the Mini RNA isolation kit (Zymo Research, Orange, CA). Caps were snapped onto tubes containing an additional 30 µL of lysis buffer. The cell lysate was vortexed vigorously, spun down, and frozen on dry ice until completion of the preparation. No effort was made to include axons or dendrites in the cell capture; therefore, each "cell" is actually a partial cell body, representing considerably less than a single neuron's cytoplasmic content. RNA was purified from lysates containing approx 90–140 partial neurons by using the Mini RNA isolation kit. RNA quality was determined using the Bioanalyzer 2100 eukaryote RNA pico assay (Agilent Technologies, Palo Alto, CA). RNA was considered to be of sufficient quantity and quality if 18S and 28S rRNA bands were visualized.

For isolation of RNA from human blood, samples from five individuals were drawn into PAXgene blood RNA tubes (PreAnalytiX, QIAGEN). Blood from six rats was collected in either PAXgene blood RNA tubes or EDTA vacutainer tubes. The PAXgene tubes were stored at –70°C until processing. RNA was isolated from the EDTA anticoagulated blood by using the QIAamp RNA blood mini kit protocol (QIAGEN). RNA from the samples collected in the PAXgene tubes was purified using the PAXgene blood RNA kit. To ensure RNA quality, 28S/18S ribosomal RNA ratios (~2 for undergraded RNA) was determined using the Agilent 2100 Bioanalyzer eukaryote total RNA nano assay.

2.2. Labeled Target Preparation and GeneChip Hybridization

For GeneChip hybridization of blood samples, 3 µg of human blood total RNA or 5 µg of rat blood total RNA was processed according to the Affymetrix eukaryotic target preparation protocol (GeneChip Expression Analysis Technical Manual, Affymetrix). As a quality check, the smear size distribution of cRNA before and after fragmentation was evaluated using the Agilent 2100 Bioanalyzer eukaryote total RNA nano assay.

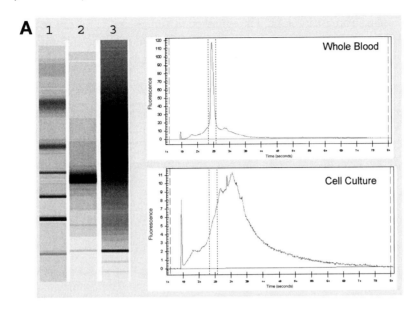

Fig. 1. Labeled cRNA target was visualized using the Agilent 2100 Bioanalyzer. *(Caption continued on next page.)*

Samples were hybridized to probe arrays for 16 h at 45°C with rotation. Human blood samples were hybridized to the HGU95Av2 GeneChip, representing approx 10,000 full-length genes, whereas rat blood samples were hybridized to the RGU34A Gene-Chip, representing approx 7000 full-length genes and 1000 expressed sequence tag clusters. Probe arrays were visualized using an antibody amplification step. Arrays were scanned and analyzed using Affymetrix Microarray Suite 5.0. GeneChips were scaled using all probe set scaling; the trimmed mean signal of each probe array was adjusted to a target intensity of 100. For a GeneChip to be included in our subsequent analysis, quality metrics (i.e., scale factor, noise, background, percentage of present calls, glyceraldehyde-3-phosphate dehydrogenase, and β-actin 3×/5× ratios) for each probe array type had to be consistent and fall within a predetermined acceptable range.

Using TaqMan quantitative real-time PCR, we estimated the LMM-LPC samples to contain 2–5 ng of RNA. The RNA from each sample was subjected to two rounds of T7-directed RNA amplification by using a protocol outlined in the Affymetrix GeneChip Eukaryotic Small Sample Target Labeling Assay, Version II Technical Note (Affymetrix, part no. 701265 rev. 3). As an experimental control, human brain total RNA was diluted, and 5 ng was amplified in parallel with the laser-microdissected samples. Samples were hybridized to the HGU133A GeneChip, representing approx 22,000 well-characterized human genes, and the data were collected and analyzed as outlined for the blood samples.

3. Results

3.1. Gene Expression Profiling of Human and Rat Whole Blood

Total RNA isolated from human and rat blood by using the PAXgene system showed no evidence of degradation. Compared with human blood, total RNA yields from rat blood were approx 10 times higher. Gel visualization of the biotin-labeled cRNA targets generated for GeneChip hybridization showed an atypical dominant band of 680–

1100 base pairs (bp) (**Fig. 1A**). This band represented >40% of the total cRNA area. Hemoglobin α and β gave the highest microarray signals for both human and rat blood and are presumably represented in the dominant cRNA target peak.

3.2. Expression Profiling of Whole Blood Vs Erythrocyte-Depleted Blood

Total RNA yields from rat blood samples prepared using the QIAamp RNA blood mini kit were 64% lower than the same samples prepared using PAXgene. RNA was intact for all samples, regardless of isolation procedure (data not shown). The QIAamp cRNA exhibited a small to absent atypical dominant band compared with labeled cRNA targets generated from whole blood PAXgene samples. GeneChips hybridized with whole blood samples (PAXgene) reported 48% fewer detectable cells compared with hybridizations in which the majority of erythrocytes were removed (QIAamp). Although the same blood was used to prepare RNA with either the QIAamp or PAXgene procedure, signal data correlated poorly between the two methods ($R^2 = 0.41$). In contrast, blood RNA isolated from separate rats by using the same isolation procedure gave a high signal correlation ($R^2 = 0.97$). To visualize both the significance and magnitude of change between whole blood and erythrocyte-depleted blood, a volcano plot was generated where the negative log of p values between the two conditions was plotted against the log 2 of -fold change (**Fig. 2**). From the volcano plot, a subset of known genes were identified that showed significant, robust signal differences between the two blood RNA isolation procedures. The signal intensity for genes encoding proteins specific to erythrocytes was much higher in the whole blood samples. Removal of the majority of erythrocytes from blood samples resulted in the detection of genes specific to leukocytes. Many of these genes were in the low-signal range and were not detectable in the samples from which erythrocytes were not depleted.

3.3. Effect of Variable Erythrocyte Contamination on Microarray Data

Analysis of biotin-labeled cRNA product with the Agilent Bioanalyzer indicates the efficiency of erythrocyte specific lysis by using the QIAamp blood RNA isolation kit varies between samples. Specifically, the percentage of the total cRNA area in the 680–1100-bp peak of the Bioanalyzer electropherograms varied among multiple RNA isolations (**Fig. 1B**). In six representative blood samples, the percentage of the total cRNA

Fig. 1. (*Continued from previous page*) (**A**) Compared with cRNA generated from a purified cell population (lane 3), the cRNA from whole blood showed an atypical dominant band of 680–1100 bp (lane 2). This band constituted >40% of the total cRNA area and represented overabundant red blood cell message. Lane 1 contained RNA marker (0.02, 0.2, 0.5, 1.0, 2.0, and 4.0 kilobases). (**B**) (*Opposite page*) Labeled cRNA target generated from whole blood where the majority of red blood cells were selectively lysed (QIAamp) showed differences in the percentage of the total cRNA area occupied by the 680- to 1100-bp band (i.e., 16–23%). This difference reflected variability in the efficiency of red blood cell lysis and resulted in different degrees of red blood cell-contaminating message. (**C**) *(See page 68.)* Probe sets whose signal showed a positive linear correlation ($R^2 = 0.9725$) with the percentage of the total cRNA area occupied by the 680- to 1100-bp band and a significant overexpression in whole blood (PAXgene) vs red blood cell-depleted samples (QIAamp) were identified as markers for the level of erythrocyte contamination. Aquaporin 1 (Aqp1) is presented as an example of a probe set that met the criteria.

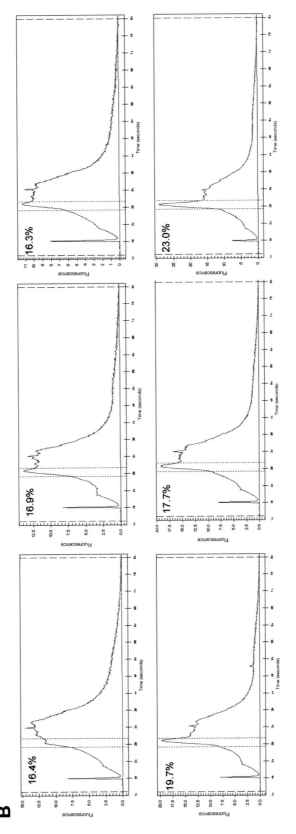

67

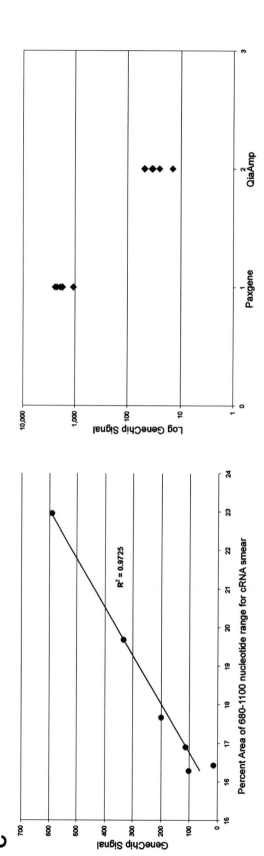

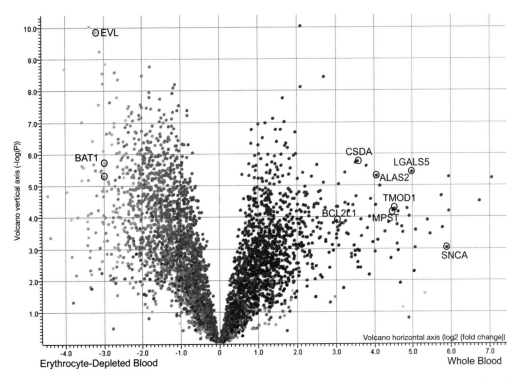

Fig. 2. To visualize both the significance and magnitude of change between whole blood (PAXgene) and erythrocyte-depleted blood (QAIamp), a volcano plot was generated where the negative log of *p* values between the two conditions was plotted against the -fold change (log 2 scale). Genes that were not detectable in any of the samples were removed from the plot. The color scale reflects the GeneChip signal intensity associated with each gene. Dark colors are associated with high-signal genes and light colors are associated with low-signal genes. Many red blood cell-specific or -enriched genes were highly overabundant in the whole blood profiled samples. The erythrocyte-depleted samples included many low-signal genes that were not detectable when whole blood was profiled.

area in this region ranged from 16 to 23%. For 184 probe sets, the GeneChip signal intensity correlated with the percentage of area of the 680- to 1100-bp region ($p < 0.05$ linear fit) (**Fig. 1C**). The intensity of 79 of the 84 probe sets was significantly increased in the whole blood vs erythrocyte-depleted samples ($p < 0.0001$ paired *t*-test) (**Fig. 1C**). An additional 12 probe sets were identified as erythrocyte specific by an annotation search. These 91 probe sets make up a list of genes associated with erythrocyte contamination in total RNA preparations from rat blood (**Table 1**).

3.4. Reproducibility of LMM-LPC Coupled to GeneChip Microarray Analysis

A series of experiments were performed to ascertain the overall reproducibility of our LMM-LPC–coupled small sample prep microarray procedure. Multiple samples of pyramidal neurons from PFC were obtained from the same brain tissue to compare the variability between samples processed in parallel or on separate days. For each sample analyzed, 80 to 125 cell captures were performed. The same operator was

Table 1
A List of 49 Known Genes Associated With the Level of Erythrocyte Contamination in Total RNA Preparations From Rat Blood[a]

Probe set ID	Gene symbol	PAXgene median signal	QIAamp median signal	Probe set ID	Gene symbol	PAXgene median signal	QIAamp median signal
L12384_at	Arf5	2,711	620.1	M58364_at[b]	Gch	1,257	112.6
D90401_g_at	Af6	706.4	117	rc_AI179576_s_at[b]	Hbb	12,053	3,685
D86297_at	Alas2	17,818	1,071	rc_AI178971_at[b]	Hba1	13,395	408.5
X67948_at	Aqp1	1,861	33.88	rc_AI012802_at	Hagh	3,982	64.7
S69383_at[b]	Alox12	12,736	474.9	X06827_at	Hmbs	634.4	50.6
rc_AA891873_f_at	Atpi	1,725	327.9	rc_AI231213_at[b]	Kai1	715	164.1
rc_AA859938_at	Bnip3l	8,774	900.8	L21711_s_at	Lgals5	13,345	421.7
S78284_s_at[b]	Bcl2l1	607.9	72.37	D50564_at	Mpst	1,879	84.25
Y07704_at[b]	Best5	4,149	276.1	X89968_g_at	Napa	1,790	414
U60578cds_s_at[b]	Ca2	3,459	64.22	S61973_at[b]	Grina	7,511	655.8
M11670_at[b]	Cat	1,087	69.78	U06099_at	Prdx2	12,437	700.5
rc_AI235585_s_at	Ctsd	590.1	203.3	U73030_at	Pttg1	1,525	134.7
X76697_at	Cd52	2,490	18.7	X55660_at[b]	Pcsk3	1,514	151.9
D16237_at	Cdc25b	8,545	347.8	AF089817_i_at	Rgs19ip1	1,877	172.4
D28557_s_at	Csda	13,364	1,118	AB015191_at[b]	Rh	3,187	132.4
rc_AA891107_at	Nudt4	2,723	166.9	rc_AA859911_g_at	Siat5	147	56.83
AJ009698_at	Emb	2,951	594.5	X89225cds_s_at	Slc3a2	535.2	119.3
rc_AI231807_at	Ftl1	13,615	2,059	J04793_at[b]	Slc4a1	2,424	100.9
AB003515_at	Gabarapl2	6,099	527.1	U39549_at	Syngr1	282	70.58
rc_AA892649_at	Gabarap	4,101	1,216	L20822_at	Stx5a	694.9	196.5
D13518_at	Gata1	379.6	23.38	AF007758_at[b]	Snca	450.5	21.1
rc_H31692_at	Gerp95	225.4	42.9	M15474cds_s_at	Tpm1	465.6	57.82
S65555_at[b]	Gclm	1,897	65.63	AB005143_s_at	Ucp2	9,914	1,038
rc_AI232783_s_at	Glns	19,077	1,375	Y00350_at	Urod	896.2	89.1
X07365_s_at	Gpx1	11,812	2,639				

[a] As described previously, these genes were identified by a positive linear correlation ($R^2 = 0.9725$) with the percentage of the total cRNA area occupied by the 680- to 1100-bp band and a significant overexpression in whole blood (PAXgene) vs red blood cell-depleted samples (QIAamp).

[b] Multiple probe sets for that gene are present in list. Only one representative probe set shown.

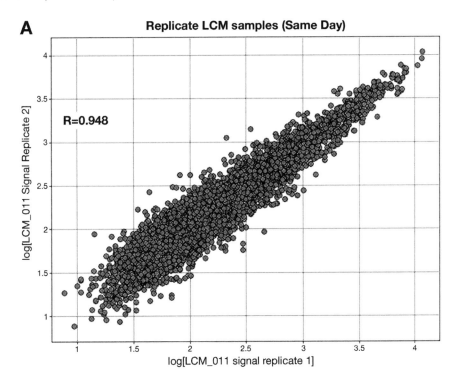

Fig. 3. The overall reproducibility of the LMM-LPC small-sample preparation procedure for microarrays was validated. Correlation values were generated using Affymetrix GeneChip signal intensity values for all probe sets called present or marginal. *(Continued on next page.)*

used throughout these experiments to minimize capturing and handling variations. Each sample was converted to labeled cRNA targets and hybridized to human U133A GeneChips. Correlation values were generated using signal intensity values for all probe sets called present or marginal. The correlation values obtained from two samples microdissected and amplified in parallel were similar to the correlation values obtained from two samples processed on different days ($R = 0.948$ and 0.932, respectively; **Fig. 3A,B**). To compare the variability of LMM-LPC with that of the amplification procedure alone, RNA purified from a large amount of tissue was diluted to 5-ng aliquots and processed in parallel by using the small-sample amplification method. Correlations for these samples were very similar to those obtained from our LMM-LPC sample replicates (**Fig. 3C**; $R = 0.954$), suggesting that the acquisition of cells via laser microdissection does not add significantly to the overall variability of the procedure.

3.5. Enrichment of Neuronal Transcripts

We performed a detailed comparison of Affymetrix GeneChip data obtained from LMM-LPC layer III BA9 cortical pyramidal neurons, an entire cortical tissue microsection before laser microdissection, and a diluted an aliquot of whole brain RNA. Over-all, our data demonstrated a reduction in sample transcript heterogeneity and an

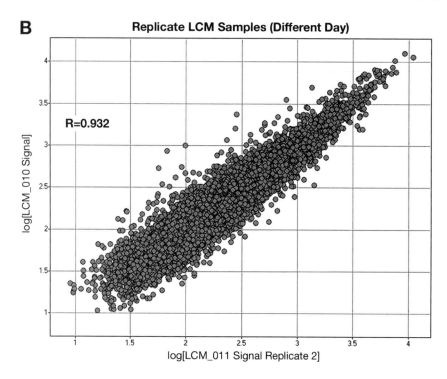

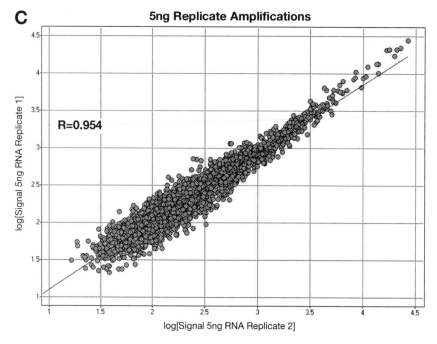

Fig. 3. *(Continued from previous page)* To determine the reproducibility of the entire LMM-LPC amplification procedure, LMM-LPC samples were processed on the same day (**B**) or on different days (**C**). Correlation values for all samples were similar, suggesting that the acquisition of cells via laser microdissection does not add immensely to the overall variability and that the small-sample preparation procedure itself is reproducible.

increase in microarray sensitivity when neurons are laser captured. Approximately 20% of the transcripts scored as present in GeneChip data obtained from whole brain were not detectable in LMM-LPC neurons. When the LMM-LPC neurons were compared with large regions of cortex layer III BA9, 13% of the transcripts that scored present in the cortex were not detectable in the neurons. Conversely, 10 and 14% of the transcripts detected in LMM-LPC neurons were not detectable in cortical tissue slice or whole brain respectively. These transcripts were detected in the neuronal samples but not in the more heterogeneous samples in which they were presumably present. This finding can by explained by an increase in their effective concentrations to the level of microarray detection because of the exclusion of competing signal noise from extraneous cell types.

A closer examination of the transcripts detected in LMM-LPC neurons, but not detected in whole brain, tissues slice, or both, was undertaken to determine whether profiling microdissected neurons indeed enriched for neuronal specific gene expression. Transcripts detected only in whole brain, tissue slices, or both also were examined to confirm that gene expression specific to neuronal support cells or noncortical neurons was absent in the microdissected neurons. The entire data set generated from these comparisons is too large to be included in this chapter; however, a sampling of the data has been tabulated to exemplify our findings **(Table 2)** *(20–58)*. For those genes reliably detected (Microarray Suite 5.0; detection *p* value < 0.05) the signal intensity is given, whereas "not detected" genes are marked as ND. All genes listed demonstrated consistent trends across replicate ($n = 3$) experiments. The majority of genes in **Table 2** show a switch in detection call from absent to present (or *vice versa*) between LCM, slice, or whole brain samples. We think these genes are the best indicators of neuronal-specific gene expression and enhanced sensitivity of the GeneChip platform because of a reduction in sample heterogeneity.

To show that a reduction in sample heterogeneity was achieved in the microdissected neuronal samples, we examined a list of well-established nonneuronal cell markers specific to neuroglia as well as cerebellum-specific neuronal markers. Included in the table are references describing the restricted expression of these genes. **Table 2** lists several known neuroglia markers such as GFAP, S100B, and SOX10 that are detected in whole brain and to a lesser extent in the PFC tissue slice but not in the LMM-LPC neurons. This observation indicates that the RNA contributions from glia and other non-neuronal support cells of the brain are excluded from the sample obtained by laser microdissection. To assess whether a reduction in sample heterogeneity was achieved on a neuronal level, several neuronal markers specific to or enriched in the cerebellum were evaluated. As expected, they were all detected in whole brain but were not detected in the LMM-LPC PFC neurons. Additionally, they were not detected in the PFC tissue slices except for ZIC1, which was detected at approx 40-fold lower levels in two of three slice samples compared with whole brain. Combined, these results strongly indicate a reduction in sample complexity has been achieved through the use of LMM-LPC.

To determine whether an increase in the detection of neuronal specific gene expression was achieved with LMM-LPC, the signal intensities and detection values for a number of genes known to be expressed in cortical neurons were compared across the LCM, slice, and whole brain samples. **Table 2** lists many examples of genes that are detected at very low levels, if at all, when macrodissected prefrontal cortex or whole

Table 2
A Reduction in Sample Heterogeneity Leads to Increased Sensitivity[a]

Probe set ID	Gene name	Gene symbol	LCM intensity	Slice intensity	Brain intensity	Reference
A. Cortical Neurons						
Neuroglia						
203540_at	Glial fibrillary acidic protein	GFAP	ND	823.5	3321	20
201162_at[b]	Insulin-like growth factor binding protein 7	IGFBP7	ND	108.4	400.3	21
201005_at	CD9 antigen; leukocyte antigen MIC3	CD9	ND	41.2	145.4	22
209686_at	S100 calcium-binding protein, β	S100B	ND	142.7	467.4	23
209842_at	SRY (sex determining region Y)-box 10	SOX10	ND	ND	373.9	24
202935_s_at[b]	SRY (sex-determining region Y)-box 9	SOX9	ND	315.8	229	25
204733_at	Kallikrein 6 preproprotein; protease M; neurosin; zyme	KLK6	ND	ND	545.2	26
209301_at	Carbonic anhydrase II	CA2	ND	53.9	169.5	27
200696_s_at	Gelsolin	GSN	ND	100.3	150.9	28
216617_s_at	Myelin-associated glycoprotein	MAG	ND	ND	104.9	29
205989_s_at[b]	Myelin oligodendrocyte glycoprotein	MOG	ND	40.3	97.4	30
Cerebellar neurons						
205747_at	Cerebellin	CBLN1	ND	ND	290.9	31
207060_at	Engrailed-2	EN2	ND	ND	115.4	32
207182_at	GABA$_A$ receptor, α6 precursor	GABRA6	ND	ND	36.6	37
206373_at	Zinc finger protein of the cerebellum 1	ZIC1	ND	79.5	3331	33
210400_at	N-methyl-D-aspartate receptor subunit 2C precursor	GRIN2C	ND	ND	195.3	34
206163_at	mab-21-like protein 1	MAB21L1	ND	ND	118.2	35
219327_s_at	G protein-coupled receptor family C, group 5	GPRC5C	ND	ND	172.2	36
B. Neuronal-specific transcripts						
Neuronal-specific ion channels						
211323_s_at[b]	inositol 1,4,5-triphosphate receptor, type 1	ITPR1	127	ND	ND	38
210123_s_at	α 7 neuronal nicotinic acetylcholine receptor	CHRNA7	111.9	ND	ND	39
214044_at[b]	ryanodine receptor 2	RYR2	307.1	195.4	ND	40

Probe ID	Gene	Description				Ref
210432_s_at	SCN3A	sodium channel, voltage-gated, type III, α	138.8	150	ND	41
205802_at[b]	TRPC1	transient receptor potential cation channel, subfamily C	207.4	178.6	ND	42
220294_at[b]	KCNV1	potassium channel Kv8.1	91.7	40.4	ND	43
206678_at	GABRA1	(GABA) A receptor, α1	934.8	508.7	ND	44
213714_at[b]	CACNB2	calcium channel, voltage-dependent, β 2 subunit	193.9	111	ND	45
206730_at[b]	GRIA3	glutamate receptor 3 isoform flop precursor	2205	1110	272.1	46
210383_at	SCN1A	sodium channel protein, brain I α-subunit	532.4	275.6	51.2	47
205358_at	GRIA2	glutamate receptor, ionotropic, AMPA 2	2106	1576	412.1	46

Neuronal-specific

G-protein coupled receptors

Probe ID	Gene	Description				Ref
206772_at	PTHR2	PTH2 receptor	124.2	97.8	ND	48
207548_at	GRM7	glutamate receptor, metabotropic 7; mGluR 7	260.7	232	ND	49
211616_s_at[b]	HTR2A	5-hydroxytryptamine (serotonin) receptor 2A	664.1	368.9	66	50

Other neuronal-specific proteins

Probe ID	Gene	Description				Ref
208320_at[b]	CABP1	calbrain	1875	721.3	398.2	51
204105_s_at	NRCAM	neuronal cell adhesion molecule	1037	933.5	188.4	52
209915_s_at[b]	NRXN1	neurexin I	366.2	215	ND	53
212990_at[b]	SYNJ1	synaptojanin 1	677.9	474.1	141.3	54
205893_at	NLGN1	neuroligin 1	89.4	50.3	22.8	55
203999_at[b]	SYT1	synaptotagmin 1	4151	3814	986.7	56
204338_s_at	RGS4	RGS4; regulator of G-protein signalling 4	220.5	82	ND	57
204715_at	PANX1	Pannexin 1	70.9	50.8	ND	58

ND, probe set not detectable.

[a] Brain reflected the most heterogeneous population of cells. Cortical slices maintained a variety of cell types, yet it contained a more focused set of transcripts. Finally, LMM-LPC captured layer III cortical neurons represented a homogeneous cell population and would be expected to lack transcripts present in the more heterogeneous samples. A sampling of probe sets was chosen to highlight differential expression between brain, cortical slices, and LMM-LPC cortical neurons. (A) Validating the homogeneity of the LMM-LPC sample for cortical neurons only, transcripts associated with nonneuronal cell types (i.e., neuroglia) or cerebellar neurons were undetectable. (B) The detection of neuronal-specific transcripts was enhanced in neurons obtained by LMM-LPC. Some neuronal specific genes were only detectable in the LMM-LPC samples (i.e., ITPR1 and CHRNA7), whereas for the majority, expression was more strongly detected in the absence of transcripts from nonneuronal cell types.

[b] Multiple probe sets for that gene are present, showing equivalent intensity patterns. Data from one representative probe set is shown..

brain is profiled. However, in the LMM-LPC samples, the signal intensity observed for these genes was increased compared with either whole brain or PFC tissue slice. Many of these examples are in gene classes such as ion channels and G protein-coupled receptors, which are generally difficult to detect using microarrays. Our localization results are consistent with published reports using other techniques such as reverse transcription-PCR, *in situ* hybridization, or immunohistochemistry (*see* **Table 2**, reference), and they reinforce the claim that with the use of microdissection, increased microarray sensitivity is achieved.

Discussion

Both examples detailed in this chapter demonstrate that a reduction in sample heterogeneity increases microarray detection sensitivity. Isolating RNA from defined cell populations with minimal cellular disruption led to more biologically meaningful expression profiles. In the first example, the detection sensitivity for genes specific to leukoctyes was increased when the majority of erythrocytes were removed. Removing the more abundant erythrocytes significantly reduced the noise and complexity of the microarray data. In the second example, microarray detection sensitivity for genes specific to pyramidal neurons was increased by laser microdissection of those specific cells from heterogeneous brain tissue. This technique, when combined with small-scale RNA isolation and amplification, generated a robust gene expression profile from a distinct population of approx 150 neurons.

To identify biomarkers from whole blood, the benefits of immediate RNA stabilization in which all cells, including erythrocytes, are profiled must be weighed against the improvement in microarray sensitivity when sample heterogeneity is reduced. For effective profiling of the leukocyte population, immature erythrocytes (reticulocytes) need to be removed (mature erythrocytes are not a concern because they do not contain RNA). Although reticulocytes constitute only 0.5–1.5% of the erythrocyte population *(2)*, their contribution to the RNA pool is significant because of the overwhelming prevalence of erythrocytes compared with other cell types. The impact of erythrocytes to the total RNA population is evident from the distinctive peak present in biotin-labeled cRNA targets generated from whole blood. This peak constitutes approx 40% of the total cRNA population and consists of erythroctye-specific messages, including TMOD1 *(59)*, MPST *(60)*, LGALS5 *(61)*, ALAS2 *(62)*, and α and β globin, the most abundant signals in profiled whole blood (**Fig. 2**). DNA microarray detection sensitivity is reduced by the overwhelming prevalence of reticulocyte message, which dilutes the contribution of leukocyte mRNA to the overall expression profile.

Sample heterogeneity was reduced by selectively lysing the majority of erythrocytes; this raised leukocyte specific messages to detectable levels. Several leukocyte-specific genes such as BAT1 *(63)* and EVL *(64)* were only detectable in the erythrocyte-lysed samples (**Fig. 2**), verifiying enrichment. Although the overall gene expression profile from multiple-lysed samples was well maintained ($R^2 = 0.97$), the efficiency of erythrocyte removal is not absolute and can introduce variability. In one of our studies, significant expression differences were identified that resulted solely from differences in erythrocyte contamination between the samples. To detect this problem, a list of genes associated with rat erythrocytes was created (**Table 1**). Tracking changes in the expression of these genes provides a quality control measure for erythrocyte contamination. By removing these genes from subsequent analyses, the introduction of erythrocyte expression artifacts is avoided.

Because many blood biomarker discovery projects begin in rodent model systems, a secondary aspect of our study was to adapt human blood RNA isolation procedures for rat. For whole blood, both human and rat exhibited similar cRNA target profiles with a prominent peak representative of erythrocyte-specific message. In an analogous study using human samples, the expression profiles from whole blood were compared with those from erythrocyte-depleted blood (Affymetrix Technical Note: An Analysis of Blood Processing Methods to Prepare Samples for GeneChip Expression Profiling). The results of this study agreed with those from the rat study; an overabundance of erythrocyte-specific message in profiled whole blood was accompanied by a reduction in microarray sensitivity. Despite the differences in species and in data analysis methods, several genes were identified as erythrocyte specific in both human and rat, including SNCA, BCL2L1, and CSDA.

To improve microarray sensitivity from whole blood while maintaining the benefits of immediate RNA stabilization, Affymetrix has developed a Globin Reduction Protocol (Affymetrix Technical Note: Globin Reduction Protocol: A Method for Processing Whole Blood RNA Samples for Improved Array Results). Oligonucleotides that hybridize to human globin mRNAs (α1, α2, and β) are added to whole blood total RNA. The RNA to DNA duplex is then digested with RNase H, resulting in cleavage of the poly-A tail from the globin mRNA, rendering it unamplifiable. GeneChip detection was greatly improved, providing further support that overabundant erythrocyte messages are responsible for microarray sensitivity loss in whole blood profiled samples.

The challenge in the second example was much greater: isolating a subpopulation of neurons defined by their morphology and anatomical location within complex post-mortem brain tissue. For this task, a different technology was used, laser microdissection. Specifically, LMM-LPC was used to isolate layer III BA9 cortical neurons. By limiting cellular disruption during tissue freezing and fixation, RNA from the multitude of neighboring neurons and nonneuronal support cells was stabilized. Thus, gene expression data from these captured cells represent a more accurate physiological "snapshot." For our method to be useful in practice, many facets of the LMM-LPC, RNA isolation, and amplification procedure were examined and optimized for small-scale samples. For example, several methods for preparing RNA from small numbers of cells were tested. RNA isolation methods using glass fiber spin columns performed better than protocols where cells were lysed and RNA was precipitated from solution. The Zymo Research Mini RNA preparation kit was finally chosen because it gave the best RNA yields from small cell numbers. RNA was reliably recovered from as few as 40 cells (data not shown).

Given the unusually small sample size, the reproducibility of the technique was critical to evaluate. We wanted to determine whether starting amounts of RNA as low as 1 ng coupled with two rounds of T7 RNA polymerase based linear amplification would significantly add to the inherent microarray variability commonly observed by our laboratory and others (65). Data obtained from replicate LMM-LPC–amplified samples showed a high degree of correlation, indicating that the method was highly reproducible (**Fig. 3**). These correlations were very close to those obtained from replicate amplifications by using diluted aliquots of a control RNA, suggesting that the LMM-LPC procedure itself does not add significant variability to the overall process. Note, our study did not address potential skewing of expression levels because of multiple rounds of amplification, a problem customarily associated with this technique *(66)*. We did show that expression data were consistent when generated from RNA samples amplified by the same method.

Once our system was validated, we explored how microarray sensitivity is affected by a reduction in sample heterogeneity. We found that 20 and 13% of the transcripts profiled were detected in whole brain and tissue slices, respectively, but not in LMM-LPC neurons. A detailed evaluation confirmed that these differences were largely attributable to genes with expression restricted to nonneuronal cell types such as GFAP and SOX10 (**Table 2**). This finding verified the specificity of our LMM-LPC technique to capture only the cortical pyramidal neurons of interest. Conversely, 10 and 14% of the transcripts detected in LMM-LPC neurons were not detectable in cortical tissue slice or whole brain, respectively. These transcripts were confirmed as neuronal specific by others using techniques such as reverse transcription-PCR, *in situ* hybridization, or immunohistochemistry (*see* **Table 2**, reference). Although presumably still present in the broader tissue regions, their effective concentrations are raised to the level of detection in the LMM-LPC neurons that lack competing signal noise from extraneous cell types. More importantly, many of these transcripts were members of the ion channel and G protein-coupled receptor gene families that are typically difficult to detect via microarrays.

Capturing individual neurons is a laborious and time-consuming process. Most experiments that couple laser microdissection with microarrays, use hundreds or thousands of neurons *(12,67)*. Recent studies have shown that capture of single neurons can be successfully coupled to transcript profiling by using glass cDNA arrays and fluorescently labeled targets *(12)*. We have not yet established the lower limit of microdissected cells required for use in our system. It is our intention to further drive down the number of neurons required for use with the Affymetrix GeneChip platform to increase our throughput. This approach will reveal additional issues, such as the diversity of gene expression because of biological variability between tissue donors.

In this chapter, we detail two experiments in which we physically separated cells from their milieu to produce a catalog of their gene expression. In each of the experiments, efforts were made to maintain the native expression profile while reducing sample complexity. Our intent is to further develop and evaluate methods that will parse the cell populations with greater precision, to identify those cells that will yield greater insight into CNS disease.

References

1. Greenberg, S. A. (2001) DNA microarray gene expression analysis technology and its application to neurological disorders. *Neurology* **57,** 755–761.
2. Bergman, R. A., Afifi, A. K., and Heidger, P. M., eds. (1999) *Atlas of Microscopic Anatomy, Section 4: Blood.* www.anatomyatlases.org.
3. Hartel, C., Bein, G., Muller-Steinhardt, M., and Kluter, H. (2001) Ex vivo induction of cytokine mRNA expression in human blood samples. *J. Immunol. Methods* **249,** 63–71.
4. Braide, M., and Bjursten, L. M. (1986) Optimized density gradient separation of leukocyte fractions from whole blood by adjustment of osmolarity. *J. Immunol. Methods* **93,** 183–191.
5. Riches, P., Gooding, R., Millar, B. C., and Rowbottom, A. W. (1992) Influence of collection and separation of blood samples on plasma IL-1, IL-6 and TNF-α concentrations. *J. Immunol. Methods* **153,** 125–131.

6. Freeman, R., Wheeler, J., Robertson, H., Paes, M. L., and Laidler, J. (1990) In-vitro production of TNF-α in blood samples. *Lancet* **336**, 312–313.

7. Tanner, M. A., Berk, L. S., Felten, D. L., Blidy, A. D., Bit, S. L., and Ruff, D. W. (2002) Substantial changes in gene expression level due to the storage temperature and storage duration of human whole blood. *Clin. Lab. Haematol.* **24**, 337–341.

8. Stordeur, P., Zhou, L., and Goldman, M. (2002) Analysis of spontaneous mRNA cytokine production in peripheral blood. *J. Immunol. Methods* **261**, 195–197.

9. Rainen, L., Oelmueller, U., Jurgensen, S., et al. (2002) Stabilization of mRNA expression in whole blood samples. *Clin. Chem.* **48**, 1883–1890.

10. Pahl, A., and Brune, K. (2002) Stabilization of gene expression profiles in blood after phlebotomy. *Clin. Chem.* **48**, 2251–2253.

11. Insel, T. R. (2004) Prologue: expression profiling within the central nervous system, part II. *Neurochem. Res.* **29**, i–ii.

12. Ginsberg, S. D., Elarova, I., Ruben, M., et al. (2004) Single-cell gene expression analysis: implications for neurodegenerative and neuropsychiatric disorders. *Neurochem. Res.* **29**, 1053–1064.

13. Evans, S. J., Datson, N. A., Kabbaj, M., et al. (2002) Evaluation of Affymetrix Gene Chip sensitivity in rat hippocampal tissue using SAGE analysis. Serial analysis of gene expression. *Eur. J. Neurosci.* **16**, 409–413.

14. Bonner, R. F., Emmert-Buck, M., Cole, K., et al. (1997) Laser capture microdissection: molecular analysis of tissue. *Science* **278**, 1481–1483.

15. Luo, J., Isaacs, W. B., Trent, J. M., and Duggan, D. J. (2003) Looking beyond morphology: cancer gene expression profiling using DNA microarrays. *Cancer Invest.* **21**, 937–949.

16. Schneider, B. G. (2004) Principles and practical applications of laser capture microdissection. *J. Histotechnol.* **27**, 69–78.

17. Jin, L., Ruebel, K. H., Bayliss, J. M., Kobayashi, I., and Lloyd, R. V. (2003) Immunophenotyping combined with laser capture microdissection. *Acta Histochem. Cytochem.* **36**, 9–13.

18. Kehr, J. (2003) Single cell technology. *Curr. Opin. Plant Biol.* **6**, 617–621.

19. Trillo-Pazos, G., Diamanturos, A., Rislove, L., et al. (2003) Detection of HIV-1 DNA in microglia/macrophages, astrocytes and neurons isolated from brain tissue with HIV-1 encephalitis by laser capture microdissection. *Brain Pathol.* **13**, 144–154.

20. Brownell, E., Lee, A. S., Pekar, S. K., Pravtcheva, D., Ruddle, F. H., and Bayney, R. M. (1991) Glial fibrillary acid protein, an astrocytic-specific marker, maps to human chromosome 17. *Genomics* **10**, 1087–1089.

21. Degeorges, A., Wang, F., Frierson, H. F., Jr., Seth, A., and Sikes, R. A. (2000) Distribution of IGFBP-rP1 in normal human tissues. *J. Histochem. Cytochem.* **48**, 747–754.

22. Banerjee, S. A., and Patterson, P. H. (1995) Schwann cell CD9 expression is regulated by axons. *Mol. Cell Neurosci.* **6**, 462–473.

23. Leal, R. B., Frizzo, J. K., Tramontina, F., Fieuw-Makaroff, S., Bobrovskaya, L., Dunkley, P. R., and Goncalves, C. A. (2004) S100B protein stimulates calcineurin activity. *Neuroreport* **15**, 317–320.

24. Kuhlbrodt, K., Herbarth, B., Sock, E., Hermans-Borgmeyer, I., and Wegner, M. (1998) Sox10, a novel transcriptional modulator in glial cells. *J. Neurosci.* **18**, 237–250.

25. Pompolo, S., and Harley, V. R. (2001) Localisation of the SRY-related HMG box protein, SOX9, in rodent brain. *Brain Res.* **906**, 143–148.

26. Yamanaka, H., He, X., Matsumoto, K., Shiosaka, S., and Yoshida, S. (1999) Protease M/neurosin mRNA is expressed in mature oligodendrocytes. *Brain Res. Mol. Brain Res.* **71**, 217–224.

27. Agnati, L. F., Tinner, B., Staines, W. A., Vaananen, K., and Fuxe, K. (1995) On the cellular localization and distribution of carbonic anhydrase II immunoreactivity in the rat brain. *Brain Res.* **676**, 10–24.

28. Lena, J. Y., Legrand, C., Faivre-Sarrailh, C., Sarlieve, L. L., Ferraz, C., and Rabie, A. (1994) High gelsolin content of developing oligodendrocytes. *Int. J. Dev. Neurosci.* **12,** 375–386.

29. Favilla, J. T., Frail, D. E., Palkovits, C. G., Stoner, G. L., Braun, P. E., and Webster, H. D. (1984) Myelin-associated glycoprotein (MAG) distribution in human central nervous tissue studied immunocytochemically with monoclonal antibody. *J. Neuroimmunol.* **6,** 19–30.

30. Pham-Dinh, D., Mattei, M. G., Nussbaum, J. L., et al. (1993) Myelin/oligodendrocyte gly-coprotein is a member of a subset of the immunoglobulin superfamily encoded within the major histocompatibility complex. *Proc. Natl. Acad. Sci. USA* **90,** 7990–7994.

31. Pang, Z., Zuo, J., and Morgan, J. I. (2000) Cbln3, a novel member of the precerebellin family that binds specifically to Cbln1. *J. Neurosci.* **20,** 6333–6339.

32. Zec, N., Rowitch, D. H., Bitgood, M. J., and Kinney, H. C. (1997) Expression of the homeobox-containing genes EN1 and EN2 in human fetal midgestational medulla and cer-ebellum. *J. Neuropathol. Exp. Neurol.* **56,** 236–242.

33. Aruga, J., Yokota, N., Hashimoto, M., Furuichi, T., Fukuda, M., and Mikoshiba, K. (1994) A novel zinc finger protein, zic, is involved in neurogenesis, especially in the cell lineage of cerebellar granule cells. *J. Neurochem.* **63,** 1880–1890.

34. Lin, Y. J., Bovetto, S., Carver, J. M., and Giordano, T. (1996) Cloning of the cDNA for the human NMDA receptor NR2C subunit and its expression in the central nervous system and periphery. *Brain Res. Mol. Brain Res.* **43,** 57–64.

35. Mariani, M., Baldessari, D., Francisconi, S., et al. (1999) Two murine and human homologs of mab-21, a cell fate determination gene involved in Caenorhabditis elegans neural develop-ment. *Hum. Mol. Genet.* **8,** 2397–2406.

36. Robbins, M. J., Michalovich, D., Hill, J., et al. (2000) Molecular cloning and characteriza-tion of two novel retinoic acid-inducible orphan G-protein-coupled receptors (GPRC5B and GPRC5C). *Genomics* **67,** 8–18.

37. Merlo, F., Balduzzi, R., Cupello, A., and Robello, M. (2004) Immunocytochemical study by two photon fluorescence microscopy of the distribution of GABA(A) receptor subunits in rat cerebellar granule cells in culture. *Amino Acids* **26,** 77–84.

38. Sharp, A. H., Nucifora, F. C., Jr., Blondel, O., et al. (1999) Differential cellular expression of isoforms of inositol 1,4,5-triphosphate receptors in neurons and glia in brain. *J. Comp Neurol.* **406,** 207–220.

39. Agulhon, C., Abitbol, M., Bertrand, D., and Malafosse, A. (1999) Localization of mRNA for CHRNA7 in human fetal brain. *Neuroreport* **10,** 2223–2227.

40. Martin, C., Chapman, K. E., Seckl, J. R., and Ashley, R. H. (1998) Partial cloning and differential expression of ryanodine receptor/calcium-release channel genes in human tis-sues including the hippocampus and cerebellum. *Neuroscience* **85,** 205–216.

41. Whitaker, W. R., Faull, R. L., Dragunow, M., Mee, E. W., Emson, P. C., and Clare, J. J. (2001) Changes in the mRNAs encoding voltage-gated sodium channel types II and III in human epileptic hippocampus. *Neuroscience* **106,** 275–285.

42. Strubing, C., Krapivinsky, G., Krapivinsky, L., and Clapham, D. E. (2001) TRPC1 and TRPC5 form a novel cation channel in mammalian brain. *Neuron* **29,** 645–655.

43. Hugnot, J. P., Salinas, M., Lesage, F., et al. (1996) Kv8.1, a new neuronal potassium chan-nel subunit with specific inhibitory properties towards Shab and Shaw channels. *EMBO J.* **15,** 3322–3331.

44. Hevers, W., and Luddens, H. (1998) The diversity of GABAA receptors. Pharmacological and electrophysiological properties of GABAA channel subtypes. *Mol. Neurobiol.* **18,** 35–86.

45. Tanaka, O., Sakagami, H., and Kondo, H. (1995) Localization of mRNAs of voltage-depen-dent Ca(2+)-channels: four subtypes of α 1- and β-subunits in developing and mature rat brain. *Brain Res. Mol. Brain Res.* **30,** 1–16.

46. Breese, C. R., Logel, J., Adams, C., and Leonard, S. S. (1996) Regional gene expression of the glutamate receptor subtypes GluR1, GluR2, and GluR3 in human postmortem brain. *J. Mol. Neurosci.* **7,** 277–289.

47. Lossin, C., Rhodes, T. H., Desai, R. R., et al. (2003) Epilepsy-associated dysfunction in the voltage-gated neuronal sodium channel SCN1A. *J. Neurosci.* **23,** 11,289–11,295.

48. Wang, T., Palkovits, M., Rusnak, M., Mezey, E., and Usdin, T. B. (2000) Distribution of parathyroid hormone-2 receptor-like immunoreactivity and messenger RNA in the rat nervous system. *Neuroscience* **100,** 629–649.

49. Makoff, A., Pilling, C., Harrington, K., and Emson, P. (1996) Human metabotropic glutamate receptor type 7: molecular cloning and mRNA distribution in the CNS. *Brain Res. Mol. Brain Res.* **40,** 165–170.

50. Chen, K., Yang, W., Grimsby, J., and Shih, J. C. (1992) The human 5-HT2 receptor is encoded by a multiple intron-exon gene. *Brain Res. Mol. Brain Res.* **14,** 20–26.

51. Yamaguchi, K., Yamaguchi, F., Miyamoto, O., Sugimoto, K., Konishi, R., Hatase, O., and Tokuda, M. (1999) Calbrain, a novel two EF-hand calcium-binding protein that suppresses Ca2+/calmodulin-dependent protein kinase II activity in the brain. *J. Biol. Chem.* **274,** 3610–3616.

52. Davis, J. Q., Lambert, S., and Bennett, V. (1996) Molecular composition of the node of Ranvier: identification of ankyrin-binding cell adhesion molecules neurofascin (mucin+/ third FNIII domain–) and NrCAM at nodal axon segments. *J. Cell Biol.* **135,** 1355–1367.

53. Ushkaryov, Y. A., Petrenko, A. G., Geppert, M., and Sudhof, T. C. (1992) Neurexins: synaptic cell surface proteins related to the α-latrotoxin receptor and laminin. *Science* **257,** 50–56.

54. Haffner, C., Takei, K., Chen, H., et al. (1997) Synaptojanin 1: localization on coated endocytic intermediates in nerve terminals and interaction of its 170 kDa isoform with Eps15. *FEBS Lett.* **419,** 175–180.

55. Ichtchenko, K., Hata, Y., Nguyen, T., et al. (1995) Neuroligin 1: a splice site-specific ligand for β-neurexins. *Cell* **81,** 435–443.

56. Brose, N., Petrenko, A. G., Sudhof, T. C., and Jahn, R. (1992) Synaptotagmin: a calcium sensor on the synaptic vesicle surface. *Science* **256,** 1021–1025.

57. Ingi, T., and Aoki, Y. (2002) Expression of RGS2, RGS4 and RGS7 in the developing postnatal brain. *Eur. J. Neurosci.* **15,** 929–936.

58. Bruzzone, R., Hormuzdi, S. G., Barbe, M. T., Herb, A., and Monyer, H. (2003) Pannexins, a family of gap junction proteins expressed in brain. *Proc. Natl. Acad. Sci. USA* **100,** 13,644–13,649.

59. Fowler, V. M. (1987) Identification and purification of a novel Mr 43,000 tropomyosin-binding protein from human erythrocyte membranes. *J. Biol. Chem.* **262,** 12,792–12,800.

60. Wlodek, L., and Ostrowski, W. S. (1982) 3-Mercaptopyruvate sulphurtransferase from rat erythrocytes. *Acta Biochim. Pol.* **29,** 121–133.

61. Gitt, M. A., Wiser, M. F., Leffler, H., et al. (1995) Sequence and mapping of galectin-5, a β-galactoside-binding lectin, found in rat erythrocytes. *J. Biol. Chem.* **270,** 5032–5038.

62. Sadlon, T. J., Dell'Oso, T., Surinya, K. H., and May, B. K. (1999) Regulation of erythroid 5-aminolevulinate synthase expression during erythropoiesis. *Int. J. Biochem. Cell Biol.* **31,** 1153–1167.

63. Allcock, R. J., Price, P., Gaudieri, S., Leelayuwat, C., Witt, C. S., and Dawkins, R. L. (1999) Characterisation of the human central MHC gene, BAT1: genomic structure and expression. *Exp. Clin. Immunogenet.* **16,** 98–106.

64. Lambrechts, A., Kwiatkowski, A. V., Lanier, L. M., et al. (2000) cAMP-dependent protein kinase phosphorylation of EVL, a Mena/VASP relative, regulates its interaction with actin and SH3 domains. *J. Biol. Chem.* **275,** 36,143–36,151.

65. Guo, Q. M. (2003) DNA microarray and cancer. *Curr. Opin. Oncol.* **15,** 36–43.

66. Galvin, J. E. (2004) Neurodegenerative diseases: pathology and the advantage of single cell profiling. *Neurochem. Res.* **29,** 1041–1051.
67. Pierce, A. and Small, S. A. (2004) Combining brain imaging with microarray: isolating molecules underlying the physiological disorders of the brain. *Neurochem. Res.* **29,** 1145–1152.

6

Genomics to Identify Biomarkers of Normal Brain Aging

Loubna Erraji-Benchekroun, Victoria Arango, J. John Mann, and Mark D. Underwood

Summary

Aging of the brain can lead to impairments in cognitive and motor skills and is a major risk factor for several common neurological and psychiatric disorders, such as Alzheimer's disease and Parkinson disease. A better understanding of the molecular effects of brain aging may help to reveal processes that lead to age-related brain dysfunction. With the need for tissue-specific aging biomarkers, several studies have used DNA microarray analysis to elucidate gene expression changes during aging in rodents and very recently in humans. The use of microarray chips allows the assessment of thousands of genes simultaneously, and the identification of new biomarkers involved in aging that may be potential therapeutic targets for pathological aging. Different quality parameters need to be examined mainly at the level of tissue collection, RNA extraction, and sample preparation and processing. Moreover, the data sets resulting from microarray chips experiments are usually complex and require increasingly powerful and refined computational competences as well as new approaches and tools of analysis. Genomic studies usually follow a pattern of analysis ranging from data extraction and statistical analysis, to gene selection and classification into specific pathways.

Key Words: Aging; brain; chips; gene expression; human; microarray.

1. Introduction

Although progress has been made understanding some of the cellular and molecular mechanisms of neurodegenerative diseases, the mechanisms of normal aging are still poorly understood and less subject to molecular analysis. To understand and measure the aging process, several molecular studies have reported on short-lived organisms such as *Drosophila melanogaster (1)*, and *Caenorhabditis elegans (2)*, and in nonhuman mammalians such as mice *(3)* and even monkeys *(4,5)*. However, molecular characterization of aging in humans, especially in the CNS, has been lacking and complicated by limited availability of high-quality human samples.

Normal brain aging of the CNS in diverse mammalian species shares many features, such as atrophy of pyramidal neurons, synaptic atrophy, decrease of striatal dopamine receptors, cytoskeletal abnormalities, reactive astrocytes and microglia *(6)*, and deterioration in memory functions and personality disorders *(7,8)*. Aging of the brain also has been associated with instability of nuclear and mitochondrial genomes *(9,10)* and leads to an increase of neuronal injury, reinnervation, and neurite extension and sprout-

From: *Bioarrays: From Basics to Diagnostics*
Edited by: K. Appasani © Humana Press Inc., Totowa, NJ

ing *(11)*. Aging also leads to increases in inflammation-mediated neuronal damage *(12)*; oxidative damage to DNA, proteins, and lipids *(13–15)*, and a stress response characterized by the induction of heat-shock factors and other oxidative stress-induced transcripts *(16)*. Neuroimaging studies have reported age-related decreases in brain volume and have suggested that the frontal lobes are the part of the brain most profoundly affected by the aging process *(17,18)*. It seems, moreover, that the disproportionate tissue loss in the frontal cortex strongly supports the frontal theory of cognitive aging *(19,20)*, which suggests that changes in frontal cortex structure, function, or both are responsible for cognitive problems often seen in older people.

Research on age-related transcript alterations in each cell type is important in clarifying mechanisms of aging. Studies of gene expression profiling have the potential for assessing thousands of genes simultaneously and for identifying components that may be potential therapeutic targets for pathological aging. However, the data sets resulting from microarray chips experiments are complex and require increasingly powerful and refined computational competences as well as new approaches and tools of analysis. The model of genomic studies analysis usually ranges from data extraction and statistics, to gene selection and classification into specific pathways.

With the need for tissue-specific aging biomarkers, several studies have used DNA microarray analysis to elucidate gene expression changes during aging in rodents *(11,16,21)* and very recently in humans *(14,15)*.

2. Molecular Characterization of Normal Aging

2.2. Oligonucleotide DNA Microarrays: Assets and Limitations

In a study of normal aging, we performed a postmortem molecular characterization of two areas of the human prefrontal cortex, dorsolateral and orbital (BA9 and BA47) of 40 subjects ranging from 13 to 79 yr old by using U133A microarray chips. DNA microarrays allow the simultaneous analysis of gene expression patterns of whole genomes *(22)* and provide a means to analyze relatively small amounts of total RNA. However, microarray chips also present some limitations, including a decreased sensitivity to detect genes with low expression levels and the inability to measure posttranslational modifications. The heterogeneity of the tissue studied (e.g., brain) is also an issue; thus, it is very important to obtain high-quality samples.

Such arrays are synthesized based on sequence information and provide the possibility of linking gene expression patterns and functional information available in public genomic databases, thereby permitting a systematic investigation of gene involvement and function in biological systems. Microarray technology relies on the semiquantitative comparison of RNA abundance between samples, which are assumed *a priori* to represent changes in gene expression or activity of the cells. Accordingly, efforts in genome sequencing and functional gene annotations are shifting the focus to a more integrated and functional view of biological mechanisms. The large amount of data generated, however, represent a considerable analytical challenge. New microarray analytical tools are being developed and genomic information gets periodically updated, making the structure of genomic data sets complex and dynamic. Currently, a large proportion of the human genome can be surveyed on a single microarray (~47,000 genes and expressed sequenced tags; U133 plus 2.0 Affymetrix GeneChip oligonucleotide DNA microarray). On a single chip, each gene is probed by 16 to 20 probe pairs known as a probe set, consisting of 25-base pair oligonucleotides corresponding to different parts of the gene sequence. In a probe pair, a perfect match oligonucleotide corresponds to the exact gene sequence, whereas the mismatch oligonucleotide differs from the perfect match by a single base in the center of the sequence. The use of probe pair redundancy to assess the expression level of a specific transcript improves the signal-to-noise ratio (efficiencies of hybridization are averaged over multiple probes), increases the accuracy of RNA quantification (removal of outlier data), and reduces the rate of false positives. The intensity information from these probes can be combined in many ways to get an overall intensity measurement for each gene, but there is currently no consensus as to which approach yields more reliable results.

2.2. Assessment of RNA Quality and Postmortem Factors

Studies including human subjects are often difficult to control compared with other organisms or cultured cell lines, because of their environment and their medical history. For example, before starting any genomic study, all the major sources of variation in gene expression in postmortem brain samples should be characterized. A critical variable in the analysis of gene expression in human pathology lies in the quality of the tissue samples examined. The RNA extraction, microarray sample preparation, and quality control procedures are performed according to the manufacturer's protocol (http://www.affymetrix.com). The integrity and purity of mRNA in the samples can be assessed by optical densitometry, by gel electrophoresis, and by the ratio of hybridization signal that is obtained between the 3' and 5' mRNA ends for control genes (3'/5' ratio for actin and glyceraldehyde-3-phosphate dehydrogenase on oligonucleotide microarrays). A ratio close to 1 indicates low or absent mRNA damage. The use of the Agilent Bioanalyzer is also a fast and easy way to obtain reliable information about RNA quality (www.agilent.com/chem/labonchip). Microarray quality parameters include control of the noise (Raw Q) to be less than 5, a background signal less than 100, a consistent scaling factor, and a consistent number of genes detected as present across arrays.

We have sought to restrict our studies only to well-characterized samples in which the subjects were psychologically characterized, with negative neuropathology and toxicology. A neuropathological examination, including thioflavine S or immunohistochemical stains for senile plaques and neurofibrillary tangles, was performed on fixed tissue samples: several cases contained plaques or neurofibrillary tangles, but never in sufficient numbers to indicate a diagnosis of Alzheimer's disease. The brain samples examined also were limited to sudden death cases to eliminate or reduce the potential confound of the agonal state, and they were analyzed for pH to obtain an index for qualifying RNA integrity. No correlation between sample variability and brain pH and postmorten interval (PMI) indicated a high RNA integrity and quality. The PMI was kept to less than 24 h. Microarray studies performed on human subjects with agonal conditions such as coma, hypoxia, hypoglycemia, and the ingestion of neurotoxic substances at time of death, reported a significant effect on RNA integrity and gene expression profiles *(23)*. Another study showed that samples with low pH displayed decreased expression of genes implicated in energy metabolism and proteolytic activities, and increased expression of genes relating to stress response and transcription factors *(24)*.

Likewise, a population-based sample will reflect the racial constitution of the region from which the sample is collected. Our sample was variably made up of Caucasians, African Americans, Hispanics, and Asians. Males were included or excluded with the same criteria as females: males ($n = 30$) did not differ significantly from females ($n = 9$) on age, race, PMI, or brain pH. PMI, brain pH, and race did not correlate with RNA quality and gene expression *(25)*. The cause of death in the sample included suicide, which in itself could be associated with gene expression abnormalities. However, we found no evidence for molecular differences that correlated with depression or suicide *(26)*. Therefore, the effect of age on gene expression was analyzed across all samples combined in one group, thereby increasing our analytical power *(15)*.

2.3. Data Extraction: Use of Sex Chromosome Genes as Internal Controls

To extract and combine probe information, alternative algorithms have been recently described. However, the reliability of these approaches has been limited to analysis based on few synthetic internal control genes. Once gene expression levels have been determined, the issue of multiple comparisons in statistical testing of large number of genes (thousands of genes) in a comparatively small number of samples (generally from two to less than a hundred) arises. One approach commonly used to avoid this issue sets empirical statistical thresholds for expression level, fold change between samples, and significance levels, based on a small number of internal controls that are added either during processing or before hybridization of the samples onto microarrays. To assess microarray data extraction procedures and develop specific and sensitive statistical analysis, we developed the use of sex chromosome genes as biological internal controls *(25)* to compare alternate probe-level data extraction algorithms. Sexual dimorphism originates in the differential expression of X- and Y-chromosome—linked genes. The expression of Y-chromosome genes is not restricted to the testes, and several genes are expressed in the male CNS *(27)*, although their function outside the testes is unknown. In our analyses of gene expression, we have compared three analytical tools to assess microarray data quality and to establish statistical guidelines for analyzing large-scale gene expression: Microarray Suite 5.0 (MAS 5.0) Statistical Algorithm from Affymetrix (Santa Clara, CA), Model Based Expression Index *(28)*, and robust multiarray average (RMA) *(29)*. The three methods were tested on our brain genomic data set by using transcripts from Y-chromosome genes as internal controls for reliability and sensitivity of signal detection. The results identified probe sets with significant sex effect in both brain areas and showed that RMA-generated gene expression values were markedly less variable and more reliable than MAS 5.0 and Model Based Expression Index-derived values *(25)*.

2.4. Statistical Analysis and Gene Selection

In microarray studies, the different steps include probe intensity extraction and statistical and bioinformatics analyses. Our human brain samples were hybridized onto Affymetrix U133A microarrays (22,283 probe sets). Signal intensities were extracted with the RMA algorithm *(29)*.

Univariate statistical tests were used to assess correlations between gene expression levels and age of subjects. To further reduce and refine the genes' list for analysis, we removed genes with low expression (i.e., expressed in 10% or fewer samples) or with a coefficient of variation below 2% (based on a log 2 scale). These criteria were selected based on the argument that genes with low expression or low variability cannot yield relationships with the age variable of interest or contribute to the group variability. In our study of aging genes, this still left 11,546 genes for statistical testing. We computed significance values for Pearson correlation coefficients, measuring linear [log 2(age) and log 2(signal)] and exponential [age and log 2(signal)] relationships between age and gene expression. The *p* values were adjusted for multiple testing by the Benjamini–Hochberg method for controlling the false discovery rate *(30)* with an experiment-wise false discovery rate at 5%. A multifactorial analysis including demographic (age, sex, and race), clinical (psychiatric diagnostic), sample (pH and PMI), and array parameters also was performed on the 20 most affected probe sets per brain area to confirm the strong effect of age. Not surprisingly, age explained more than 50% of the variation, whereas other variables (e.g., pH and PMI) explained only 1–2%, array parameters explained approximately 1–7% of the variation depending on the brain area. Our gene selection led to a list of 588 age-affected probe sets. They were clustered using Cluster and Treeview software *(31)*, and a consistent effect of aging throughout the prefrontal cortex was observed *(15)*.

2.5. Gene Expression Validation

Validation of altered gene expression detected by microarray can be performed by real-time PCR or *in situ* hybridization to look at transcription level changes, and by Western blot analysis, receptor autoradiography, or immunohistochemistry for protein expression confirmation.

Real-time PCR is a very sensitive technique that allows the detection of product amplification during the PCR by measuring the online incorporation of fluorescence that is either incorporated or released by an internal probe. In the linear range of amplification, the amount of PCR products is directly correlated with relative levels of mRNA and can therefore be used to compare expression levels either between different genes in the same sample or between same genes across different samples. *In situ* hybridization allows the cellular localization of the corresponding RNA: brains are cut on a cryostat and mounted onto glass slides and stored at –80°C until hybridized with a specific radioactive probe.

Western blot analysis permits the quantification of the corresponding protein; however, it does not give information about its localization. Quantitative receptor autoradiography can be used to determine the presence and to quantify the selected corresponding protein. Immunohistochemistry allows cellular localization of the selected protein from fixed tissue samples.

Finally, another alternative to study gene expression alterations at the protein level is proteomics. Protein microarrays include high-resolution 2D electrophoresis and mass spectrometry and are performed to isolate the proteins in a first step and then to identify them and determine their modifications.

2.6. BioInformatics

Computational methods have increasing impact because data are more complex, and large-scale methods for data generation become more prominent. To answer some of the biological questions and improve the overall understanding of cellular and molecu-

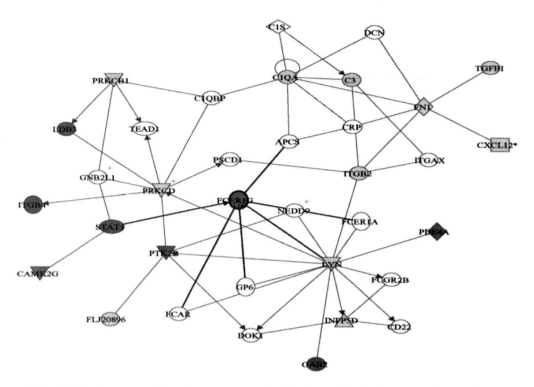

Fig. 1. Visual representation of a gene network using Ingenuity Pathways Analysis software.

lar mechanisms, refined models and softwares are developed. Genes and gene products do not function independently but instead are interrelated into networks and molecular systems. For example, it is more consistent to analyze groups or families of genes than to analyze genes individually.

One of the methods applied for this purpose is the gene "functional class scoring" method that examines the statistical distribution of individual gene scores among all genes in the gene ontology class *(32)* and does not involve an initial gene selection step *(33)*. This method assigns scores to classes or groups of genes and allows determination of the effect of age on these groups or families of genes.

Another way of classification into cellular and molecular pathways is provided by a few types of software such as the Pathway Assist software (Iobion Informatics LLC, La Jolla, CA), the Ingenuity Pathways Analysis software (Ingenuity Systems, Inc., Redwood City, CA) **(Fig. 1)**, or the geWorkBench software (http://amdec-bioinfo.cu-genome.org) to look at biological association networks and biological interactions.

3. Aging Is a Continuous and Specific Process Throughout Adult Life

3.1. Continuous Effect of Age

Following the analytical steps described above, we identified progressive changes in expression of many genes. Using our approach, we confirmed gene expression alterations at the transcriptome level in two areas of the prefrontal cortex in a large cohort of human subjects *(15)*. Our cross-sectional study in subjects covered a 66-yr age range (13–79 yr) and suggested a continuous process of aging in the brain as reflected by changes in gene transcriptome during adult life. Age-related transcriptional changes were robust and highlighted by many genes being similarly affected across two prefrontal cortex areas in a heterogeneous group of human subjects *(15)*. We found that the most age-affected gene family reported to the structure and function of nervous system involving genes implicated in synaptic transmission and signal transduction. Another well-represented gene family was related to cellular defenses and included genes implicated in inflammatory response, oxidative stress, and response to injury. Increased expression in genes and gene families associated with inflammation and oxidative stress, and decreased expression in genes involved in synapse function and integrity support the notion of loss in the functional capacity of neurons with age *(15)*. Our results confirmed studies reporting early onset and continuous rates of morphological changes *(34)* and subtle cognitive decline *(35,36)* across the life span. They also were consistent with gene expression profiling studies of the aging hippocampus in rodents, describing progressive transcriptional changes between young, middle-aged, and old animals *(37)*. The transcriptional changes in the aging human prefrontal cortex that we found confirm other studies done at the level of the transcriptome *(14,38)* and show an unbiased report of the molecular and cellular mechanisms during aging. Unlike transcriptome analyses, protein microarrays are more limited in the number of peptides identified, and they do not offer overviews of cellular functions. Nonetheless, although RNA levels may not always correlate with protein levels, numerous studies confirm our RNA results during aging at the protein or function levels, including, among others, astrocyte markers (glial fibrillary acidic protein [GFAP]; *6*), S100B *(39)*, neurotransmitter receptors (HTR2A *[40–43]*, ADRA2A *[44]*), Ca^{2+}-binding proteins (CALB1/2; *45*), enzymes (monoamine oxidase B [MAOB] *[46]*, CA4/10 *[47]*), and trophic factors (insulin-like growth factor 1, reviewed in **ref. 48**, CLU *[49]*; NTRK3 *[50]*). Gene expression alterations associated with cell signaling and transduction showed increased expression of growth factors, and decreased expression of most genes implicated in Ca^{2+} homeostasis and regulation, such as calbindin, the calcium channels, decreased expression of channels such as GABRA5, potassium and sodium channels, decreased expression of G protein-coupled receptors, neuropeptides, kinases, and phosphatases. Upregulated genes were linked to increased proinflammatory state (GFAP, complement components, and cytokines), increased oxidative stress and DNA damage (MAOB and GPX3), and antiapoptotic effort in the aging brain (BCL2, CLU, and BAD; *see* **Table 1**).

4. Summary and Conclusions

We have developed statistical and analytical approaches for microarray analysis. We demonstrated the use of sex genes as biological internal controls for genomic analysis of complex tissues. We also suggested analytical guidelines for testing alternate oligonucleotide microarray data extraction protocols and for adjusting multiple statistical analyses of differentially expressed genes.

We have applied these analytical approaches to study gene expression alterations in the human brain during aging, and we reported data suggesting an extensive and selective reorganization of glial and neuronal functions during aging, for which we provide

Table 1
Age-Related Gene Expression Alterations [a]

Affymetrix probe set	Gene symbol	Gene name	FC old vs young-BA9	FC old vs young-BA47
Structure and function of neuron/glia and synapse				
205625_s_at	CALB1	Calbindin 1, (28 kDa)	**−2.28**	**−2.40**
215531_s_at	GABRA5	γ-aminobutyric acid (GABA)$_A$ receptor, α5	**−1.86**	**−1.73**
208482_at	SSTR1	Somatostatin receptor 1	**−1.49**	**−1.34**
209372_x_at	TUBB	Tubulin, β polypeptide	**−1.43**	**−1.23**
211616_s_at	HTR2A	5HT2$_A$ receptor	**−1.43**	**−1.69**
205823_at	RGS12	Regulator of G protein signalling 12	**−1.41**	**−1.35**
209869_at	ADRA2A	Adrenergic, α-2A-, receptor	**−1.40**	**−1.38**
200951_s_at	CCND2	Cyclin D2	**−1.36**	**−1.24**
221730_at	COL5A2	Collagen, type V, α2	**−1.33**	**−1.53**
207767_s_at	EGR4	Early growth response 4	**−1.33**	**−1.22**
211651_s_at	LAMB1	Laminin, β 1	**−1.27**	**−1.23**
202154_x_at	TUBB4	Tubulin, β, 4	**−1.22**	−1.05
204685_s_at	ATP2B2	ATPase, Ca^{2+} transporting, plasma membrane 2	−1.17	**−1.26**
211577_s_at	IGF1	Insulin-like growth factor 1 (somatomedin C)	−1.17	−1.16
211535_s_at	FGFR1	Fibroblast growth factor receptor 1	**1.24**	**1.26**
212205_at	H2AV	Histone H2A.F/Z variant	**1.27**	1.03
204041_at	MAOB	Monoamine oxidase B	**1.27**	**1.28**
201302_at	ANXA4	Annexin A4	**1.32**	**1.33**
212244_at	GRINL1A	Glutamate receptor, ionotropic, *N*-methyl-D-asparate-like 1A	**1.33**	1.15
205117_at	FGF1	Fibroblast growth factor 1 (acidic)	**1.38**	**1.49**
Cellular defenses				
217767_at	C3	Complement component 3	**−1.77**	**−1.46**
203104_at	CSF1R	Colony-stimulating factor 1 receptor (Macrophage)	**−1.61**	**−1.58**
1861_at	BAD	BCL2-antagonist of cell death	**−1.33**	−1.13
203414_at	MMD	Monocyte to macrophage differentiation-associated	**−1.31**	**−1.33**
219825_at	P450RAI-2	Cytochrome P450 retinoid metabolizing protein	**−1.27**	**−1.51**
204897_at	PTGER4	Prostaglandin E receptor 4 (subtype EP4)	**−1.22**	**−1.26**
218336_at	PFDN2	Prefoldin 2	1.01	1.15
204440_at	CD83	CD83 antigen	1.16	1.08
201896_s_at	DDA3	p53-regulated DDA3	1.20	**1.26**
214428_x_at	C4A	Complement component 4A	**1.24**	**1.57**
222043_at	CLU	Clusterin (complement lysis inhibitor)	**1.30**	**1.31**
203685_at	BCL2	B-cell CLL/lymphoma 2	**1.32**	**1.48**
206118_at	STAT4	Signal transducer and activator of transcription 4	**1.37**	**1.36**
205830_at	CLGN	Calmegin	**1.42**	**1.22**
203540_at	GFAP	Glial fibrillary acidic protein	**1.77**	**1.88**

[a] Increased or decreased expression of some selected genes related to the "structure and function of neuron/glia and synapse" and "cellular defenses" families during aging. Significant chnages are shown in bold for the two areas of the prefrontal cortex.

BA9, BA47: Brodmann areas 9 and 47; FC, -fold change old vs young subjects.

putative molecular markers. The transcriptional correlates of aging seem to implicate less than 10% of all genes in the brain. This number could be explained by mRNA differences restricted to small number of cells being considerably diluted and impossible to differentiate in the overall RNA pool extracted from whole brain areas. Although we did not assess changes in protein levels, modifications, functions, or their combinations, our results support the notion that aging correlates with very specific molecular changes in the brain, as opposed to widespread and nonspecific alterations. The age-related transcriptional changes begin early in adulthood, are continuous throughout adult life, and suggest the possibility of early identification of mechanisms that are either preventive or detrimental to age-related brain functions.

Large-scale RNA monitoring methodologies have reached a level that make them amenable to the study of complex diseases in the CNS; however, functional annotations, bioinformatic analysis, and other developing experimental platforms such as protein arrays are rapidly expanding the scope of current genomic studies.

Acknowledgments

We thank Hanga Galfalvy, Paul Pavlidis, Etienne Sibille, and Peggy Smyrniotopoulos for participating in the different aspects of the project. This publication was made possible by grants 5T32MH20004, F32MH63559, R01-MH40210, and MH64168 from the National Institute of Mental Health, and National Institute of Mental Health Conte Center for the Neuroscience of Mental Disorders grant MH62185.

References

1. Helfand, S. L., and Inouye, S. K. (2002) Rejuvenating views of the ageing process. *Nat. Rev. Genet.* **3,** 149–153.
2. Hill, A. A., Hunter, C. P., Tsung, B. T., Tucker-Kellogg, G., and Brown, E. L. (2000) Genomic analysis of gene expression in *C. elegans. Science* **290,** 809–812.
3. Jiang, C. H., Tsien, J. Z., Schultz, P. G., and Hu, Y. (2001) The effects of aging on gene expression in the hypothalamus and cortex of mice. *Proc. Natl. Acad. Sci. USA* **98,** 1930–1934.
4. Kayo, T., Allison, D. B., Weindruch, R., and Prolla, T. A. (2001) Influences of aging and caloric restriction on the transcriptional profile of skeletal muscle from rhesus monkeys. *Proc. Natl. Acad. Sci. USA* **98,** 5093–5098.
5. Uddin, M., Wildman, D. E., Liu, G., et al. (2004) Sister grouping of chimpanzees and humans as revealed by genome-wide phylogenetic analysis of brain gene expression profiles. *Proc. Natl. Acad. Sci. USA* **101,** 2957–2962.
6. Finch, C. E. (2003) Neurons, glia, and plasticity in normal brain aging. *Neurobiol. Aging* **24(Suppl. 1),** S123–S127.
7. Burt, T., Prudic, J., Peyser, S., Clark, J., and Sackeim, H. A. (2000) Learning and memory in bipolar and unipolar major depression: effects of aging. *Neuropsychiatry Neuropsychol. Behav. Neurol.* **13,** 246–253.
8. Devanand, D. P., Turret, N., Moody, B. J., et al. (2000) Personality disorders in elderly patients with dysthymic disorder. *Am. J. Geriatr. Psychiatry* **8,** 188–195.
9. Gaubatz, J. W. (1995). *Molecular Basis of Aging,* CRC, Boca Raton, FL, pp. 71–82.
10. Lee, C. M., Weindruch, R., and Aiken, J. M. (1997) Age-associated alterations of the mitochondrial genome. *Free Radic. Biol. Med.* **22,** 1259–1269.
11. Lee, C. K., Weindruch, R., and Prolla, T. A. (2000) Gene-expression profile of the ageing brain in mice. *Nat. Genet.* **25,** 294–297.

12. Blumenthal, H. T. (1997) Fidelity assurance mechanisms of the brain with special reference to its immunogenic CNS compartment: their role in aging and aging-associated neurological disease. *J. Gerontol. A Biol. Sci. Med. Sci.* **52,** B1–B9.
13. Sohal, R. S., Mockett, R. J., and Orr, W. C. (2002) Mechanisms of aging: an appraisal of the oxidative stress hypothesis. *Free Radic. Biol. Med.* **33,** 575–586.
14. Lu, T., Pan, Y., Kao, S. Y., et al. (2004) Gene regulation and DNA damage in the ageing human brain. *Nature* **429,** 883–891.
15. Erraji-Benchekroun, L., Underwood, M. D., Arango, V., et al. (2005) Molecular aging in human prefrontal cortex is selective and continuous throughout adult life. *Biol. Psychiatry* **57,** 549–558.
16. Weindruch, R., Kayo, T., Lee, C. K., and Prolla, T. A. (2002) Gene expression profiling of aging using DNA microarrays. *Mech. Ageing Dev.* **123,** 177–193.
17. Salat, D. H., Kaye, J. A., and Janowsky, J. S. (1999) Prefrontal gray and white matter volumes in healthy aging and Alzheimer disease. *Arch. Neurol.* **56,** 338–344.
18. Tisserand, D. J., Bosma, H., Van Boxtel, M. P., and Jolles, J. (2001) Head size and cognitive ability in nondemented older adults are related. *Neurology* **56,** 969–971.
19. West, R. L. (1996) An application of prefrontal cortex function theory to cognitive aging. *Psychol. Bull.* **120,** 272–292.
20. MacPherson, S. E., Phillips, L. H., and Della, S. S. (2002) Age, executive function, and social decision making: a dorsolateral prefrontal theory of cognitive aging. *Psychol. Aging* **17,** 598–609.
21. Lee, C. K., Allison, D. B., Brand, J., Weindruch, R., and Prolla, T. A. (2002) Transcriptional profiles associated with aging and middle age-onset caloric restriction in mouse hearts. *Proc. Natl. Acad. Sci. USA* **99,** 14,988–14,993.
22. Lander, E. S. (1999) Array of hope. *Nat. Genet.* **21,** 3-4.
23. Tomita, H., Vawter, M. P., Walsh, D. M., et al. (2004) Effect of agonal and postmortem factors on gene expression profile: quality control in microarray analyses of postmortem human brain. *Biol. Psychiatry* **55,** 346–352.
24. Li, J. Z., Vawter, M. P., Walsh, D. M., et al. (2004) Systematic changes in gene expression in postmortem human brains associated with tissue pH and terminal medical conditions. *Hum. Mol. Genet.* **13,** 609–616.
25. Galfalvy, H. C., Erraji-Benchekroun, L., Smyrniotopoulos, P., et al. (2003) Sex genes for genomic analysis in human brain: internal controls for comparison of probe level data extraction. *BMC Bioinformatics* **4,** 37.
26. Sibille, E., Arango, V., Galfalvy, H. C., et al. (2004) Gene expression profiling of depression and suicide in human prefrontal cortex. *Neuropsychopharmacology* **29,** 351–361.
27. Xu, J., Burgoyne, P. S., and Arnold, A. P. (2002) Sex differences in sex chromosome gene expression in mouse brain. *Hum. Mol. Genet.* **11,** 1409–1419.
28. Li, C., and Wong, W. H. (2001) Model-based analysis of oligonucleotide arrays: expression index computation and outlier detection. *Proc. Natl. Acad. Sci. USA* **98,** 31–36.
29. Irizarry, R. A., Bolstad, B. M., Collin, F., Cope, L. M. , Hobbs, B., and Speed, T. P. (2003) Summaries of Affymetrix GeneChip probe level data. *Nucleic Acids Res.* **31,** e15.
30. Benjamini, Y., and Hochberg, Y. (1995) Controlling the false discovery rate: a practical and powerful approach to multiple testing. *J. R.. Statist. Soc. B* **57,** 289–300.
31. Eisen, M. B., Spellman, P. T., Brown, P. O., and Botstein, D. (1998) Cluster analysis and display of genome-wide expression patterns. *Proc. Natl. Acad. Sci. USA* **95,** 14,863–14,868.
32. Ashburner, M., Ball, C. A., Blake, J. A., et al. (2000) Gene ontology: tool for the unification of biology. The Gene Ontology Consortium. *Nat. Genet.* **25,** 25–29.

33. Pavlidis, P., Qin, J., Arango, V., Mann, J. J., and Sibille, E. (2004) Using the gene ontology for microarray data mining: a comparison of methods and application to age effects in human prefrontal cortex. *Neurochem. Res.* **29,** 1213–1222.

34. Resnick, S. M., Pham, D. L., Kraut, M. A., Zonderman, A. B., and Davatzikos, C. (2003) Longitudinal magnetic resonance imaging studies of older adults: a shrinking brain. *J. Neurosci.* **23,** 3295–3301.

35. Park, D. C., Lautenschlager, G., Hedden, T., Davidson, N. S., Smith, A. D., and Smith, P. K. (2002) Models of visuospatial and verbal memory across the adult life span. *Psychol. Aging* **17,** 299–320.

36. Fozard, J. R. (1983) Functional correlates of 5-HT$_1$ recognition sites. *Trends Pharmacol. Sci.* **4,** 288–289.

37. Verbitsky, M., Yonan, A. L., Malleret, G., Kandel, E. R., Gilliam, T. C., and Pavlidis, P. (2004) Altered hippocampal transcript profile accompanies an age-related spatial memory deficit in mice. *Learn. Mem.* **11,** 253–260.

38. Weindruch, R., and Prolla, T. A. (2002) Gene expression profile of the aging brain. *Arch. Neurol.* **59,** 1712–1714.

39. Kato, K., Suzuki, F., Morishita, R., Asano, T., and Sato, T. (1990) Selective increase in S-100 beta protein by aging in rat cerebral cortex. *J. Neurochem.* **54,** 1269–1274.

40. Arango, V., Ernsberger, P., Marzuk, P. M., et al. (1990) Autoradiographic demonstration of increased serotonin 5-HT$_2$ and β-adrenergic receptor binding sites in the brain of suicide victims. *Arch. Gen. Psychiatry* **47,** 1038–1047.

41. Meltzer, C. C., Smith, G., Price, J. C., et al. (1998) Reduced binding of [^{18}F]altanserin to serotonin type 2A receptors in aging: persistence of effect after partial volume correction. *Brain Res.* **813,** 167–171.

42. Sheline, Y. I., Mintun, M. A., Moerlein, S. M., and Snyder, A. Z. (2002) Greater loss of 5-HT(2A) receptors in midlife than in late life. *Am. J. Psychiatry* **159,** 430–435.

43. Weissmann, D., Mach, E., Oberlander, C., Demassey, Y., and Pujol, J. F. (1986) Evidence for hyperdensity of 5HT1B binding sites in the substantia nigra of the rat after 5,7-dihydroxytryptamine intraventricular injection. *Neurochem. Int.* **9,** 191–200.

44. Sastre, M., and Garcia-Sevilla, J. A. (1994) Density of alpha-2A adrenoceptors and Gi proteins in the human brain: ratio of high-affinity agonist sites to antagonist sites and effect of age. *J. Pharmacol. Exp. Ther.* **269,** 1062–1072.

45. Bu, J., Sathyendra, V., Nagykery, N., and Geula, C. (2003) Age-related changes in calbindin-D28k, calretinin, and parvalbumin-immunoreactive neurons in the human cerebral cortex. *Exp. Neurol.* **182,** 220–231.

46. Sastre, M., and Garcia-Sevilla, J. A. (1993) Opposite age-dependent changes of alpha 2A-adrenoceptors and nonadrenoceptor [^{3}H]idazoxan binding sites (I2-imidazoline sites) in the human brain: strong correlation of I2 with monoamine oxidase-B sites. *J. Neurochem.* **61,** 881–889.

47. Sun, M. K., and Alkon, D. L. (2002) Carbonic anhydrase gating of attention: memory therapy and enhancement. *Trends Pharmacol. Sci.* **23,** 83–89.

48. Ghigo, E., Arvat, E., Gianotti, L., et al. (1996) Human aging and the GH-IGF-I axis. *J. Pediatr. Endocrinol. Metab.* **9(Suppl. 3),** 271–278.

49. Trougakos, I. P., and Gonos, E. S. (2002) Clusterin/apolipoprotein J in human aging and cancer. *Int. J. Biochem. Cell Biol.* **34,** 1430–1448.

50. Torres, G. (1995) Fenfluramine-induced c-*fos* in the striatum and hypothalamus: a tract-tracing study. *Neuroreport* **6,** 1679–1683.

7

Gene Expression Profiling for Biomarker Discovery

Kazuhiko Uchida

Summary

The DNA microarray is a powerful method used to detect global expression of genes under-stand the physical status of cells. Since this technology was established, it has been applied to many fields of medical investigation. Many types of tumors have been analyzed, and correlations have been found between gene expression profiles and biological characteristics such as invasion metastasis, and prognosis. Quantitative analysis of tumor-specific gene expression has revealed that altered gene expression is associated with the pathology and the altered biological function of cancer cells. This chapter describes the clinical application of microarrays (bioarrays) for the iden-tification of potential diagnostic markers for cancer by measuring tumor-specific expression of thousands of genes. Expression profile analysis using a microarray followed by protein expression analysis is useful for the development of molecular biomarkers for cancer diagnosis.

Key Words: Biomarker; gene expression; microarray; quantitative PCR; protein expression; proteome.

1. Bioarray and Clinical Applications

1.1. Technology Platform

Since microarrays have been developed, several technological modifications and novel technology platforms have been reported. In the late 1990s, the cDNA microarray and Affymetrix GeneChip (Affymetrix, Santa Clara, CA) were two major microarray platforms. The cDNA microarray consists of more than 10,000 cDNAs, including expressed sequence tag clones. A GeneChip is prepared by on-chip generation of oligonucleotides. Completion of the Human Genome Project accelerated the develop-ment of a new generation of microarray and the application of this powerful technol-ogy to many research fields. Although nearly 3,000 research publications have been documented, the clinical application of this technology to diagnosis has just launched. In an era of personalized medicine, microarray technology will contribute to advances in diagnosis as well as prognosis prediction. Additionally, microarray will contribute to the development of new therapies, because it will aid in the understanding of dis-eases at the molecular level.

From: *Bioarrays: From Basics to Diagnostics*
Edited by: K. Appasani © Humana Press Inc., Totowa, NJ

1.1.1. Bioarray for Transcriptome

1.1.1.1. OLIGONUCLEOTIDE MICROARRAYS

Oligonucleotide microarrays are the most conventional type of microarray. The five major microarray suppliers are Affymetrix (Santa Clara, CA) (http://www.affymetrix.com/) *(1)*; Agilent Technologies (Palo Alto, CA) (http://www.chem.agilent.com/) *(2)*; CodeLink™, GE Healthcare (Little Chalfont, Buckinghamshire, United Kingdom) (http://www4.amershambiosciences.com/APTRIX/upp01077.nsf/Content/codelink_bioarray_system) *(3)*; Illumina, Inc. (San Diego, CA) (http://www.illumina.com/) *(4)*; and NimbleGen Systems Inc. (Madison, WI) (http://www.nimblegen.com/) *(5)*. Affymetrix and NimbleGen Systems Inc. are on-chip synthetic oligonucleotide arrays. The NimbleGen array is used only for custom analysis service. Illumina, Inc. provides Oligator® oligonucleotide manufacturing and BeadArray™ platform technologies, and Agilent and CodeLink are glass array on which synthetic oligonucleotides are spotted.

1.1.2. cDNA Microarrays

A cDNA microarray is made by spotting cDNA PCR product on a glass slide or membrane. In earlier work on gene expression analysis by microarray, cDNA microarray is used. For example, these in-house microarrays are made as follows *(6,7)*. To produce these microarrays, human cDNA is obtained from a cDNA supplier, such as Research Genetics (Invitrogen, Carlsbad, CA). The cDNA insert is amplified by standard PCR protocols by using the primer sets. The amplified clones are purified and then spotted onto nylon membranes or slide glass by an arrayer.

At the beginning of microarray research, commercial oligonucleotide arrays were expensive for general use in research. Although the cDNA microarray can be prepared in the laboratory, the preparation and handling of cDNA is labor-intensive and the total cost is expensive.

1.1.3. DNA Analysis

1.1.3.1. OLIGONUCLEOTIDE MICROARRAYS

Oligonucleotide microarrays make it possible to genotype many thousands of single-nucleotide polymorphisms in large numbers of individuals. Recently, hybridization-based genotyping has enabled researchers to identify genes linked to several diseases. A single hybridization can analyze more than 100,000 single-nucleotide polymorphisms distributed across the human genome *(8)*.

The chromatin immunoprecipitation (ChIP) DNA microarray Chip also has been recently used for the analysis of transcription regulation sites *(9)*. By ChIP followed by PCR, DNA that coprecipitates with transcription factors is amplified, and this transcription factor-binding DNA is hybridized with an oligonucleotide array. For a comprehensive analysis of the transcription regulating sites, a tiling array that covers all of the nucleotide sequence of human genome is required.

1.1.3.2. DNA MICROARRAYS

The amplification of oncogenes, deletion of tumor suppressor genes, or both, together with gene dysfunctions caused by point mutations, are the main causes of cancer *(10,11)*. Alterations in DNA copy number, which occur when genes are deleted

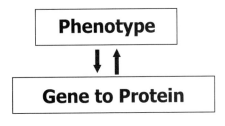

Fig. 1. Genotype and phenotype correlation. Genotype includes DNA sequence variation, DNA copy number alteration (gene amplification and deletion), gene expression, and protein expression. These genetic and genomic changes resulting protein expression and function lead phenotypic changes in cells.

or amplified, can be assayed by arrayed bacterial artificial chromosome clones or DNA oligonucleotide arrays *(12)*. Recently, the GeneChip® Mapping Assay (Affymetrix) is applied for linkage analysis, population genetics, and chromosomal copy number changes during cancer progression *(13)*.

1.2. Clinical Applications in Cancer Treatment and Diagnosis

As shown in **Fig. 1**, the purpose of biological and biomedical research is to explain the phenotype of a cell or organism. What determines the fate of a cell to die or stay alive after exposure to genotoxic agents? What is the mechanism of transformation that is responsible for the regulation of cell proliferation, histology, and the malignant potential for metastasis and invasion? Approaches using technology platforms based on the genome, transcriptome, and proteome try to address these questions.

Cancer phenotype analysis has been performed by several groups *(2,6,14)*. The capacity of cancer cells for invasion and metastasis to other tissues is an important characteristic that can be used to determine appropriate cancer therapies as well as patient prognosis. Reports of profiling have found correlations to invasion and metastasis by using microarrays in colorectal cancer *(15)*, lung cancer *(14)*, and breast cancer *(2)*. Renal cell carcinoma (RCC) shows various clinical behaviors; currently, surgical modalities are the only effective therapy against this cancer. Global expression profiling of RCC has been analyzed with cDNA microarray technology *(16,17)*. Wilhelm et al. *(18)* reported that array-based comparative genomic hybridization was capable of differentially diagnosing an RCC from a benign renal tumor on the basis of their genetic profiles *(18)*. Ami et al. *(7)* showed that several genes were up-regulated in cell lines from metastatic lesions. Transgelin, which is reportedly involved in cell proliferation and migration, was upregulated in a metastatic lesion, SKRC-52, with a unique spindle-shaped morphology. These unique profiling patterns of gene expression clearly correlated with cell morphology and metastatic potential.

2. Microarrays for Molecular Biomarker Discovery

2.1. Biomarkers for Diagnosis

2.1.1. Circulating Cancer Cells in Peripheral Blood

Tumor-specific molecular markers that identify circulating cancer cells in peripheral blood or tumor-associated proteins in serum are useful for early diagnosis, prognosis prediction, or both. Sensitive diagnostic methods that can detect cancer in an early stage (when it is difficult to detect with imaging diagnostic devices such as computed tomography and ultrasonography) improve the curability of cancer and have a major impact on the choice of therapeutic strategies.

Many types of tumor markers have been identified. Some of these markers, such as α-fetoprotein, carcinoembryonic antigen, CA19-9, CA125, and prostate-specific antigen (PSA), have become clinically useful. These tumor markers are probably not tumor specific, tissue specific, or both, and some of them have a broad spectrum in cancers (19). Although the combination of tumor marker examination and clinical examination has improved the diagnosis of cancers, no potential tumor markers that identify cancer in the early stage have been identified. To identify patients with cancer in the early stage, more sensitive and reliable tumor markers are required.

PCR and reverse transcription-PCR (RT-PCR) for tumor-specific proteins, tissue-specific proteins, or both to detect circulating cancer cells in peripheral blood have been attempted in patients with leukemias (20), prostate cancer (21), and hepatocellular carcinoma (22). Recently, a relationship between prognosis and circulating cancer cells has been reported (23). The identification of circulating cancer cells in peripheral blood from patients in an early clinical stage is an ideal diagnostic method because of its high sensitivity.

In spite of many trials, useful tumor-specific transcripts have not been identified. PSA expression was formerly thought to be prostate specific. The PSA level in serum has become one of the most useful molecular biomarkers. However, *PSA* mRNA has been detected in nonprostate cells, including normal blood cells as well as various tumors such as breast cancer, lung cancer, and ovarian cancer (24). The enhancement of sensitivity compromises specificity. So, the identification of "real" tissue specificity, tumor specificity, or both is potentially important. Screening tumor-specific gene(s) by microarray to establish the molecular diagnosis of cancer at an early stage requires the identification of genes that are expressed in cancer cells, but not in normal cells or noncancerous tissues (**Fig. 2**).

2.1.2. Diagnostic Molecular Marker for Thyroid Cancer

2.1.2.1. Gene Expression of Thyroid Cancer

Papillary thyroid carcinoma (PTC) is the most common type of malignant thyroid tumor, representing 80–90% of all thyroid malignancies. Papillary, follicular, and anaplastic thyroid carcinomas arise from follicular cells; medullary thyroid carcinomas

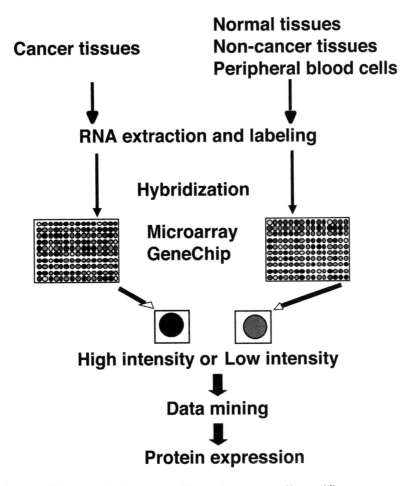

Fig. 2. Identification of tissue-specific and cancer cell-specific gene expression by microarray.

arise from the parafollicular epithelium of the thyroid. PTC is usually well differentiated; however, the clinical behavior of PTC varies widely. For example, incidental microcarcinomas grow very slowly and are noninvasive or minimally invasive. In contrast, invasive PTC with metastasis can be lethal. PTC often recurs many years after surgical removal. The prognosis for PTC is often favorable; however, approximately 20% of PTC tumors recur *(25)*, and some reach advanced stages. Postoperative follow-up for diagnosing recurrence is important for a favorable outcome. Serum thyroglobulin has been monitored by immunoassay to detect the recurrence of differentiated thyroid carcinoma. However, measurement of serum thyroglobulin is sometimes hindered by the presence of circulating factors and residual normal thyroid gland tissue producing thyroglobulin *(6)*. Thus, a reliable diagnostic molecular marker for PTC would be extremely valuable for improving cancer detection and the prognosis of patients with PTC.

2.1.2.2. Microarray and Quantitative PCR of PTC

A common approach for identifying circulating cancer cells in peripheral blood is the RT-PCR method for amplifying tumor-specific mRNA *(20–22)*. Another possible approach is to measure abnormal or ectopic tissue-specific gene expression in peripheral blood. Although RT-PCR detection of a marker for circulating cancer cells in peripheral blood would be a powerful noninvasive method, a reliable marker for PTC has not yet been identified.

cDNA microarrays provide a powerful method to quantitatively analyze cancer-specific gene expression. These microarrays can detect altered gene expression associated with the pathology or altered biology of cancer cells. As shown in **Fig. 3**, gene expression profiles clearly discriminate PTC to normal thyroid and peripheral blood lymphocytes. Most of the PTC samples are clustered in one group, and the normal thyroid samples are clustered in another group. A cDNA microarray can be used to identify potential diagnostic markers for cancer by measuring tumor-specific expression of thousands of genes in hundreds of tumors. Yano et al. *(6)* showed that candidate genes for diagnostic markers also could be characterized by analyzing the gene expression profiles of a small number of cancer tissues in combination with additional large-scale immunohistochemical analysis of protein expression *(6)*. **Figure 4** shows genes overexpressed in PTC but not in normal thyroid and peripheral blood cells. The potential biomarker candidates were included in these genes. The upregulated genes in PTC compared with normal thyroid are summarized in **Table 1**.

2.1.2.3. Immunohistchemical Confirmation of Biomarkers

By analysis of cancer-specific protein expression in PTC, basic fibroblast growth factor (bFGF) and platelet-derived growth factor (PDGF) were identified as potential diagnostic tools for cancer. Before the use of bFGF and PDGF as diagnostic molecular markers, the expression of these proteins needed to be immunohistochemically examined in thyroid tissues. **Table 2** summarizes the results of an immunohistochemical analysis of protein expression in 55 differentiated PTCs, 4 follicular variant PTCs, 6 follicular thyroid carcinomas (FTCs), 5 hyperfunctioning thyroid tissues, 11 benign thyroid neoplasms, and 10 normal thyroids. These samples included the tissues analyzed by microarray. Thirty-four of 59 (54%) PTCs were positive for bFGF, whereas none of the 10 normal thyroids was positive for bFGF.

Fig. 3. *(opposite page)* Clustering diagram for human 3968 genes in thyroid cancer and normal or noncancerous diseases of the thyroid. Expression profiles were analyzed for seven Papillary thyroid carcinoma (PTC) samples and seven normal thyroid samples, including three pairs of normal/tumoral thyroid tissues. The cutoff for differential expression associated with PTC was set at twofold. Clustering analysis was performed using the model-based approach with GeneSpring™ software (Agilent Technologies). Genes with similar expression patterns are clustered together. Each horizontal block represents a single gene. The color scale correlates color with relative expression. The basal expression of each gene is represented in yellow. A shift toward red indicates an increase in expression, whereas a shift toward blue signifies a decrease in expression.

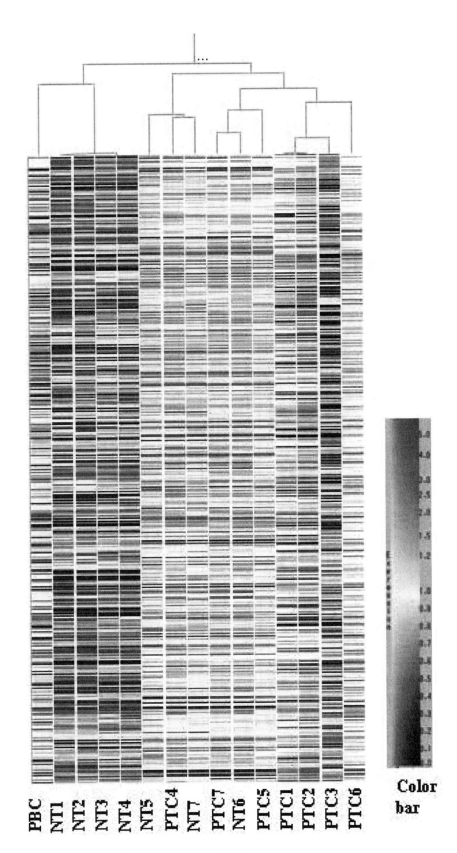

PBC NT1 NT2 NT3 NT4 NT5 PTC4 NT7 PTC7 NT6 PTC5 PTC1 PTC2 PTC3 PTC6

Color
bar

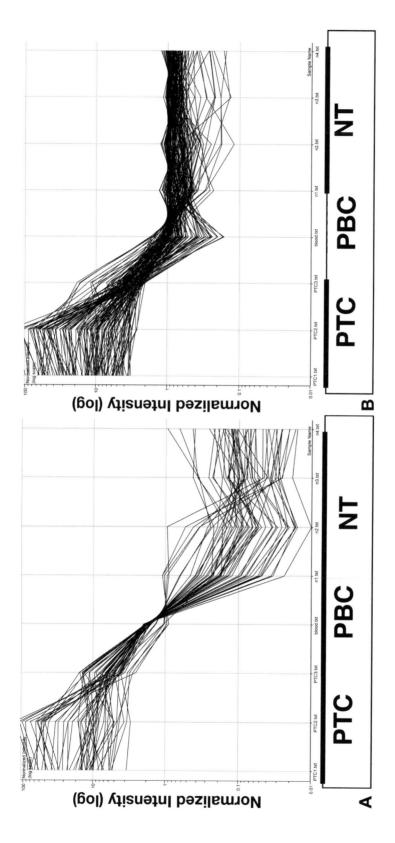

Fig. 4. Genes overexpressed in papillary thyroid carcinoma (PTC), but not in normal thyroid and peripheral blood cells. (**A**) Genes highly expressed in PTC and downregulated in normal thyroid (NT). (**B**) Genes highly expressed in PTC but not in NT or peripheral blood cells (PBC).

Table 1
Upregulated Genes in Papillary Thyroid Carcinoma
Compared With Normal Thyroid[a]

Cell cycle control proteins	Tetracycline transporter-like protein
	Cyclin H, cyclin D1, SPHAR gene for cyclin-related protein
	Heme oxygenase (decycling) 1
Oncogene	v-raf, putative oncogene, vav2
	V-rel avian reticuloendotheliosis viral oncogene
Growth factor	FGF, VEGR-B, PDGF, placental growth factor
	BMP5, TGF, β-induced
	FHF-1, hepatocyte growth factor-like protein
Growth factor receptor	bFGF receptor, FGF 4 receptor
	Insulin-like growth factor 1 recetor, growth factor bound protein
Phosphatase	Inositol polyphosphate phosphatase-like protein 1
	Nuclear dual-specificity phosphatase
	MAP kinase phosphatase, M-phase inducer phosphatase 2
	Inositol polyphosphate 4-phosphatase
	Protein phosphatase-1 inhibitor, protein-yrosine phosphatase
	L-3-phosphoserine-phosphatase

BMP, bone morphogenic protein; FGF, firbroblast growth facotr; MAP, mitogen-activated protein; PDGF, platelet-derived growth factor; TNF, tumor necrosis factor; VEGF, vascular endothelial growth factor.

[a] List of genes is not exhaustive.

Table 2
Overexpression of Basic Fibroblast Growth Factor (bFGF)
and Platelet-Derived Growth Factor (PDGF) in Thyroid Tissues

Tissue	bFGF	(%)	PDGF	(%)
Normal thryoid	0/10[a]	(0)	0/10	(0)
Hyperfunctioning tissue	3/5	(60)	1/5	(20)
Multinodular goiter	2/6	(33)	0/6	(0)
Adenoma	2/5	(40)	0/5	(0)
Papillary carcinoma (well-differentiated)	30/55	(55)	45/55	(82)
Papillary caricinoma (follicular variant)	2/4	(50)	3/4	(75)
Follicular carcinoma	4/6	(67)	6/6	(100)

[a] Number of samples with positive staining/total number of samples, including the seven cases studies by DNA array analysis.

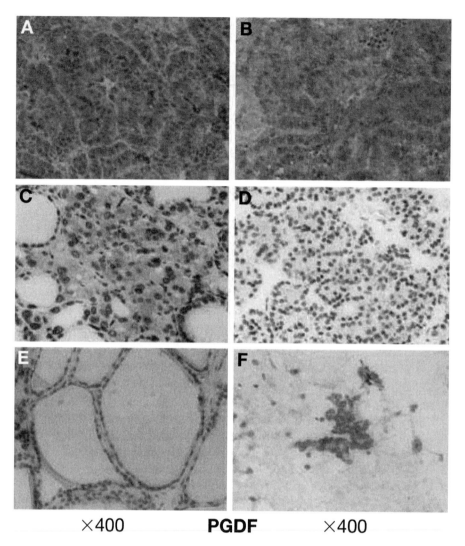

×400   **PGDF**   ×400

Fig. 5. Overexpression of PDGF in papillary thyroid carcinoma (PTC). Strong staining for anti-platelet-derived growth factor (PDGF) antibody is observed only in tumor cells of PTC (**A,B**). The parafollicular cells of hyperfunctioning thyroid tissue showed a weakly positive immunoreaction (**C**). Follicular adenoma (**D**) and normal thyroid (**E**) are negative for PDGF immunostaining. Positive staining for anti-PDGF antibody is observed in cytological specimens from PTC obtained by fine needle aspirating biopsy (**F**). Original magnification, 3400.

Positive staining for bFGF also was observed in hyperfunctioning thyroid tissues and benign neoplasms. As shown in **Fig. 5**, thyroid neoplasms PTC and FTC showed increased proportions of positive immunostaining for PDGF; 48 of 59 (81%) cases of PTC, and 6 of 6 (100%) cases of FTC were positive for PDGF. In contrast, follicular cells in normal thyroid tissue from multinodular goiters and adenomas were negative for PDGF immunostaining. Immunochemical analysis also showed that neoplastic cells retrieved by fine needle aspiration biopsy were immunoreactive for PDGF.

These studies demonstrate that microarray technology can be used to develop molecular markers for cancer diagnosis. Some, not all, of the genes shown to have strong expression might be used as serum tumor markers for cancer.

3. Prospects for Gene Expression Profiling in Clinical Use

Accumulating data indicate the importance of global examination of gene expression and protein expression. The protein microarray as well as DNA microarray may become potential diagnostic tools *(26)*. In addition to transcriptome, proteome, and peptidome analyses with technological breakthroughs may lead to the successful isolation of a tumor marker with high sensitivity and high specificity. These data also may improve our understanding of carcinogenesis. In the near future, expression profiling diagnosis may be used clinically for several types of diseases.

References

1. Fodor, S. P., Rava, R. P., Huang, X. C., Pease, A. C., Holmes, C. P., and Adams, C. L. (1993) Multiplexed biochemical assays with biological chips. *Nature* **364**, 555–556.
2. van't Veer, L. J., Dai, H., van de Vijver, M. J., et al. (2002) Gene expression profiling predicts clinical outcome of breast cancer. *Nature* **415**, 530–536.
3. Ramakrishnan, R., Dorris, D., Lublinsky, A., et al. (2002) An assessment of Motorola CodeLink microarray performance for gene expression profiling applications. *Nucleic Acids Res.* **30**, e30.
4. Ferguson, J. A., Boles, T. C., Adams, C. P., and Walt, D. R. (1996) A fiber-optic DNA biosensor microarray for the analysis of gene expression. *Nat. Biotechnol.* **14**, 1681–1684.
5. Singh-Gasson, S., Green, R. D., Yue, Y., et al. (1999) Maskless fabrication of light-directed oligonucleotide microarrays using a digital micromirror array. *Nat. Biotechnol.* **17**, 974–978.
6. Yano, Y., Uematsu, N., Yashiro, T., et al. (2004) Gene expression profiling identifies platelet-derived growth factor as a diagnostic molecular marker for papillary thyroid carcinoma. *Clin. Cancer Res.* **10**, 2035–2043.
7. Ami, Y., Shimazui, T., Akaza, H., et al. (2005) Gene expression profiles correlate with the morphology and metastasis characteristics of renal cell carcinoma cells. *Oncol. Rep.* **13**, 75–80.
8. Kennedy, G. C., Matsuzaki, H., Dong, S., et al. (2003) Large-scale genotyping of complex DNA. *Nat. Biotechnol.* **21**, 1233–1237.
9. Buck, M. J. and Lieb, J. D. (2004) ChIP-chip: considerations for the design, analysis, and application of genome-wide chromatin immunoprecipitation experiments. *Genomics* **83**, 349–360.
10. Uchida, K. (2003) Gene amplification and cancer. *Nat. Encyclopedia Hum. Genome* **2**, 593–598.
11. Kashiwagi, H., and Uchida, K. (2000) Genome-wide profiling of gene amplification and deletion in cancer. *Hum. Cell* **13**, 135–141.
12. Pinkel, D., Segraves, R., Sudar, D., et al. (1998) High resolution analysis of DNA copy number variation using comparative genomic hybridization to microarrays. *Nat. Genet.* **20**, 207–211.
13. Zhou, X., Cole, S. W., Hu, S, and Wong, D. T. (2004) Detection of DNA copy number abnormality by microarray expression analysis. *Hum. Genet.* **114**, 464–467.
14. Chen, J. J., Peck, K., Hong, T. M., et al. (2001) Global analysis of gene expression in invasion by a lung cancer model. *Cancer Res.* **61**, 5223–5230.

15. Hegde, P., Qi, R., Gaspard, R., et al. (2001) Identification of tumor markers in models of human colorectal cancer using a 19,200-element complementary DNA microarray. *Cancer Res.* **61,** 7792–7797.
16. Boer, J. M., Huber, W. K., Sultmann, H., et al. (2001) Identification and classification of differentially expressed genes in renal cell carcinoma by expression profiling on a global human 31,500-element cDNA array. *Genome Res.* **11,** 1861–1870.
17. Takahashi, M., Rhodes, D. R., Furge, K. A., et al. (2001) Gene expression profiling of clear cell renal cell carcinoma: gene identification and prognostic classification. *Proc. Natl. Acad. Sci. USA* **98,** 9754–9759.
18. Wilhelm, M., Veltman, J. A., Olshen, A. B., et al. (2002) Array-based comparative genomic hybridization for the differential diagnosis of renal cell cancer. *Cancer Res.* **62,** 957–960.
19. Ishikawa, T., Kashiwagi, H., Iwakami, Y., et al. (1998) Expression of alpha-fetoprotein and prostate-specific antigen genes in several tissues and detection of mRNAs in normal circulating blood by reverse transcriptase-polymerase chain reaction. *Jpn. J. Clin. Oncol.* **28,** 723–728.
20. Downing, J. R., Shurtleff, S. A., Zielenska, M., et al. (1995) Molecular detection of the (2;5) translocation of non-Hodgkin's lymphoma by reverse transcriptase-polymerase chain reaction. *Blood* **85,** 3416–3422.
21. Kawakami, M., Okaneya, T., Furihata, K., Nishizawa, O., and Katsuyama, T. (1997) Detection of prostate cancer cells circulating in peripheral blood by reverse transcription-PCR for hKLK2. *Cancer Res.* **57,** 4167–4170.
22. Kar, S., and Carr, B. I. (1995) Detection of liver cells in peripheral blood of patients with advanced-stage hepatocellular carcinoma. *Hepatology* **21,** 403–407.
23. Cristofanilli, M., Budd, G. T., Ellis, M. J., et al. (2004) Circulating tumor cells, disease progression, and survival in metastatic breast cancer. *N. Engl. J. Med.* **351,** 781–791.
24. Yu, H., Diamandis, E. P., Levesque, M., Asa, S. L., Monne, M., and Croce, C. M. (1995) Expression of the prostate-specific antigen gene by a primary ovarian carcinoma. *Cancer Res.* **55,** 1603–1606.
25. Loh, K. C., Greenspan, F. S., Gee, L., Miller, T. R., and Yeo, P. P. (1997) Pathological tumor-node-metastasis (pTNM) staging for papillary and follicular thyroid carcinomas: a retrospective analysis of 700 patients. *J. Clin. Endocrinol. Metab.* **82,** 3553–3562.
26. Liotta, L. A., Espina, V., Mehta, A. I., et al. (2003) Protein microarrays: meeting analytical challenges for clinical applications. *Cancer Cell* **3,** 317–325.

8

Array-Based Comparative Genomic Hybridization

Applications in Cancer and Tuberculosis

Murali D. Bashyam and Seyed E. Hasnain

Summary

There has been a huge increase in DNA sequence data during the past decade from various biological systems. Most notably, completion of human and several pathogen genomes has enabled us to apply several high-throughput technological innovations to understand the human disease process. This chapter deals with one such technology, i.e., array-based comparative genomic hybridization (aCGH). Genomic alterations have long been implicated in several disease processes, including cancer. Earlier techniques such as conventional karyotyping, G-banding, FISH, and so on, either suffered from a lower resolution or were prohibitively expensive for whole genome coverage. The comparative genomic hybridization technique was the first step towards whole genome profiling of genomic amplifications and deletions; however, it could at best offer a resolution of approx 10–20 Megabases (Mb). The advent of the microarray technology during the later part of the 1990s has enabled the high-resolution mapping of genomic alterations at a high resolution (< 1 Mb). The present chapter discusses the aCGH technology and its use in studying cancer and *Mycobacterium tuberculosis* infection.

Key Words: Array-based comparative genomic hybridization; aCGH; microarray; gene amplification; homozygous deletion; oncogene; tumor suppressor gene.

1. Introduction

In the recent history of modern biology, technological inventions and innovations have revolutionized the pace and quality of biomedical research. Automation of DNA sequencing has enabled the complete sequencing of the genomes of >260 archaeal, bacterial, and eukaryotic organisms during the past decade, and several more genome projects are nearing completion (www.tigr.org, http://www.cbs.dtu.dk/services/ GenomeAtlas and www.ncbi.nlm.nih.gov/genomes/index.html). It is important to use this vast information to answer questions related to basic cellular processes such as growth, differentiation, and response to environmental changes; infections and genetic diseases and disorders; and aging; drug therapy, and molecular medicine. Thus, it is not only important to determine the function of all genes that are discovered from the genome sequence but also to understand how various genes perform these functions involving the coordination of various cellular pathways. Microarray technology has revolutionized studies in the aforementioned areas during the past 8–10 yr. Details of microarray technology and uses of gene expression microarrays in several areas of research are covered in other chapters in this book.

From: *Bioarrays: From Basics to Diagnostics*
Edited by: K. Appasani © Humana Press Inc., Totowa, NJ

A typical array experiment yields data on expression levels of thousands of genes or expressed sequence tags for several samples. Statistical algorithms have been developed to correlate gene expression signatures with the specific biological phenomenon being studied. However, it is not always a simple exercise because 1) the gene expression microarray experiment may not yield a significant differential gene expression pattern, 2) the same data set may yield different signatures when addressed by different laboratories, 3) the observed change in expression profile may be a result of sample handling itself, and 4) it may be difficult to determine whether the changes observed are causes or an effect of the phenomenon being studied. To deal with these problems associated with whole genome gene expression analyses, one interesting alternative that has become increasingly popular is to characterize changes in the genome at the "DNA" level. Changes in DNA are stable and are not linked to "sample handling," and it is easier to establish whether the change is a cause or an affect of the phenomenon being studied.

It is now well established that genetic differences arising out of changes in the genome underlie the basis for several important developmental defects; diseases such as cancer; variation in response to drugs; and several issues relating to infectious diseases, including variation in virulence and drug resistance. Genome diversity may result from single-nucleotide polymorphisms, gene duplications, amplification or deletion of regions of the genome, recombination mechanisms, and changes in repetitive DNA elements. Changes in the genome can occur at three levels: 1) the nucleotide level, including base substitutions and microsatellite instability; 2) the subchromosomal level, including localized amplifications and deletions; and 3) the chromosomal level, including aneuploidy and chromosomal translocations. Several cytogenetic techniques have been developed to identify these abnormalities. A technique called comparative genomic hybridization (CGH) was developed to identify and map subchromosomal changes *(1)*. To improve its resolution, several array-based CGH techniques have been developed. In this chapter, we highlight the important achievements in the fields of cancer and geographic genomics of *Mycobacterium tuberculosis* that have been achieved by using the array-based CGH (aCGH) approach.

2. Historical Aspects of CGH

The fundamental discovery that human cells harbor 23 pairs of chromosomes was made in 1956 *(2)*. Further studies uncovered deviations from this rule. The advent of karyotyping in 1950s made it possible to identify various kinds of chromosomal abnormalities. Karyotyping led to the establishment of a link between specific chromosomal abnormalities and cytogenetic disorders such as Down syndrome, Turner's syndrome, Klinefelter's syndrome, and so on. Further advances led to the development of "banding techniques" that enabled the identification of subchromosomal structural abnormalities, including translocations and large deletions and inversions. During the past 2–3 decades, several techniques have been developed that can identify chromosomal abnormalities at the molecular level; these techniques include fluorescence *in situ* hybridization (FISH) *(3)*, spectral karyotyping, chromosome painting, and so on. In 1992, Kallioniemi and colleagues *(1)* introduced CGH for identification of DNA amplifications and deletions. In CGH, DNA from the test and reference genomes are differentially labeled using fluorescent dyes and hybridized to a normal metaphase spread. The ratio of the fluorescence at different regions along the chromosomes provides the measure of copy number changes for each chromosome, at an approximate resolution of 20 megabases (Mb). CGH has mainly been used to detect chromosomal changes in various forms of cancers and to a lesser extent in genetic disorders *(4)*. Although it is a whole genome technique, the important lacuna has been a lack of resolution *(5)*; changes of the order of 5 Mb or less cannot be identified *(6)*, and it is difficult to accurately map the boundaries of the changes detected. Currently, FISH is the only technique with a higher resolution, but it requires knowledge of the sequence of the region, and it is not a whole genome technique.

The introduction of aCGH has made it possible to detect DNA lesions at submegabase levels *(7,8)*. aCGH is a direct amalgamation of CGH and microarrays. There are two popular aCGH formats; one format uses bacterial artificial chromosome (BAC) clones to generate the array, whereas the other format uses cDNA clones or oligonucleotides. The microarray is made of cloned DNA fragments (500–2 kilobases [Kb] on cDNA array and 100–250 Kb in BAC arrays) whose exact map positions are known. An additional advantage of cDNA arrays is that each cDNA element on the array represents a functional gene; therefore, the genomic changes detected are directly mapped at a gene-by-gene resolution. The resolution of aCGH depends on the size of the arrayed DNA elements and the distance between elements that lie adjacent to each other on the chromosome. In this respect, it is easier to understand why cDNA arrays provide a higher resolution than the BAC arrays. The first array-based CGH approach was documented by Solinas-Toldo and colleagues *(9)* followed by Pinkel and co-workers *(7)*. Later, Pollack and co-workers *(8)* showed the usefulness of the first genome-wide array-based CGH. As with CGH, the main application of aCGH has been in cancer research.

3. aCGH and Cancer

Chromosomal abnormalities are not only a hallmark for blood malignancies but also for solid tumors. Initially, detection of chromosomal abnormalities by using classical cytogenetics was more common in blood malignancies, because it was easier to obtain good-quality metaphases in large numbers. The advent of CGH made the analyses of copy number changes in solid tumors a reality, because the dependence on mitoses was deemed unnecessary. However, because of its low resolution, it renders the identification of "driver" genes within amplicons a difficult task. The advent of aCGH has alleviated this lacuna. The initial reports of aCGH were based on arrays made from BAC clones *(7,9)*. Using BAC clones specific for chromosome 20 (six clones for the p arm and 16 clones for the q arm with an average spacing of 3 Mb), Pinkel and colleagues *(7)* could identify localized changes mainly restricted to 20q. The use of cDNA arrays for CGH in subsequent years by the group led by Patrick O. Brown (Stanford University, Stanford, CA) has improved the resolution and sensitivity of the technique. In 1999, Pollack and co-workers used a cDNA array fabricated at Stanford University to assess copy number changes in breast cancer cell lines and tumor samples. They were able to successfully identify two subpeaks of an amplicon localized to 17q22-24 that was previously identified as one peak by using conventional CGH *(8)*. The current cDNA arrays available from The Stanford Functional Genomics Facility (SFGF) provide very good coverage of the human genome. The array is mainly based on clone sets belonging to the Integrated Molecular Analysis of Genomes and their Expression consortium,

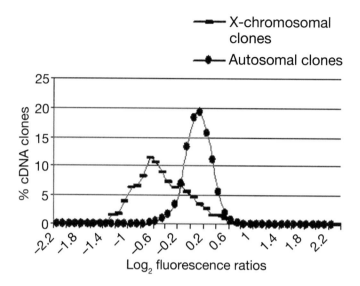

Fig. 1. Measurement of single copy loss. Genomic DNAs from normal female (labeled with Cy3) and normal male (labeled with Cy5) were compared using aCGH on the human microarray from the Stanford Functional Genomics Facility. The comparison includes 20,760 autosomal clones and 662 X-chromosomal clones. The graph compares the fluorescence ratios of autosomal clones and X-chromosomal clones; the fluorescence ratio values (Cy5/Cy3) are plotted on the *x*-axis and the percentage of clones on the *y*-axes. The fluorescence values are binned by intervals of 0.2.

thereby making it easier to localize map positions for each clone. Moreover, even if mapping information for a particular clone is incorrect, the results obtained for a particular clone will be correct. Thus, aCGH result can actually be used to validate the map position of genes! The SFGF has recently developed oligonucleotide-based microarrays.

The biggest advantage of using cDNA arrays for CGH is the feasibility of performing gene expression arrays in parallel. The DNA copy number and gene expression levels for each element on the array can be determined using the same array platform, thereby facilitating the identification of candidate oncogenes within amplicons *(10)*. This advantage is lost when using BAC arrays for CGH. Another difficulty associated with BAC arrays is their maintenance; BACs are usually single copy vectors and therefore result in low yield of DNA. In addition, their size presents with a minor problem during purification from bacterial transformants. Because BAC clones represent genomic DNA, they usually contain a variety of repeat sequences, which presents with another problem when analyzing ratio values after hybridization. Recently, repeat-free regions from genes have been selectively amplified using PCR and used to generate locus specific-arrays *(11)*. However, this is more labor-intensive and expensive than the other strategies.

Although it is widely thought that BAC arrays provide a more accurate measurement of copy number changes, recent work on pancreatic cancer has shown that cDNA array-based CGH can provide highly accurate copy number values *(12)*. The sensitivity of cDNA arrays for aCGH can be gauged from **Fig. 1**. aCGH was performed between DNA isolated from a normal female (labeled with Cy3 and a normal male (labeled with Cy5) by using the arrays from The Stanford Functional Genomics Facility. The mean fluorescence ratio for autosomes was close to 1 (thereby indicating no change), whereas that for X-chromosome was 0.6 (thereby indicating a single copy loss). Previous work using conventional CGH had revealed certain broad regions of gains and losses in pancreatic cancer cell lines. With an aim of identification of highly localized regions of DNA amplifications and homozygous deletions, aCGH analysis on pancreatic cancer cells lines has been carried out, by using a BAC array platform *(13)*, or a cDNA array platform *(12,14)*, or a combination of both *(15)*. Although several changes identified were common to two or more studies, some changes were identified only by one group, indicating the varying resolution or different methods for identifying statistically significant gains and losses. The aCGH work done by the Stanford group was supplemented by a parallel gene expression analysis by using the same Stanford array platform. This study has revealed 14 high-level localized amplicons and 15 localized homozygous deletions *(12)*. An important use of the aCGH technique was discovered serendipitously; the aCGH profiles of two cell lines, namely, AsPC1 and MPanc-96, both obtained from the American Type Culture Collection (Manassas, VA), were nearly identical **(Fig. 2)**. Because the profile was similar to that reported earlier for AsPC1 in another study *(16)* and AsPC1 was established earlier than MPanc-96, it was fair to assume that the profile belonged to AsPC1. On the basis of these results, American Type Culture Collection has since removed the MPanc-96 cell line from their collection. The higher resolution of aCGH is clearly indicated in a comparison of CGH profiles obtained from conventional CGH **(Fig. 3A)** and from aCGH **(Fig. 3B)** for the AsPC1 cell line. The CGH study could identify a broad amplification located at 7q; however, the molecular boundaries for the amplicon could not be identified and it could not be ascertained which genes were resident within the amplicon **(Fig 3A)**. However, aCGH has not only identified the exact boundaries of the amplicon but also revealed the genes located within the amplicon **(Figs. 3B** and **4)**. Gene expression arrays performed in parallel provide the unique ability to measure expression levels for each gene within the amplicon, thereby facilitating identification of putative "oncogenes" **(Fig 4)**. **Figure 4** depicts the fluorescence ratios (log$_2$ scale) for DNA and RNA for the pancreatic cancer cell line AsPC1; genes residing within the 7q21.2-21.3 amplicon are indicated. Therefore, this is a good example of the power of cDNA arrays to detect novel cancer-specific genes. The aCGH and gene expression results should be validated by independent techniques (FISH and reverse transcription-PCR). The genes identified by this approach can be further characterized by performing functional studies in cell lines and animal models to determine their role in tumorigenesis.

The usefulness of aCGH mainly depends on the coverage offered by the arrays being used; if the coverage is low (thereby resulting in a low resolution), several localized amplicons and homozygous deletions are likely to be missed. The study carried out by the Stanford group has identified a few highly localized deletions spanning <50 Kb. The previous aCGH reports on pancreatic cancer have been unable to identify such small homozygous deletions. When analyzing amplifications identified through aCGH, it is important to pick high-level changes, because they are more likely to harbor genes central to development and progression of the tumor, especially for aggressive tumors such as cancer of the pancreas.

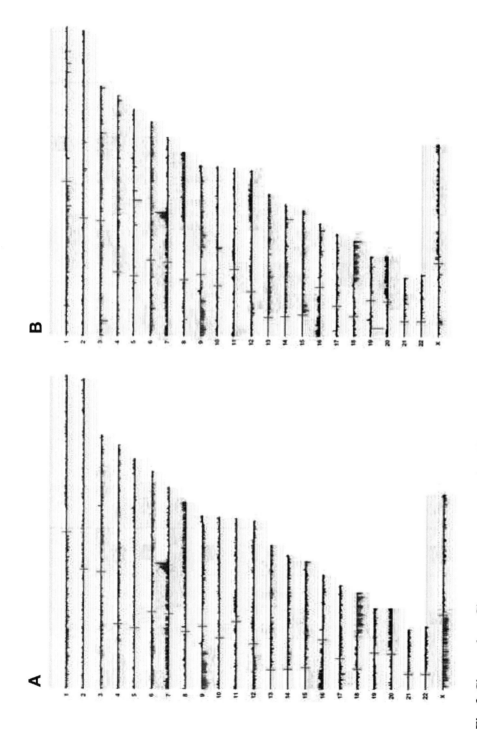

Fig. 2. Cluster along Chromosomes representation (CLAC) of aCGH results for pancreatic cancer cell lines AsPC1 (**A**) and M-Panc96 (**B**). aCGH was performed as described in the text. The gains and losses were identified using CLAC (**48**). Each chromosome is depicted as a dark black horizontal line; the centromere is depicted by the vertical line with the p arm on the left. Fluorescence ratios (log_{10} scale) are plotted as a moving average of five adjacent clones according to their chromosomal map positions; positive and negative ratios are depicted by vertical lines above and below the horizontal line, respectively.

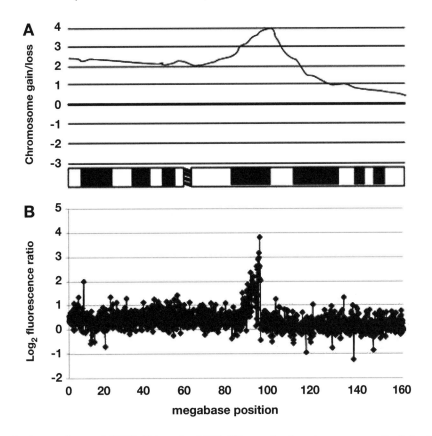

Fig. 3. Comparison of CGH (**A**) and aCGH (**B**) results for chromosome 7 of the human pancreatic cancer cell line AsPC1. The CGH result (**A**) is modified with permission from Ghadimi et al. *(16)*. The chromosomal gain/loss is plotted on the *y*-axis, and the chromosomal position is plotted on the *x*-axis. The aCGH result is depicted in **B**; log 2 fluorescence ratios are plotted on the *y*-axis, and the chromosomal position is plotted on the *x*-axis.

To achieve a higher level of sensitivity and resolution in aCGH, it is important to use oligonucleotide arrays. Several groups are currently working on developing whole genome oligonucleotide arrays; these arrays will provide the highest-resolution arrays to date *(17–19)*.

Recently, aCGH was used for prognostication in stomach cancer *(20)* and in several other cancers *(4,21,22)*. Although more popular for cancer research, aCGH has been successfully shown to detect chromosomal abnormalities associated with several genetic disorders such as detection of microdeletions and microduplications associated with mental retardation *(23)*, congenital aural atresia *(24)*, cardiofaciocutaneous syndrome *(25)*, and so on. aCGH also has been used successfully to identify prominent gene copy number changes that highlight the molecular differences in the great ape lineages *(26)*. In **Subheading 4.**, we highlight the outstanding findings in tuberculosis (Tb) research in relation to geographic genomics and strain variation of the etiological agent *M. tuberculosis (27)*.

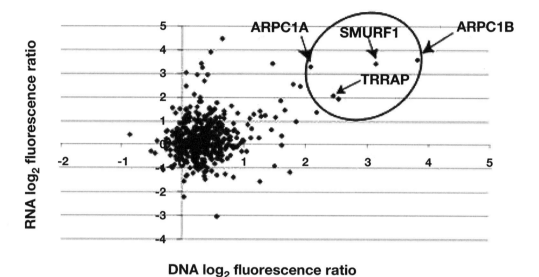

Fig. 4. Graphical representation of DNA and RNA ratios (log 2 scale) for genes or expressed sequence tags located on chromosome 7 for the pancreatic cancer cell line AsPC1. The *x*-axis denotes DNA ratios, and the *y*-axis denotes RNA ratios. Genes resident within the amplicon located at 7q21.2-21.3 are indicated.

4. aCGH and Tuberculosis

Tb has been the scourge of humankind for generations. Considering the long history of Tb, *M. tuberculosis* is probably the most successful human pathogen. *Mycobacterium tuberculosis* is the cause of the largest number of deaths from any single infectious agent and approximately one-third of the world's population is suspected to be asymptomatically infected with the pathogen. Although Tb was prevalent mainly in developing countries, much has been written and discussed about the resurgence of Tb during the past 1–2 decades in the developed countries, mainly due to its "evil nexus" with AIDS *(28)* and the emergence of multidrug-resistant strains of *M. tuberculosis* (MDR-TB) *(29)*. Tuberculosis has been declared by World Health Organization as a worldwide emergency *(30)*. The importance of basic and applied research to develop new strategies for diagnosis, treatment, and prevention of Tb cannot be overemphasized. Understanding the mechanisms responsible for 1) establishment of the disease, 2) resistance to drugs, and 3) dormancy constitute some of the important prerequisites for achieving these goals. MDR-TB has become a global threat, and it is important to understand its molecular basis so that better treatment regimens can be developed. The unraveling of the genomic sequence of *M. tuberculosis* and use of high-throughput technologies has facilitated several concerted approaches toward eradication of Tb.

The microarray technology has been routinely used in the past to study gene expression profiles to understand the pathogenesis, drug resistance, and basic biology of *M. tuberculosis (31–33)*. However, the study of genomic changes is also important in this context. Tb exhibits a wide spectrum of disease states; only about 10% of infected population show symptoms of the disease, and even among the infected population, there is wide variation in clinical symptoms. Although this variation can be attributed in part to environmental and host factors, the genetic variation in pathogen strains also is increasingly being recognized as an important cause. Use of aCGH to identify genomic changes (mainly deletions) has been particularly useful to understand pathogen evolution.

One of the first applications of aCGH in tuberculosis research was the evaluation of various Bacillus Calmette-Guerin (BCG) strains vis-à-vis the pathogenic *M. tuberculosis* H37Rv *(34)*. Strain differences are important to determine efficacy of vaccines in different populations, assuming that different efficacies are the result of adaptation to different populations. The main objective of the study by Behr and colleagues *(34)* was to determine whether the differential efficacies of various BCG vaccine strains could be correlated with presence of gene duplications, deletions, or a combination. This work has provided a benchmark for further studies on pathogen strain variation and evolution. DNA from *M. tuberculosis* H37Rv and BCG were differentially labeled and hybridized on to a whole genome PCR-based *M. tuberculosis* microarray. Because the genomic sequences of *M. tuberculosis* and BCG are almost entirely similar, it was fair to assume that a fluorescence ratio (ratio of fluorescence exhibited by BCG DNA to *M. tuberculosis* DNA) of 1 would mean equal hybridization from both samples (thereby indicating that there was no change in either genomes for that gene/loci); a lower fluorescence ratio would indicate a deletion of that particular gene in the BCG genome and *vice versa*. The study also developed a method to map the fluorescence ratios along the bacterial genome. This work revealed 16 deleted regions in the BCG strains compared with *M. tuberculosis* (ranging in size from 2 to ~13 Kb). Of these 16 regions, nine regions (that included 61 open reading frames [ORFs]) were absent from BCG as well as from virulent strains of *Mycobacterium bovis*. Some of these ORFs might play a role in the increased person-to-person spread of *M. tuberculosis* compared with virulent *M. bovis*. More importantly, the ORFs that are deleted from all BCG strains can be used to design better diagnostics to differentiate *M. tuberculosis* from BCG in a BCG-vaccinated population. Based on their findings, Behr and colleagues have been able to construct a historical time line of deletions that have resulted during the repeated subculturing of BCG vaccine strains across the world.

This study has spurred similar work in other countries to determine the genetic identities of BCG vaccines being administered. BCG Sofia was tested and found to be indistinguishable from BCG Russia *(35)*. In another study, 13 BCG vaccine strains were tested to identify strain-specific deletions, using the Affymetrix platform *(36)*. This study has revealed that none of the present BCG vaccine strains are identical to the original BCG vaccine. In total, nine deleted regions were identified; some deletions were specific to a particular strain, whereas others were more common. In comparison with *M. bovis*, each strain was missing an average of 19 ORFs. Surprisingly, during the period that BCG was subjected to continuous passaging (from its discovery in 1908 until it was lyophilized in 1961), the strains have undergone a much higher proportion of deletions compared with clinical strains of *M. tuberculosis* over a much longer period. These deletions indicate that genomic content required for a pathogenic strain might be higher than for a laboratory nonpathogenic strain *(36)*. However, it is possible that the deletions detected in *M. tuberculosis* clinical strains could be an underestimation, because this comparison is only dependent on work done on *M. tuberculosis* clinical strains isolated from infected patients from San Francisco.

The aforementioned studies, mainly restricted to analysis of evolution of BCG strains, were extended to analysis of *M. tuberculosis* clinical strains by Kato-Maeda and colleagues *(37)*. Their protocol was based on previous work that had shown the use of the Affymetrix platform to carry out such analyses *(38)*. This study showed that it was possible not only to detect deletions but also to define the molecular boundaries of the deletions, to the extent possible. However, the Affymetrix raw data could not reveal the deletions, and specific algorithms had to be used to "clean" the data to reveal the deletions. Usually, this problem is not encountered when using the cDNA array platform, as evident in the work on pancreatic cancer where even single copy loss could be easily detected.

Kato-Maeda and colleagues *(37)* analyzed 15 clinical isolates, each with a distinct genotype, collected from patients over a period of 7 yr (1991–1998), from the San Francisco area. Complete clinical data as well as data on healthy contacts were collected. They used an Affymetrix gene chip exclusively designed for the *M. tuberculosis* genome and compared fluorescence intensities of hybridization of DNA from each strain with that of *M. tuberculosis* H37Rv strain separately. There was no difference detected between the genomes of *M. tuberculosis* H37Rv and H37Ra (an avirulent derivative of H37Rv). This study has revealed a distinct pattern of deleted regions spread throughout the genome whereby the deletions were not randomly distributed. In total, 25 different deleted sequences were identified, which included 93 ORFs and 22 intergenic regions. Insertion sequences and prophages were more commonly present in deleted regions than was expected by chance. In five instances, the deleted regions were directly replaced by IS6110 elements. In some other cases, the deletions were the result of complex genomic rearrangements. Therefore, it seems that the *M. tuberculosis* genome is highly "active" and particularly the IS6110 element may be responsible for part of the genomic rearrangements and deletions *(39–41)*. *M. tuberculosis* also contains approx 40 other mobile genetic elements that may participate in generating deletions. Among the deleted ORFs, there were four genes belonging to the PE/PPE family and 15 that could play putative role in pathogenesis, latency, or both. The study also revealed a good correlation between percentage of genome deleted and the proportion of infected patients having pulmonary cavities; although no significant correlation between the amount of genome deletion and transmission and pathogenicity of the strains could be established.

Two larger studies were recently reported by Tsolaki and colleagues *(42,43)*. The studies included 100 well-characterized strains, chosen based on the IS6110 banding pattern to select strains unique to San Francisco. Again, the deletions were more or less confined to certain regions of the genome rather than being randomly distributed. Many deletions were common in closely related isolates, whereas some others were common in unrelated isolates. Because all deletions were present in clinical strains, it is safe to assume that none of the deleted regions will be involved in pathogenesis; rather, pathogenesis could be attributed to host interactions and response to antibiotics. The study revealed a total of 68 different deletions that included 224 genes *(42)*. Among the 68 different deletions, only three deletions occurred in different lineages, whereas all others came under the same lineage. The magnitude of deletions observed was comparable with that observed between *M. tuberculosis* and *M. bovis*. As reported previously, mobile genetic elements were found to be deleted more frequently than by chance.

More surprising was the high frequency of deletion of genes involved in intermediary metabolism and respiration; this high frequency may be an effect of selection pressure of immune system *(42)*. The study has attempted to correlate the deletions with specific host population, in order to establish pathogen–host relationship. This kind of association is based on the premise that genomic changes (single-nucleotide polymorphisms and genomic deletions) are usually one-time events (and are irreversible considering that horizontal gene transfer is an extremely rare event in mycobacteria; *[43]*); therefore, they are stable markers for construction of phylogeny of *M. tuberculosis* strains. The strains that did not cluster with other strains isolated in San Francisco were the ones isolated from patients who arrived in San Francisco from other countries, so the infection presumably occurred in their native countries *(44)*. Apparently, a particular strain of *M. tuberculosis* establishes stable association with a particular host population or a given niche *(45)*, and this association remains stable for a long time. It may be possible, therefore, to predict the *M. tuberculosis* genotype based on the place where the person contracted the disease *(44)*! This is a tempting proposition; however, it is based only on one study from San Francisco. Authors of this study acknowledge that this study includes no samples from the Indian subcontinent. Several such studies would have to be conducted on strains from other parts of the world to arrive at a concrete conclusion.

All aCGH studies have used *M. tuberculosis* H37Rv as the reference genome to profile deletions in BCG strains as well as in clinical strains of *M. tuberculosis*. This would make it impossible to detect deletions in regions that are already lost in H37Rv. Therefore, to obtain a complete coverage of genomes of all *M. tuberculosis* strains, it is imperative to design a microarray that can query all deletions. For this purpose, it may be worthwhile to sequence a few clinical strains of the pathogen from different geographic regions. This approach would aid in development of a more comprehensive tuberculosis array.

Several alternative strategies have been used for genotyping strains from different geographical regions. An exhaustive and comprehensive study was reported by Ahmed and colleagues *(46)*. Their work was based on use of fluorescent amplified fragment length polymorphism (FAFLP) to study pathogen evolution. Because *M. tuberculosis* harbored a lower proportion of single-nucleotide polymorphisms compared with other bacteria, it had been proposed that it could be a "recent" pathogen *(47)*. The study by Ahmed and colleagues suggests that deletions of genes, acquisition of genes, or both might be more important than point mutations in establishing the genetic diversity of the pathogen strains. Results from the FAFLP data actually indicate that *M. tuberculosis* might harbor a higher proportion of single-nucleotide polymorphisms than previously reported. Deletions identified using the FAFLP approach seem to be random in the 150 strains analyzed, contrary to what was proposed previously *(40)*. This contradiction can be resolved by mapping deletions in a large number of strains from different geographical regions. The FAFLP and other molecular typing studies also have revealed that strains from different geographic regions cluster together, indicating that they have coevolved with the genotype of the local population *(45,46)*, corroborating the results obtained from the aCGH studies.

Several other genome studies on pathogen strains have revealed interesting results. The work by Brosch and colleagues *(40)* has shown a specific pattern of deletions that distinguish strains belonging to the *M. tuberculosis* complex from each other. The study has elucidated the interesting proposition that the human pathogen *M. tuberculosis* might not have evolved from its bovine counterpart *M. bovis*, but rather both might have evolved from a common progenitor strain that was already a human pathogen. Again, this study was based on analyses of deletion patterns from different strains. The CGH studies have revealed a relatively less genomic variation in *M. tuberculosis* compared with other bacteria such as *Escherichia coli*. However, given the ability of *M. tuberculosis* to adapt to different environments and to cause disease with varying degree of virulence in various populations, it is possible that the pathogen might harbor a wider range of genomic changes than reported in this and other aCGH studies. Therefore, it is important to conduct a large study, including clinical strains from various geographic regions to establish concrete correlations with pathogenicity, drug resistance, and pathogen evolution. The FAFLP results have provided a backbone for future studies on MDR-TB as well as geographic genomics of the Tb pathogen. It will now be possible to embark on a comprehensive program of Tb genomics to correlate genomic changes as well as gene expression changes with variations in pathogenesis and MDR status of various clinical strains of *M. tuberculosis*. On the basis of the large strain collection available in India, it should be possible to correlate pathogen evolution with population diversity as well as multidrug resistance. This work should therefore help to identify novel drug targets, explain variation in virulence of different clinical strains, and also delineate the molecular basis for MDR-TB.

References

1. Kallioniemi, A., Kallioniemi, O. P., Sudar, D., et al. (1992). Comparative genomic hybridization for molecular cytogenetic analysis of solid tumors. *Science* **258**, 818–821.
2. Longo, L. D. (1978). Classic pages in obstetrics and gynecology. In: The chromosome number in man (J. Hin Tjio and A. Levan. Hereditas, eds.), vol. 42, pp. 1–6, 1956. *Am J Obstet Gynecol* **130**, 722.
3. Van Prooijen-Knegt, A. C., and Van der Ploeg, M. (1982). Localization of specific DNA sequences in cell nuclei and human metaphase chromosomes by fluorescence microscopy. *Cell Biol. Int. Rep.* **6**, 653.
4. Gebhart, E. (2004). Comparative genomic hybridization (CGH): ten years of substantial progress in human solid tumor molecular cytogenetics. *Cytogenet. Genome Res.* **104**, 352–358.
5. Kallioniemi, O. P., Kallioniemi, A., Piper, J., et al. (1994). Optimizing comparative genomic hybridization for analysis of DNA sequence copy number changes in solid tumors. *Genes Chromosomes Cancer* **10**, 231–243.
6. Forozan, F., Karhu, R., Kononen, J., Kallioniemi, A., and Kallioniemi, O. P. (1997). Genome screening by comparative genomic hybridization. *Trends Genet.* **13**, 405–409.
7. Pinkel, D., Segraves, R., Sudar, D., et al. (1998). High resolution analysis of DNA copy number variation using comparative genomic hybridization to microarrays. *Nat. Genet.* **20**, 207–211.
8. Pollack, J. R., Perou, C. M., Alizadeh, A. A., et al. (1999). Genome-wide analysis of DNA copy-number changes using cDNA microarrays. *Nat. Genet.* **23**, 41–46.
9. Solinas-Toldo, S., Lampel, S., Stilgenbauer, S., et al. (1997). Matrix-based comparative genomic hybridization: biochips to screen for genomic imbalances. *Genes Chromosomes Cancer* **20**, 399–407.

10. Pollack, J. R., Sorlie, T., Perou, C. M., et al. (2002). Microarray analysis reveals a major direct role of DNA copy number alteration in the transcriptional program of human breast tumors. *Proc. Natl. Acad. Sci. USA* **99**, 12,963–12,968.

11. Mantripragada, K. K., Buckley, P. G., Jarbo, C., Menzel, U., and Dumanski, J. P. (2003). Development of NF2 gene specific, strictly sequence defined diagnostic microarray for deletion detection. *J. Mol. Med.* **81**, 443–451.

12. Bashyam, M. D., Bair, R., Kim, Y. H., et al. (2005). Array-based comparative genomic hybridization identifies localized DNA amplifications and homozygous deletions in pancreatic cancer. *Neoplasia* **7**, 556–562.

13. Holzmann, K., Kohlhammer, H., Schwaenen, C., et al. (2004). Genomic DNA-chip hybridization reveals a higher incidence of genomic amplifications in pancreatic cancer than conventional comparative genomic hybridization and leads to the identification of novel candidate genes. *Cancer Res.* **64**, 4428–4433.

14. Aguirre, A. J., Brennan, C., Bailey, G., et al. (2004). High-resolution characterization of the pancreatic adenocarcinoma genome. *Proc. Natl. Acad. Sci. USA* **101**, 9067–9072.

15. Heidenblad, M., Schoenmakers, E. F., Jonson, T., et al. (2004). Genome-wide array-based comparative genomic hybridization reveals multiple amplification targets and novel homozygous deletions in pancreatic carcinoma cell lines. *Cancer Res.* **64**, 3052–3059.

16. Ghadimi, B. M., Schrock, E., Walker, R. L., et al. (1999). Specific chromosomal aberrations and amplification of the AIB1 nuclear receptor coactivator gene in pancreatic carcinomas. *Am. J. Pathol.* **154**, 525–536.

17. Barrett, M. T., Scheffer, A., Ben-Dor, A., et al. (2004). Comparative genomic hybridization using oligonucleotide microarrays and total genomic DNA. *Proc. Natl. Acad. Sci. USA* **101**, 17,765–17,770.

18. Brennan, C., Zhang, Y., Leo, C., et al. (2004). High-resolution global profiling of genomic alterations with long oligonucleotide microarray. *Cancer Res.* **64**, 4744–4748.

19. Carvalho, B., Ouwerkerk, E., Meijer, G. A., and Ylstra, B. (2004). High resolution microarray comparative genomic hybridisation analysis using spotted oligonucleotides. *J. Clin. Pathol.* **57**, 644–646.

20. Weiss, M. M., Kuipers, E. J., Postma, C., et al. (2003). Genome wide array comparative genomic hybridisation analysis of premalignant lesions of the stomach. *Mol. Pathol.* **56**, 293–298.

21. Oostlander, A. E., Meijer, G. A., and Ylstra, B. (2004). Microarray-based comparative genomic hybridization and its applications in human genetics. *Clin. Genet.* **66**, 488–495.

22. Albertson, D. G., and Pinkel, D. (2003). Genomic microarrays in human genetic disease and cancer. *Hum. Mol. Genet.* **12(Spec. no. 2)**, R145–R152.

23. Vissers, L. E., de Vries, B. B., Osoegawa, K., et al. (2003). Array-based comparative genomic hybridization for the genomewide detection of submicroscopic chromosomal abnormalities. *Am. J. Hum. Genet.* **73**, 1261–1270.

24. Veltman, J. A., Jonkers, Y., Nuijten, I., et al. (2003). Definition of a critical region on chromosome 18 for congenital aural atresia by array CGH. *Am. J. Hum. Genet.* **72**, 1578–1584.

25. Rauen, K. A., Albertson, D. G., Pinkel, D., and Cotter, P. D. (2002). Additional patient with del(12)(q21.2q22): further evidence for a candidate region for cardio-facio-cutaneous syndrome. *Am. J. Med. Genet.* **110**, 51–56.

26. Fortna, A., Kim, Y., MacLaren, E., et al. (2004). Lineage-specific gene duplication and loss in human and great ape evolution. *PLoS Biol.* **2**, E207.

27. Majeed, A. A., Ahmed, N., Rao, K. R., et al. (2004). AmpliBASE MT: a *Mycobacterium tuberculosis* diversity knowledgebase. *Bioinformatics* **20**, 989–992.

28. MacDougall, D. S. (1999). TB & HIV: the deadly intersection. *J. Int. Assoc. Physicians AIDS Care* **5**, 20–27.

29. Chakhaiyar, P., and Hasnain, S. E. (2004). Defining the mandate of tuberculosis research in a postgenomic era. *Med Princ. Pract.* **13**, 177–184.

30. Chakhaiyar, P., and Hasnain, S. E. (1995). Tuberculosis epidemic poses international threat. World Health Organization. *AIDS Wkly Plus*, pp. 24–25.

31. Kendall, S. L., Rison, S. C., Movahedzadeh, F., Frita, R., and Stoker, N. G. (2004). What do microarrays really tell us about M. tuberculosis? *Trends Microbiol.* **12**, 537–544.

32. Waddell, S. J., Stabler, R. A., Laing, K., Kremer, L., Reynolds, R. C., and Besra, G. S. (2004). The use of microarray analysis to determine the gene expression profiles of *Mycobacterium tuberculosis* in response to anti-bacterial compounds. *Tuberculosis* **84**, 263–274.

33. Boshoff, H. I., Myers, T. G., Copp, B. R., et al. (2004). The transcriptional responses of *Mycobacterium tuberculosis* to inhibitors of metabolism: novel insights into drug mechanisms of action. *J. Biol. Chem.* **279**, 40,174–40,184.

34. Behr, M. A., Wilson, M. A., Gill, W. P., et al. (1999). Comparative genomics of BCG vaccines by whole-genome DNA microarray. *Science* **284**, 1520–1523.

35. Stefanova, T., Chouchkova, M., Hinds, J., et al. (2003). Genetic composition of Mycobacterium bovis BCG substrain Sofia. *J. Clin. Microbiol.* **41**, 5349.

36. Mostowy, S., Tsolaki, A. G., Small, P. M., and Behr, M. A. (2003). The in vitro evolution of BCG vaccines. *Vaccine* **21**, 4270–4274.

37. Kato-Maeda, M., Rhee, J. T., Gingeras, T. R., Salamon, H., Drenkow, J., Smittipat, N., and Small, P. M. (2001). Comparing genomes within the species *Mycobacterium tuberculosis*. *Genome Res.* **11**, 547–554.

38. Salamon, H., Kato-Maeda, M., Small, P. M., Drenkow, J., and Gingeras, T. R. (2000). Detection of deleted genomic DNA using a semiautomated computational analysis of GeneChip data. *Genome Res.* **10**, 2044–2054.

39. Fang, Z., Doig, C., Kenna, D. T., et al. (1999). IS6110-mediated deletions of wild-type chromosomes of *Mycobacterium tuberculosis*. *J. Bacteriol.* **181**, 1014–1020.

40. Brosch, R., Philipp, W. J., Stavropoulos, E., Colston, M. J., Cole, S. T., and Gordon, S. V. (1999). Genomic analysis reveals variation between *Mycobacterium tuberculosis* H37Rv and the attenuated M. tuberculosis H37Ra strain. *Infect. Immun.* **67**, 5768–5774.

41. Ho, T. B., Robertson, B. D., Taylor, G. M., Shaw, R. J., and Young, D. B. (2000). Comparison of *Mycobacterium tuberculosis* genomes reveals frequent deletions in a 20 Kb variable region in clinical isolates. *Yeast* **17**, 272–282.

42. Tsolaki, A. G., Hirsh, A. E., DeRiemer, K., et al. (2004). Functional and evolutionary genomics of *Mycobacterium tuberculosis*: insights from genomic deletions in 100 strains. *Proc. Natl. Acad. Sci. USA* **101**, 4865–4870.

43. Cole, S. T., Brosch, R., Parkhill, J., et al. (1998). Deciphering the biology of *Mycobacterium tuberculosis* from the complete genome sequence. *Nature* **393**, 537–544.

44. Hirsh, A. E., Tsolaki, A. G., DeRiemer, K., Feldman, M. W., and Small, P. M. (2004). Stable association between strains of *Mycobacterium tuberculosis* and their human host populations. *Proc. Natl. Acad. Sci. USA* **101**, 4871–4876.

45. Ahmed, N., Caviedes, L., Alam, M., et al. (2003). Distinctiveness of *Mycobacterium tuberculosis* genotypes from human immunodeficiency virus type 1-seropositive and -seronegative patients in Lima, Peru. *J. Clin. Microbiol.* **41**, 1712–1716.

46. Ahmed, N., Alam, M., Rao, K. R., et al. (2004). Molecular genotyping of a large, multicentric collection of tubercle bacilli indicates geographical partitioning of strain variation and has implications for global epidemiology of *Mycobacterium tuberculosis*. *J. Clin. Microbiol.* **42**, 3240–3247.

47. Sreevatsan, S., Pan, X., Stockbauer, K. E., et al. (1997). Restricted structural gene polymorphism in the *Mycobacterium tuberculosis* complex indicates evolutionarily recent global dissemination. *Proc. Natl. Acad. Sci. USA* **94,** 9869–9874.
48. Wang, P., Kim, Y., Pollack, J., Narasimhan, B., and Tibshirani, R. (2005). A method for calling gains and losses in array CGH data. *Biostatistics* **6,** 45–58.

9

Regional Specialization of Endothelial Cells as Revealed by Genomic Analysis

Jen-Tsan Ashley Chi, Zhen Wang, and Anil Potti

Summary

The vascular system is locally specialized to accommodate widely varying needs of individual tissues. The regional specialization of vascular structure is closely linked to the topographic differentiation of endothelial cells (ECs). The gene expression programs that characterize specific ECs define their physiological specialization and their role in the development of vascular channels and epithelial organs. Our understanding of EC regional differentiation is very limited. To assess the heterogeneity of ECs on a global scale, we used DNA microarrays to obtain the global gene expression profiles of more than 50 cultured ECs purified from 14 different anatomic locations. We found that ECs from different blood vessels and microvascular ECs from different tissues have distinct and characteristic gene expression profiles. Pervasive differences in gene expression patterns distinguish the ECs of large vessels from microvascular ECs. We identified groups of genes characteristic of arterial and venous endothelium. Hey2, the human homolog of the zebrafish gene *gridlock*, was expressed only in arterial ECs and could trigger arterial-specific gene expression programs when introduced into venous ECs. Several genes critical in the establishment of left–right asymmetry were expressed preferentially in venous ECs, suggesting a surprising link between vascular differentiation and body plan development. Tissue-specific expression patterns in different tissue microvascular ECs suggest they are distinct differentiated cell types that play roles in the local physiology of their respective organs and tissues. Therefore, ECs from different anatomical locations constitute many distinct, differentiated cell types that carry out unique genetic programs to specify the site-specific design and functions of blood vessels to control internal body compartmentalization, regulate the trafficking of circulating cells, and shape the vascular development. In this chapter, we discuss these findings and their implications in different aspects of vascular biology during development and vascular diseases.

Key Words: Differentiation; endothelial cell; Hey2; microarray.

1. Introduction

Endothelial cells (ECs) line the inside of all blood and lymphatic vessels, forming a structurally and functionally heterogeneous population of cells in a large network of vascular channels. Their complexity and diversity have long been recognized; yet, very little is known about the molecules and regulatory mechanisms that mediate the heterogeneity of different EC populations. The constitutive organ- and microenvironment-specific phenotype of ECs controls internal body compartmentalization, regulating the trafficking of circulating cells to distinct vascular beds and plays an important role in shaping the vascular development during development as well as tissue repair and

From: *Bioarrays: From Basics to Diagnostics*
Edited by: K. Appasani © Humana Press Inc., Totowa, NJ

remodeling. Furthermore, ECs lining different vascular channels need to adapt to the differences in the hemodynamics and oxygen tensions of the blood flow they carry as well as the local microenvironments and adjacent cells. These differences also may play a critical role in pathological conditions, including inflammation, tumor angiogenesis, and wound healing.

The sequence of the entire human genome promised the possibility of acquiring genomic level knowledge on different aspects of biology. One important advance is the development of microarrays in which the expression levels of tens of thousands of genes are assessed in a parallel manner with one single experiment. This technology has triggered the explosion of expression data in a variety of biological systems. This immense amount of data requires the development and implementation of different bioinformatics tools and database infrastructures to explore and derive biological information. This combination of genomic expression tools and bioinformatics has led us to a greater understanding of many biological and medical dilemmas on a global and systemic level. For example, microarrays have allowed the detailed molecular portraits of human cancers and revealed the unexpected complexities from classical surgical pathology and leads to the revolution of molecular classification with the identification of gene signatures associated with different significant clinical outcomes *(1)*. These studies provide important insight into the biology of these cancers and guide the associated clinical decisions during the treatment to benefit the patients in more a specific manner, thus realizing the vision of personalized medicine. Microarrays are also powerful tools to assess the heterogeneity of different cell types, because of their ability to assess the expression levels of tens of thousands of genes simultaneously to arrive at the understanding of the extent and nature of heterogeneity on a global scale and by dissecting the molecular pathways leading to the heterogeneity. We have used DNA microarrays to explore the diversity of ECs in different types of blood vessels and different anatomical locations as reflected in the gene expression programs when these ECs are cultured under identical conditions *(2)*. Our analysis shows that ECs from different blood vessels or anatomical sites have intrinsic characteristic gene expression programs with in vitro passages and should be considered as distinct and specialized cell types. Our analysis also reveals the molecular pathways leading to the endothelial specialization in vivo.

2. Overview of Gene Expression Patterns

In our experiments, 52 purified EC samples, representing 14 distinct locations, were propagated in identical culture conditions. This sample set included ECs from five arteries (aorta, coronary artery, pulmonary artery, iliac artery, and umbilical artery), two veins (umbilical vein and saphenous vein), and seven tissues (skin, lung, intestine, uterus myometrium, nasal polyps, bladder, and myocardium). All ECs displayed a "cobblestone" appearance and were free of contamination by spindle-shaped cells. The purity of EC lineage is further assessed using flow cytometry or staining with CD31 antibody. We analyzed mRNA from 53 different cultured EC samples, including two samples from one coronary artery EC culture that was sampled twice at successive passages (coronary artery 2a and 2b) with DNA microarrays containing 43,000 elements. The first significant result was a striking order and consistency in the expression patterns, reflecting the sites of origins of the cultured ECs. Unsupervised hierarchical clustering *(3)* of the gene expression patterns from all 53 samples produced a consistent grouping of the cells according to their sites of origin (**Fig. 1A**). This finding suggests that ECs from different locations have distinct and characteristic expression patterns that persist with in vitro culture. The majority of ECs had expression patterns that clustered into discrete groups of ECs from the same location. Overall, the samples were divided into two major branches: I and II. Branch I was composed of three subgroups: 1) an "artery" group consisting of all ECs cultured from arteries, including aorta, coronary artery, pulmonary artery, iliac artery, and umbilical artery; 2) a "vein" group, consisting of ECs cultured from umbilical vein and saphenous vein; and 3) a group we called "tissue type II" consisting of ECs cultured from nasal polyps, bladder, and myocardium (**Fig. 1A**). Branch II of the samples, which we designated as "tissue type I" (**Fig. 1A**) contained all the microvascular ECs from skin, lung, intestine, and myometrium. The most prominent differences between the two branches are defined by two large groups of genes that we labeled as the "large vessel cluster" and the "microvascular cluster," respectively (**Fig 1B**).

3. Gene Expression Patterns Between Macrovascular and Microvascular ECs

To identify genes with the most consistent different levels of expression between ECs from large vessels and microvascular ECs (28 large vessel ECs and 25 microvascular ECs), we used a Wilcoxon rank sum test *(4)*, and a *p* value of 0.005 was considered significant. We selected 521 large vessel EC-specific genes and 2521 microvascular EC-specific genes for analysis (**Fig 1C**).

The distinct gene expression patterns seen are likely to be related to the characteristic differences in physiological functions and the immediate microenvironments of these vascular channels. It is also important to note that the differentially expressed genes play diverse roles in endothelial biology, including the biosynthesis of and interaction with extracellular matrix (ECM), neuronal signaling and migration, angiogenesis, and lipid metabolism. Large vessel ECs differentially expressed several genes involved in the biosynthesis and remodeling of ECM, such as fibronectin, collagen 5α1 and 5α2, and osteonectin (**Fig 1C**, gene names shown in blue). These differences are probably related, in part, to the relatively thick vascular wall surrounding the endothelium of the large vessels. Incontrast, microvascular ECs express genes encoding basement membrane proteins, such as laminin, collagen 4α1 and 4α2, and collagen 4α-binding protein and ECM-interacting proteins, such as CD36, α1 integrin, α4 integrin, α9 integrin, and β4 integrin (**Fig. 1C**, gene names shown in blue), perhaps related to the intimate association of microvascular ECs with the basement membrane and ECM.

ECs present a physical barrier to both blood-borne pathogens and immune cells, which must transverse the barrier for trafficking between tissues and the bloodstream. Microvascular ECs express higher levels of transcripts encoding proteins involved in the trafficking of circulating blood cells and pathogens (**Fig. 1C**, gene names shown in blue), such as CD36, as a cellular receptor for *Plasmodium falciparum* *(5)* and CEACAM-1 (CD66a) with *Neisseria* bacteria *(6,7)*, and macrophage mannose receptors (CD206) are important for pathogen trapping and lymphocyte recruitment *(8)*. These microvascular EC-specific gene product proteins might play important roles in conferring the specificity of their migratory paths. We have confirmed the specific

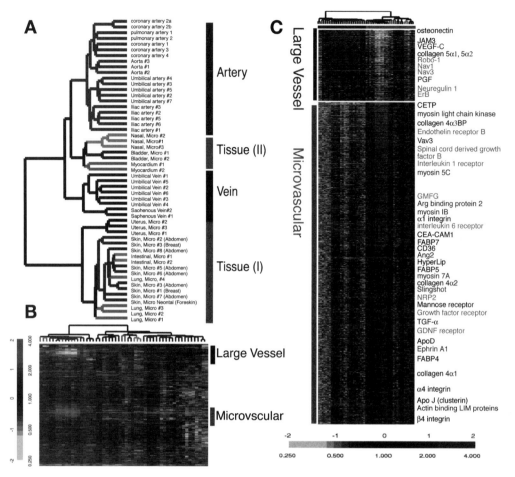

Fig. 1. Global view of endothelial cell diversity and macrovascular or microvascular differentiation. (**A**) Gene expression patterns of cultured ECs organized by unsupervised hierarchical clustering. The global gene expression patterns of 53 cultured ECs were sorted based on similarity by hierarchical clustering. Approximately 6900 genes were selected from the total data set, based on variations in expression relative to the mean expression level across all samples greater than threefold in at least two cell samples. The sites of origin of each EC culture are indicated and color coded. The anatomical origins of skin EC are indicated. The apparent order in the grouping of EC gene expression patterns is indicated to the right of the dendrogram. (**B**) Overview of gene expression patterns of all EC samples. The variations in gene expression described in **A** are shown in matrix format (*1*). The scale extends from 0.25- to 4-fold over mean (–2 to +2 in log 2 space) as is indicated on the left. Gray represents missing data. The gene clusters characteristic of large vessel and microvascular ECs are indicated on the right. (**C**) Features of large vessels and microvascular EC gene expression programs. Large vessel-specific (521) and 2521 microvascular EC-specific genes are shown in ascending order of *p* values. Genes involved ECM biosynthesis and interaction (blue), neuroglial signaling and migration (orange), angiogenesis (red), and lipid metabolism (black) are labeled by the indicated colors.

surface expression of mannose receptor and $\alpha 1$ integrin on skin microvascular ECs with flow cytometry.

The similarity of the migration paths of blood vessels and nerves has inspired interest in their interactions during development. The difference between large and microvascular ECs further reveal the divergent way in which large vessels and microvascular circulation seem designed to communicate with peripheral nerves during development and maturation. Large vessel ECs express many genes thought to be expressed in cells of neuronal lineage, such as robo-1, neuron navigator 1 and -3, neuroligin, neurogranin, and neuroregulin and its receptor ErbB (**Fig. 1C**, gene names shown in orange). Some of these proteins play important roles in neuronal migration during development. For example, robo-1 is a surface receptor for the slit proteins that act as migration signals. This role suggests the possibility that large vessel ECs may respond to some of the same guidance signals that specify the paths of neural processes through the mesenchymes of different target tissues and organs *(9)*. This hypothesis is consistent with the recent study that when nerves are severed, large vessels still migrate to their target locations, whereas microvascular circulation is greatly affected *(10)*. Our results show that microvascular ECs express many genes suggestive of paracrine signals with neuroglial cells, such as growth hormone receptor, endothelin receptor B, glycine receptor, purinergic receptor, glial cell line-derived neurotrophic factor receptor, platelet-derived growth factor receptor, interleukin-1 receptor, and interleukin-6 receptor (**Fig. 1C**, gene names shown in orange). Microvascular ECs also express secreted factors that promote the survival and differentiation of neuroglial cells, such as transforming growth factor (TGF)-α, glial maturation factor γ, stromal cell-derived factor-1, and spinal cord-derived growth factor. These proteins, uniquely expressed in small vessel ECs, suggest an intimate functional interaction between microvasculature and peripheral nerves. Because of the importance of Schwann cells in microvasculature development *(10)*, the microvascular EC-specific expression of genes involved in the EC–glial cell interaction (glial cell line derived-neurotrophic factor receptor and glial maturation factor γ) is particular noteworthy.

Large vessel ECs expressed placental growth factor, which is involved in the establishment of collateral circulation, a response to ischemia unique to large vessels *(11)*. They also express vascular endothelial growth factor-C, a growth factor essential for the growth and differentiation of lymphatic vessels *(12)* (**Fig 1C**, gene names shown in red). Microvascular networks are the main sites for angiogenesis in adults. Many genes associated with angiogenesis were expressed specifically in microvascular ECs (**Fig 1C**, gene names shown in red). Angiopoietin 2, a marker of tumor angiogenesis, modulates the remodeling of vessels during angiogenesis by reverting vessels to a more plastic state that may facilitate the vascular sprouting necessary for subsequent remodeling *(13)*. Lmo2 is a LIM-only transcription factor involved in angiogenesis. Sprouty is a fibroblast growth factor (FGF) antagonist that participates in regulating the branching pattern of insect tracheal trees *(14)*. TGF-α induces ephrin-A1 expression that is involved in tumor angiogenesis *(15)*. The expression of both TGF-α and ephrin-A1 in microvascular ECs raises the possibility of autocrine regulation. The higher level of expression of actin binding LIM protein 1, actinin-associated LIM protein, Arg-binding protein 2, Slingshot, vav3, myosin IB, myosin 5C, myosin 7A, and myosin light chain kinase in the microvascular ECs may be related to the ability of microvascular

ECs to undergo extensive cytoskeletal remodeling and migration during angiogenesis. The microvascular EC gene clusters included genes related to lipid transport and metabolism, such as Apo D, Apo J (clusterin), ApoL, cholesteryl ester transfer protein, FABP4, FABP5, and Hyperlip (**Fig. 1C**, gene names shown in black), consistent with a major role for small vessel ECs in mediating lipid transport and metabolism.

Finally, our strategy and findings will greatly enhance the validity of studies that demonstrate that systemic elements such as oxidative stress may modulate the expression and function of vascular genes, such as vascular endothelial growth factor, FGF, and platelet-derived growth factor, which play key atherogenic roles by their regulation of cell growth, differentiation, and fibroproliferative responsiveness. Appropriate gene expression profiling of ECs would delineate the changes induced by oxidative stress on the profile of growth factor gene expression in ECs and describe the impact that these modifications have on endothelial phenotype as well as on the behavior of neighboring vascular smooth muscle cells and fibroblasts, knowledge of which would be extremely useful in modifying angiogenesis and atherogenesis.

4. Gene Expression Pattern Differences Between Arterial and Venous ECs

Arteries and veins in the vertebrate circulatory system are functionally defined by the direction of blood flow relative to the heart. Recent evidence indicates that the artery–vein identities of ECs are established before blood circulation begins *(16)*. Several molecular markers specifically expressed in arteries or veins have been identified in model organisms *(16)*. In our unsupervised clustering analysis, all the arterial and venous ECs were separated into two different branches (arterial branch and venous branch in **Fig.1A**), reflecting extensive differences in their expression patterns. Using a rank sum test to identify genes with the largest, consistent differences in expression between two groups of samples: eight venous ECs (six umbilical veins and two saphenous veins) and 20 arterial ECs (three aortas, two pulmonary arteries, five coronary arteries, five umbilical arteries, and five iliac arteries), we selected 817 vein-specific genes and 59 artery-specific genes with $p < 0.005$ (**Fig 2A**).

The higher number of vein-specific genes compared with artery-specific genes may reflect the relatively low diversity of vein samples (two types, eight samples) compared with artery samples (five types, 20 samples). EphB4, reported previously to be venous specific, was among the venous EC-specific genes. Genes involved in determining left–right (L–R) asymmetry of the body plan, including smoothened, growth differentiation factor 1, lefty-1, and lefty-2 were particularly noteworthy members of the venous-specific group. During development, right-sided looping of the developing cardiac tube is the first sign of the L–R asymmetry, and the L–R determination is intimately connected to the development of the heart and vasculature. Defects in L–R asymmetry, such as situs inversus, are characteristically associated with vascular anomalies. Disruption of lefty-1 in mice leads to malpositioning of venous vessels and anomalies in the heart and its connection with major vessels *(17)*. Lefty-1 is an antagonist of Nodal through the activin-like receptor. Mutations in activin receptor-like kinase-1 result in persistent arterial-venous shunts and early loss of anatomical, molecular, and functional distinctions between arteries and veins *(18)*. The persistent expression of these genes in venous ECs suggests a molecular connection between L–R determination and the distinct differentiation programs of arterial and venous ECs.

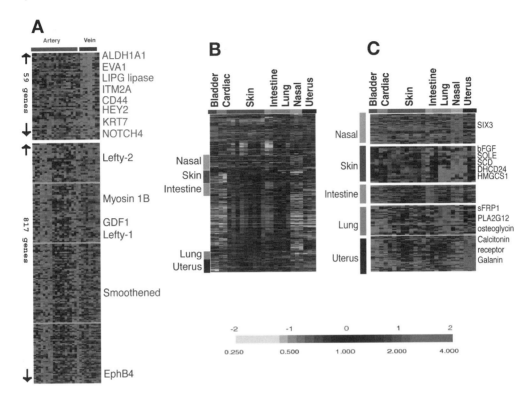

Fig. 2. Artery- or vein- and tissue-specific EC gene expression programs. (**A**) Artery- or vein-specific genes identified by a Wilcoxon rank sum test are shown in ascending order of p value within each gene list and names of select artery-specific genes (red) and vein-specific genes (blue) are shown. (**B,C**) Tissue-specific EC gene expression programs. The expression patterns of tissue specific genes as identified by multiclass SAM analysis among all the tissue microvascular ECs were shown (**B**). Clusters of genes with unique tissue expression in nasal polyps (pink), skin (brown), intestine (orange), lung (blue), and uterus (black) are marked by the indicated color and expanded on the right (**C**) with selected gene names.

The arterial EC-specific gene cluster includes cell surface proteins (Notch 4, EVA1, CD44, Ephrin-B1, and integral membrane protein 2A), metabolic enzymes (aldehyde dehydrogenase A1 and endothelial lipase), C17, keratin 7, and a transcription factor termed Hairy/Enhancer of split-related basic HLH protein 2 (Hey2). We have confirmed the arterial-specific expression of Hey2 and C17 transcript with real-time PCR and CD44 surface expression with flow cytometry. Several of these genes have already been implicated in vascular development. The Notch family of receptors and ligand Delta-like 4 show arterial expression in zebrafish and mouse and are essential for arterial cell fate determination *(19)*. Hey2, a member of the Hairy-related transcription factor family of transcription factors, is induced by Notch signaling *(20)*. The zebrafish homolog of Hey2 is the gene targeted by the *gridlock (grl)* mutation *(21)*, which leads to a localized defect in vascular patterning of dorsal aorta, consistent with specific expression of *grl* in dorsal aorta. The apparently conserved expression pattern of the Notch pathway (Notch 4 and Hey2) in human arterial ECs highlights the potential importance of this pathway in human arterial EC differentiation. This observation is further confirmed by the genetic studies with Hey2 and Delta 4 in mouse *(22)*.

5. Hey2 Activates Expression of Arterial-Specific Genes

To further examine the possible role of Hey2 in arterial EC differentiation, we tested the effect of Hey2 expression patterns on vein-derived ECs. We infected umbilical vein ECs with a retroviral vector carrying the Hey2 gene, along with a green fluorescent protein (GFP) marker, or with a control vector expressing GFP alone, and selected the infected ECs by positive GFP expression by using flow cytometry. Hey2 transduction leads to elevated expression of follistatin, an antagonist of activin, as well as several artery-specific genes identified in our previous analysis, including aldehyde dehydrogenase A1, EVA1, and keratin 7, and to downregulation of myosin I. These results suggest a pathway of arterial EC differentiation in which Hey2 turns on features of artery-specific gene expression program. Hey2 induction of follistatin also may contribute to arterial differentiation by antagonizing TGF-β family members (such as GDF, lefty-1, and lefty-2) expressed by venous ECs. This study has shown the importance of Hey 2 in driving the arterial differentiation of arterial ECs.

6. Genes Differentially Expressed in ECs from Different Tissues

Previous studies with phage display have revealed molecular heterogeneity in the vascular beds in different organs and tissues *(23)*, but little is known about the detailed origins of this heterogeneity. ECs are also implicated in the development of pancreas *(24)* and liver *(25)*, suggesting their ability to deliver specific paracrine-inductive signals during development. Our survey of different fibroblasts from diverse anatomical sites revealed intrinsic differences in gene expression programs, related to the anatomical site of origin *(9)*. We were therefore interested in exploring the possibility that there might be similar consistent differences among ECs derived from different organs and tissues. We used a permutation-based technique termed significance analysis of microarrays (SAM) *(26)* to systematically identify genes whose expression in cultured ECs varied according to the tissues of origin. SAM was used for multiclass classification because rank sum test is not able to perform multiclass classification. The analysis identified more than 2000 such genes, with an estimated false discovery rate less than 0.2% (**Fig. 2B**). Some of the identified genes varied in expression among ECs from different groups of tissues (such as the tissue type I and II branches in **Fig.1A**), and some were highly specific to ECs from a single tissue. Groups of genes specifically expressed in one or two of the tissues we examined are marked and expanded in **Fig. 2C**.

Nasal polyp ECs expressed SIX3, a homeodomain protein with forebrain-specific expression *(27)* (**Fig. 2C**). The specific expression of SIX3 in ECs from the nasal cavity suggests that developmentally patterned transcriptional programs in ECs may preserve the positional memory even after transfer to tissue culture. Diverse tissue-specific genes point to different physiological properties of ECs from each site. In contrast, the skin ECs expressed basic FGF and a set of genes involved in cholesterol biosynthesis, including squalene epoxidase, 24-dehydrocholesterol reductase, stearoyl-CoA desaturase, fatty acid desaturase, and 3-hydroxy-3-metheylglutaryl-CoA synthase 1. The lung ECs specifically expressed phospholipase A2 group XII, an enzyme involved in surfactant secretion, as well as the developmental regulators secreted frizzled related protein 1 and osteoglycin. Myometrium ECs specifically expressed the calcitonin receptor and gallanin. Calcitonin is important for the implantation of embryos *(28)*, and gallanin is a peptide hormone that stimulates the contraction of the uterine myometrium *(29)*. Their specific expression in myometrial EC may point to active roles played by ECs in the function and physiology of the uterus. Together, these results suggest that ECs from different anatomic tissues are distinct differentiated cell types, with specialized roles in the functions and physiology of the respective tissues or organs from which they were derived. It will be an important challenge to trace the steps by which angioblasts adopt a distinct differentiated fate in each different tissue or organ during development and tissue repair, especially because of participation of circulating endothelial precursor cells.

7. Conclusions

Recent studies analyzing the cultured ECs purified from postcapillary high endothelial venule ECs rapidly lost their specialized characteristics when isolated from the lymphoid tissue microenvironment compared with high endothelial venule endothelial cells *(30)*, and many surface proteins fail to be expressed in culture cells *(31)*. The microenvironment plays an important role in the expression patterns of ECs in vivo, but our studies have provided the evidence that a significant amount of location-specific gene expression still persists with in vitro passage in cell culture. These cell autonomous expression patterns suggest the existence of genetic and epigenetic mechanisms maintaining the cell type expression in the absence of the local environmental influences *(32)*. This might provide us with the "critical" window to probe and study the establishment of the endothelial specialization in different anatomical locations.

In previous studies, some lymphatic markers were present in the microvascular gene cluster, including prox-1, desmoplakin *(12)*, and neuropilin 2, a gene essential for the development of lymphatic vasculature *(33)*. These results suggest that lymphatic ECs (LECs) may have been copurified in the microvascular EC cultures. This copurification has been documented in a separate study *(34)*. Delineation of LEC vs blood EC (BEC) gene expression with microarrays has been recently reported previously *(35,36)*, and some proposed differences between LEC vs BEC do occur in the microvascular vs large vessel ECs gene lists. Most of the genes that we identified comparing large vessels vs microvascular ECs were not different between LEC and BEC in the previous studies. Thus, the difference between large vessel and microvascular EC gene expression could not be solely accounted for by LEC contamination in the microvascular EC preparations. But, it does suggest the importance of understanding EC heterogeneity that can lead to purification and identification of EC subsets with unique properties. The use of global gene expression coupled with flow cell cytometry, in which single-cell analysis is performed, gives us clues into the further division of ECs with potentially significant functional differences. A good example is the finer subdivision of hematopoeitic lineages with different regulatory and effector functions, with the further subdivision of a cell population thought to be a homogenous group, with the help of monoclonal antibody and flow cell cytometry. Similar progress can be made to identify different populations of ECs or endothelial precursor cells with the appropriate biological assays. This approach will lead to a better understanding of EC heterogeneity and how it plays a role in vascular development and tissue remodeling.

One important feature of the microarray experiment is to allow public access to the enormous amount of information we have gathered in the genomic studies. This access will allow investigators to explore and identify the differences in which they are interested to promote the advancement of the knowledge in the field. We have deposited the detailed protocol and all original data in a searchable format at http://microarray-pubs.stanford.edu/endothelial.

Acknowledgments

We are grateful to Dr. Patrick Brown and members of Brown laboratory at Stanford University (Stanford, CA) for the support and input into this study.

References

1. Alizadeh, A. A., Eisen, M. B., Davis, R. E. et al. (2000) Distinct types of diffuse large B-cell lymphoma identified by gene expression profiling. *Nature* **403**, 503–511.
2. Chi, J. T., Chang, H. Y., Haraldsen, G. et al. (2003) Endothelial cell diversity revealed by global expression profiling. *Proc. Natl. Acad. Sci. USA* **100**, 10,623–10,628.
3. Eisen, M. B., Spellman, P. T. Brown, P. O., and Botstein, D. (1998) Cluster analysis and display of genome-wide expression patterns. *Proc. Natl. Acad. Sci. USA* **95**, 14,863–14,868.
4. Troyanskaya, O. G., Garber, M. E., Brown, P. O., Botstein, D., and Altman, R. B. (2002) Nonparametric methods for identifying differentially expressed genes in microarray data. *Bioinformatics* **18**, 1454–1461.
5. Mota, M. M., Jarra, W., Hirst, E., Patnaik, P. K., and Holder, A. A. (2000) Plasmodium chabaudi-infected erythrocytes adhere to CD36 and bind to microvascular endothelial cells in an organ-specific way. *Infect. Immun.* **68**, 4135–4144.
6. Muenzner, P., Dehio, C., Fujiwara, T., Achtman, M., Meyer, T. F., and Gray-Owen, S. D. (2000) Carcinoembryonic antigen family receptor specificity of Neisseria meningitidis Opa variants influences adherence to and invasion of proinflammatory cytokine-activated endothelial cells. *Infect. Immun.* **68**, 3601–3607.
7. Muenzner, P., Naumann, M. Meyer, T. F., and Gray-Owen, S. D. (2001) Pathogenic Neisseria trigger expression of their carcinoembryonic antigen-related cellular adhesion molecule 1 (CEACAM1; previously CD66a) receptor on primary endothelial cells by activating the immediate early response transcription factor, nuclear factor-kappaB. *J Biol Chem.* **276**, 24,331–24,340.
8. Irjala, H., Johansson, E. L., Grenman, R., Alanen, K., Salmi, M., and Jalkanen, S. (2001) Mannose receptor is a novel ligand for L-selectin and mediates lymphocyte binding to lymphatic endothelium. *J. Exp. Med.* **194**, 1033–1042.
9. Chang, H. Y., Chi, J. T., Dudoit, S., et al. (2002) Diversity, topographic differentiation, and positional memory in human fibroblasts. *Proc. Natl. Acad. Sci. USA* **99**, 12,877–12,882.
10. Mukouyama, Y. S., Shin, D., Britsch, S., Taniguchi, M., and Anderson, D. J. (2002) Sensory nerves determine the pattern of arterial differentiation and blood vessel branching in the skin. *Cell* **109**, 693–705.
11. Luttun, A., Tjwa, M., Moons, L. et al. (2002) Revascularization of ischemic tissues by PlGF treatment, and inhibition of tumor angiogenesis, arthritis and atherosclerosis by anti-Flt1. *Nat. Med.* **8**, 831–840.
12. Oliver, G., and Detmar, M. (2002) The rediscovery of the lymphatic system: old and new insights into the development and biological function of the lymphatic vasculature. *Genes Dev.* **16**, 773–783.

13. Yancopoulos, G. D., Davis, S., Gale, N. W., et al. (2000) Vascular-specific growth factors and blood vessel formation. *Nature* **407,** 242–248.
14. Hacohen, N., Kramer, S., Sutherland, D., Hiromi, Y., and Krasnow, M. A. (1998) sprouty encodes a novel antagonist of FGF signaling that patterns apical branching of the *Drosophila* airways. *Cell* **92,** 253–263.
15. Cheng, N., Brantley, D. M., and Chen, J. (2002) The ephrins and Eph receptors in angiogenesis. *Cytokine Growth Factor Rev.* **13,** 75–85.
16. Lawson, N. D., and Weinstein, B. (2002) Arteries and veins: making a difference with zebrafish. *Nat. Rev. Genet.* **3,** 674–682.
17. Meno, C., Shimono, A., Saijoh, Y., et al. (1998) lefty-1 is required for left-right determination as a regulator of lefty-2 and nodal. *Cell* **94,** 287–297.
18. Urness, L. D., Sorensen, L. K., and Li, D. Y. (2000) Arteriovenous malformations in mice lacking activin receptor-like kinase-1. *Nat. Genet.* **26,** 328–331.
19. Lawson, N. D., Scheer, N., Pham, V. N., et al. (2001) Notch signaling is required for arterial-venous differentiation during embryonic vascular development. *Development* **128,** 3675–3683.
20. Nakagawa, O., McFadden, D. G., Nakagawa, M., et al. (2000) Members of the HRT family of basic helix-loop-helix proteins act as transcriptional repressors downstream of Notch signaling. *Proc. Natl. Acad. Sci. USA* **97,** 13,655–13,660.
21. Zhong, T. P., Rosenberg, M., Mohideen, M. A., Weinstein, B., and Fishman, M. C. (2000) gridlock, an HLH gene required for assembly of the aorta in zebrafish. *Science* **287,** 1820–1824.
22. Fischer, A., Schumacher, N., Maier, M., Sendtner, M., and Gessler, M. (2004) The Notch target genes Hey1 and Hey2 are required for embryonic vascular development. *Genes Dev.* **18,** 901–911.
23. Arap, W., Kolonin, M. G., Trepel, M., et al. (2002) Steps toward mapping the human vasculature by phage display. *Nat. Med.* **8,** 121–127.
24. Lammert, E., Cleaver, O., and Melton, D. (2001) Induction of pancreatic differentiation by signals from blood vessels. *Science* **294,** 564–567.
25. Matsumoto, K., Yoshitomi, H., Rossant, J., and Zaret, K. S. (2001) Liver organogenesis promoted by endothelial cells prior to vascular function. *Science* **294,** 559–563.
26. Tusher, V. G., Tibshirani, R., and Chu, G. (2001) Significance analysis of microarrays applied to the ionizing radiation response. *Proc. Natl. Acad. Sci. USA* **98,** 5116–5121.
27. Oliver, G., Mailhos, A., Wehr, R., Copeland, N. G., Jenkins, N. A., and Gruss, P. (1995) Six3, a murine homologue of the sine oculis gene, demarcates the most anterior border of the developing neural plate and is expressed during eye development. *Development* **121,** 4045–4055.
28. Zhu, L. J., Bagchi, M. K., and Bagchi, I. C. (1998) Attenuation of calcitonin gene expression in pregnant rat uterus leads to a block in embryonic implantation. *Endocrinology* **139,** 330–339.
29. Niiro, N., Nishimura, J., Hirano, K., Nakano, H., and Kanaide, H. (1998) Mechanisms of galanin-induced contraction in the rat myometrium. *Br. J. Pharmacol.* **124,** 1623–1632.
30. Lacorre, D. A., Baekkevold, E. S., Garrido, I., et al. (2004) Plasticity of endothelial cells: rapid dedifferentiation of freshly isolated high endothelial venule endothelial cells outside the lymphoid tissue microenvironment. *Blood* **103,** 4164–4172.
31. Oh, P., Li, Y., Yu, J., et al. (2004) Subtractive proteomic mapping of the endothelial surface in lung and solid tumours for tissue-specific therapy. *Nature* **429,** 629–635.
32. Gebb, S., and Stevens, T. (2004) On lung endothelial cell heterogeneity. *Microvasc. Res.* **68,** 1–12.

33. Yuan, L., Moyon, D., Pardanaud, L., et al. (2002) Abnormal lymphatic vessel development in neuropilin 2 mutant mice. *Development* **129,** 4797–4806.
34. Makinen, T., Veikkola, T., Mustjoki, S., et al. (2001) Isolated lymphatic endothelial cells transduce growth, survival and migratory signals via the VEGF-C/D receptor VEGFR-3. *EMBO J.* **20,** 4762–4773.
35. Podgrabinska, S., Braun, P., Velasco, P., et al. (2002) Molecular characterization of lymphatic endothelial cells. *Proc. Natl. Acad. Sci. USA* **99,** 16,069–16,074.
36. Hirakawa, S., Hong, Y. K., Harvey, N., et al. (2003) Identification of vascular lineage-specific genes by transcriptional profiling of isolated blood vascular and lymphatic endothelial cells. *Am. J. Pathol.* **162,** 575–586.

PART III

BIOMARKER IDENTIFICATION BY USING CLINICAL PROTEOMICS AND GLYCOMICS

Krishnarao Appasani

Identification of new biomarkers for the detection of neuroinflammatory diseases and brain tumors will help to unravel the molecular processes underlying neural pathogenesis and ultimately to develop better therapeutics for the clinical management of the neurological disorders, which is the subject of Part III. In addition, chapters also describe the applications of antibody arrays, carbohydrate arrays, and glycoprofiling in various human diseases.

Multiple sclerosis is the most prevalent chronic inflammatory disease of the central nervous system, causing severe disability in a significant proportion of patients. Although many findings suggest that multiple sclerosis is caused by an immune response to proteins expressed in the CNS, the target antigens are not known. Protein macroarray technology in combination with epitope mapping technique was used by Cepok et al. (Chapter 10) for large-scale screening of target antigens of the focused immune response in the multiple sclerosis patients. They identified a few candidate proteins as targets, and their efficacy was further validated by ELISAs. This study demonstrates the power of protein arrays in the elucidation of disease-associated antibody responses in neuroinflammatory diseases. Mature and fully differentiated brain tissue consists of three major types of cells: neurons, astrocytes, and oligodendrocytes. Most gliomas are astrocytomas and constitute >60% of the primary neurons of the CNS. The cellular origin of gliomas is incompletely understood, and their molecular classification is not properly indexed. Therefore, Chapter 11, by Srideshmukh et al., provides a summary on glioma tumor taxonomy, based on genomics (expression profiling) and proteomics (2D gel electrophoresis and mass spectrometry) approaches. In addition, this chapter describes the differential protein expression studies performed between normal tissue and high-grade astrocytomas.

Antibody-based microarrays are among the novel class of rapidly emerging proteomic technologies that will offer new opportunities for proteome-scale analyses. These arrays not only will provide high-throughput means to perform comparative proteome analyses (healthy vs diseased) but also will allow us to study biomarker discovery, disease diagnostics, differential protein expression profiling, and finally to address signaling pathways and protein interaction networks. In Chapter 12, Wingren and Borrebaeck systematically provide the framework for designing these antibody arrays and performing assays in high-throughput manner as well as a detailed overview of the work performed in cancer and allergy research. Although the glycan structures and their influence on the macromolecule are studied in detail, total glycome profiles have not been used for the diagnosis of diseases. In Chapter 13, Laroy and Contreras describe a novel method of glycoprofiling (by using DNA sequencer-aided, fluorophore-assisted carbohydrate electrophoresis) in liver diseases, and develop the GlycoCirrhoTest. They concluded that this method is not only a sensitive assay for a tiny amount of sample but also it has a throughput capacity that allows screening of many samples.

Like nucleic acids and proteins, carbohydrates are another class of essential biological molecules. Owing to their unique physicochemical properties, carbohydrates are capable of generating structural diversity, and so they are prominent in display on the surfaces of cell membranes or on the exposed regions of macromolecule. Thus, carbohydrate moieties are suitable for storing biological signals in forms that are identifiable by other biological systems. In Chapter 14, Wang et al. summarize how to develop high-throughput carbohydrate microarrays (glycan arrays) and how they could be efficiently used for the identification of immunologic targets of other microorganisms. In addition, these glycan arrays are useful for the exploration of complex carbohydrates that are differentially expressed by host cells, including stem cells. By contrast to arrays described in previous chapters, carbohydrate microarrays require preservation of the 3D conformations and topological configurations of sugar moieties on chip to permit a targeted molecular recognition by the corresponding cellular receptors. Wang et al. predict that these glycan arrays or sugar chips are not only useful in the screening of carbohydrate-based cellular receptors of microorganisms but also to probe carbohydrate-mediated molecular recognition and immune responses.

10

Identification of Target Antigens in CNS Inflammation by Protein Array Technique

Sabine Cepok, Bernhard Hemmer, and Konrad Büssow

Summary

Multiple Sclerosis (MS) is the most prevalent chronic inflammatory disease of the CNS, causing severe disability in a significant proportion of patients. Although many findings suggest that MS is caused by an immune response to proteins expressed in the CNS, the target antigens are still unknown. Among the candidates are self- and foreign proteins expressed in the CNS compartment. Here, we describe a new approach to dissect immune responses in the CNS. We applied a protein array based on a human brain cDNA library to decrypt the specificity of the local antibody response in MS. The macroarray, containing 37,000 proteins, enabled us to perform a large-scale screening for disease-associated antigens. Target proteins were further mapped to identify high-affinity ligands and possible mimics. Using this approach, we found MS-specific high-affinity antibody responses to two peptide sequences derived from Epstein–Barr virus (EBV) proteins. Several mimics with lower affinity also were identified. Subsequent analysis revealed an elevated and specific immune response in MS patients against both EBV proteins, suggesting a putative role of EBV in the pathogenesis of MS. The study demonstrates that protein arrays can be successfully applied to identify disease-associated antibody responses in neuroinflammatory diseases.

Key Words: CNS; inflammation protein array; multiple sclerosis; target antigen.

1. Introduction

The CNS was considered to be an immunologically privileged microenvironment, protected by the highly specialized blood-brain barrier, which minimizes and regulates the infiltration of cells and macromolecules. Recent reports support the concept of a routine immune surveillance of the brain (1). During CNS infection, a vigorous immune response is mounted in the brain to eliminate the infectious agent. Pathogenic antigens are released into draining lymph nodes (e.g., nuchal or cervical) where they are presented by professional antigen presenting cells (e.g., dendritic cells) to B- and T-cells (2). B- and T-cells with high affinity for these antigens are selected in the lymph node environment, become activated, clonally expand, and acquire effector functions. After release from the lymph nodes, primed B- and T-cells migrate through body compartments and accumulate at sites of CNS inflammation where they re-encounter their target antigen. Upon reactivation, T-cells mediate effector functions (e.g., cytotoxicity, cytokine release, and recruitment of macrophages), leading to clearance of the antigenic source (e.g., resolving the infectious agent). B-cells mature either in peripheral

From: *Bioarrays: From Basics to Diagnostics*
Edited by: K. Appasani © Humana Press Inc., Totowa, NJ

lymph nodes or in the inflamed tissue into plasmablast or plasma cells, and they deliver high amounts of Ig at the site of inflammation. These antibodies deposit in the tissue, but they are also found in the cerebrospinal fluid (CSF), leading to increased antibody concentration (intrathecal Ig synthesis). After CNS infection, oligoclonal IgG bands (OCBs) are seen in the CSF, which cannot be detected in the serum, arguing for a compartmentalised immune response. In acute and chronic infectious diseases of the CNS, intrathecal Ig and the OCBs target disease-associated proteins derived from the infectious agent *(3,4)*. However, in many acute and chronic inflammatory diseases of the CNS, the focus of the local immune response is still unknown. This disparity is particularly true for multiple sclerosis (MS).

MS is the most prevalent chronic inflammatory disease of the CNS, causing severe disability in a significant proportion of patients *(5)*. Although many findings suggest that MS is caused by an immune response to proteins expressed in the CNS, the target antigens are still unknown. Among the candidates are self- and foreign proteins expressed in the CNS compartment. Here, we describe a new approach to dissect immune responses in the CNS. We applied a protein array based on a human brain cDNA library to decrypt the specificity of the local antibody response in MS. The macroarray, containing 37,000 proteins, enabled us to perform a large-scale screening for disease-associated antigens. Target proteins were further mapped to identify high-affinity ligands and possible mimics. Using this approach, we found MS-specific high-affinity antibody responses to two peptide sequences derived from Epstein–Barr virus (EBV) proteins. Several mimics with lower affinity also were identified. Subsequent analysis revealed an elevated and specific immune response in MS patients against both EBV proteins, suggesting a putative role of the virus in the pathogenesis of MS. The study demonstrates that protein arrays can be successfully applied to identify disease-associated antibody responses in neuroinflammatory diseases.

2. Multiple Sclerosis

2.1. Clinical Aspects

MS is the most common chronic inflammatory and demyelinating disease of the CNS, affecting 0.05–0.15% of Caucasians *(6,7)*. The disease usually starts in young adults and occurs more often in women than in men. In 80–90% of the cases, MS starts with a relapsing, remittent disease course characterized by recurrent episodes of neurological symptoms with spontaneous remission. Over time, the number of relapses decreases, but most patients develop progressive neurological deficits that occur independently of relapses (secondary progressive phase). In a minority of patients, MS starts with a primary progressive disease course without acute relapses.

2.2. Pathogenesis of MS

It is well established that the immune system is involved in the pathogenesis of MS. Histopathological studies pointed out that besides the hallmark of demyelination, an immune response, mostly consisting of B-cells, T-cells, macrophages, and activated microglia, is seen in acute MS lesions *(8)*. T- and B-cell responses in the CNS of MS patients are highly focused. By analyzing single cells from lesions or CSF of MS patients, extensive clonal expansion of T-cells in the local compartments was demonstrated *(9,10)*. The expanded T-cells from lesions or CSF were not found in the peripheral blood, suggesting that a specific migration of these cells occurred into the CNS compartment. Serial studies on CSF samples demonstrated the persistence of expanded T-cells *(11)*. A highly focused immune response also is observed in the B-cell compartment. Almost all MS patients show an intrathecal IgG response or OCBs confined to the CSF *(12)*. Indeed, the occurrence of OCBs is still an important diagnostic marker for chronic CNS inflammation as seen in MS. Interestingly, the antibody response involves predominantly IgG1 and IgG3 and seems to be stable over long periods *(13,14)*. The dominance and persistence of an intrathecal IgG1 and IgG3 antibody response together with the locally expanded T-cells are consistent with an ongoing immune response against proteins. This assumption is supported by a stable accumulation of antigen-experienced B-cells and antibody-secreting plasmablasts in the CSF of MS patients *(15,16)*. Investigations of the B-cell receptors in the CNS and CSF of MS patients demonstrated clonotypic expansion of B-cells in the local compartment that was not (or to a lesser extent) observed in the peripheral blood *(17–20)*. Expanded B-cells in the CSF and CNS lesions were characterized by extensive replacement mutations of the B-cell receptor genes, indicating a highly focused antigen-driven immune response.

2.3. Target Antigens of Local Immune Response in MS

For many years, researchers have been searching for targets of the local immune response in MS. Because MS lesions are focused on CNS white matter, it was assumed that components of the myelin sheath, such as myelin basic protein, myelin oligodendrocyte glycoprotein, and proteolipid protein might be the target antigens in MS. The role of these antigens was supported by early reports on acute demyelinating episodes in humans after accidental immunization with myelin components (e.g., after rabies vaccination; *[21]*). After this observation, 70 yr ago, an animal model was established, called experimental autoimmune encephalomyelitis *(22)*. In experimental autoimmune encephalomyelitis, an acute or chronic inflammatory disease of the CNS with variable degree of demyelination is induced by immunization with myelin antigens and Freund's adjuvant *(23)*.

Several laboratories tried to prove this concept by characterization of the immune responses to myelin antigens in MS patients *(24–26)*. T-cells specific for many of the myelin antigens were isolated not only from MS patients but also from healthy donors, demonstrating that autoreactive T-cells are part of the normal repertoire and not necessarily harmful *(24–26)*. Investigations on antigen recognition of myelin-specific T-cells demonstrated that these cells can potentially crossreact with foreign antigens such as viral peptides *(27,28)*. The cross-recognition, termed molecular mimicry, has provided an attractive model to explain how autoreactive T-cells could become activated by infectious agents to mediate an autoimmune process *(29)*.

Although higher antibody titers to some myelin antigens are observed in the CNS and serum of MS patients, these autoantibodies are of low affinity and also found in healthy controls *(30–32)*. A recent report demonstrated that the occurrence of antimyelin antibodies is associated with progression in early MS patients, although these results warrant confirmation *(33)*. An autoimmune response to myelin antigens would explain many of the features, but experimental evidence for this concept is still missing.

Because inflammation is observed in most infectious disorders of the CNS, the immune response seen in MS also could be a secondary event. Indeed, the idea of MS

being caused by an infectious agent has been controversially discussed for almost a
century *(34)*. The hypothesis is still attractive in view of other infectious diseases of the
CNS that cause inflammation and demyelination in humans, among them measles-
associated subacute sclerosing panencephalitis or human T-cell lymphotropic virus type
I-associated myelopathy *(35)*. In these disorders, a comparable chronic immune response
is observed in the CNS, including the occurrence of OCBs. Interestingly, the intrathecal
Ig response and the OCBs contain antibodies specific for the causative agent. Similar
to autoantibody studies, many researchers have reported on elevated antibody titers to
a broad range of pathogens in MS patients, although frequently those studies were not
confirmed by others. Nevertheless, evidence is lacking that any microbe plays a role in
the pathogenesis of MS.

2.4. Methods to Decrypt Antigen Specificity in MS

In the past decade, new techniques were developed to identify the targets of the
focused immune response in the CNS of MS patients. Studies on T-cells are impaired
by some technical difficulties, because these cells recognize processed peptide antigens only in the context of appropriate major histocompatibility complex molecules.
However, initial studies have demonstrated that it is possible to decrypt target antigens
of T-cells with unknown specificity by using expression or combinatorial peptide
libraries *(28,36,37)*.

Antibodies are much easier to handle than T-cells, although the amount of IgG available from the CSF or brain of MS patients is limited. Nevertheless, these antibodies
have been used to screen for binding of target antigens. With increasing advances in
molecular biology several new techniques have been developed to identify the specificity of antibodies in MS. Because CNS-resident proteins might be the target of the
antibody response, the development of cDNA expression libraries generated from MS
brain lesions was performed. But application of this technique to CSF antibodies has
not yet led to the discovery of new MS target antigens *(38)*.

Large libraries containing random short peptides displayed on phage surface enabled
large-scale screening of epitopes for subsequent database search *(39–41)*. Using this technology, one group identified a five-amino acid consensus sequence present in EBV
nuclear antigen and the αβ-crystallin heat-shock protein *(41)*. These candidates were
not analyzed in a large patient cohort to investigate MS-specific reactivities. Also, another
group using this approach identified several peptides, but they could not confirm MS-specific reactivities when comparing antibody responses to these peptides in MS patients
and controls *(42)*. This approach, however, has several drawbacks. The technique requires
large amounts of antibody that is usually only obtained by pooling samples from several
patients. Random phage display libraries will only detect mimotopes that may mimic
the native antigen, possibly without sharing sequence homology.

Recently, antibody–phage display libraries expressing recombinant antibodies of target tissue were constructed. Libraries generated from antibody genes in the affected tissue were probed on a chosen antigenic preparation (e.g., brain tissue) *(43)*. Such a
recombinant antibody library by using subacute sclerosing panencephalitis brain was
successfully applied to identify disease-related measles virus antigens from infected brain
tissue *(44)*. Because this strategy might miss proteins with low abundance, it may fail in
diseases when target proteins are not expressed at high levels in the affected tissue.

3. Dissection of Antibody Specificity in MS by Using Human Brain cDNA Protein Macroarrays

A novel protein array based on a cDNA expression library of human brain enabled us to perform a large-scale analysis of antibody specificity in MS. We describe here the properties of the protein arrays and the identification of target antigens in MS.

3.1. Development of Human Brain Protein Macroarrays

Dot blots and colony blots are the predecessors of today's protein microarrays. The first examples of protein arrays with large numbers of protein spots generated by automated spotting were produced on large protein-binding membranes. A technology was developed to circumvent the need to produce the elements of the array individually before spotting. Instead of arraying proteins directly, protein-expressing bacterial clones are arrayed on membranes that then produce the proteins of the array *(45)*.

The earliest high-density DNA arrays were developed by the group of Hans Lehrach in the early 1990s at the Imperial Cancer Research Fund in London, England. Hundreds of thousands of bacterial clones of cDNA and genomic libraries on large nylon membranes were arrayed. Bacteria were grown on the membranes and lysed, leading to fixation of plasmid DNA as distinct spots on the membrane. These DNA arrays were screened by radioactive or fluorescent hybridization probes in gene mapping and cloning experiments *(46)*.

Using the same arraying technology, we developed protein arrays by arraying clones of protein expression cDNA libraries. Bacteria were grown on protein-binding polyvinylidene difluoride membranes, protein expression was induced, and finally cells were lysed and the cellular protein, including the expression products of the cloned cDNA fragments, was bound as distinct spots on the array (**Fig. 1**).

After preliminary tests, a large cDNA library was constructed from human fetal brain tissue. The library was cloned in a bacterial protein expression vector featuring a hexahistidine affinity tag (His-tag); 150,000 clones were picked and arrayed in duplicate on square membranes of 22×22 cm^2 at a density of 27,648 clones per membrane *(47)*. Because of its size, this array format was later called a macroarray to distinguish it from the glass slide microarray format.

It had been shown that clones containing expression plasmids with the His-tag could be tested for protein expression by a colony blot technique by using an antibody against the affinity tag. This approach was applied to the macroarrays to identify clones that did express their cDNA insert as a His-tag fusion protein. Twenty percent of clones were detected by the antibody. Random sampling showed that the clones that were detected by the antibody mostly produced recombinant protein and mostly consisted of cDNA insert that were cloned in the correct reading frame in respect to the vector-encoded start codon *(48)*. The antibody detection led to a fivefold enrichment of productive clones. The productive clones were combined into a new, smaller library. This library consists of 37,831 clones and is being arrayed on two macroarrays at standard clone density or on one array at a slightly higher density.

Presently, macroarrays of the human fetal brain and other libraries are produced and sold by the RZPD German Resource Centre. The RZPD is actively developing the technology of the macroarrays and has now generated expression libraries and macroarrays from a set of other human tissues. Currently, the sequences of approx 3000

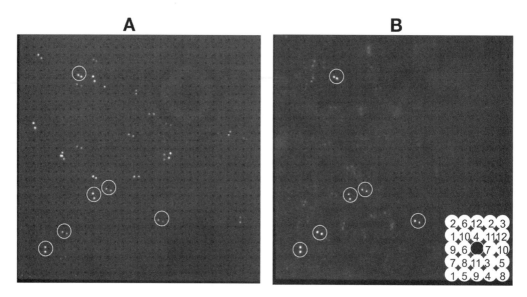

Fig.1. Development of the protein macroarray technology from DNA array technology. DNA (**A**) and protein (**B**) arrays of the same cDNA library were prepared by automated spotting. The arrays were screened for cDNA clones of the GAPDH gene in parallel by DNA hybridization (**A**) and with an anti-GAPDH antibody (**B**). The inset demonstrates the duplicate clone spotting pattern.

clones of the fetal brain library are publicly available *(49)*. The RZPD is continuously generating more sequence information for their expression library clones.

Today, researchers are using macroarrays mainly to discover antigens of antibodies from serum or CSF. However, the arrays also have been probed successfully with enzymes and binding protein to discover substrates of methyl transferases *(50)* and kinases *(51)* as well as protein interaction partners *(45)*.

3.2. Properties of Human Brain Protein Macroarrays

This novel protein array technology provides some indispensable features for identifying target antigens in MS.

1. Because inflammation is not observed in other tissues, MS-associated antigens are probably expressed in the CNS. Therefore, the human brain cDNA protein expression library represents an appropriate set of proteins to be screened.
2. The library comprises 37,831 expression clones, including full-length proteins in the right reading frame but also proteins that are not in the correct frame, resulting in the expression of polypeptides that do not occur in nature. This outcome is an advantage for a broad screening method when antibodies with unknown specificity are used.
3. The screening for antibody specificity is relatively easy to perform. The expression clones are arrayed on polyvinylidene difluoride membranes, allowing experimental techniques similar to a dot blot. Several rounds of panning required for phage display screening are not necessary. Results can be obtained in less than 16 h.
4. The assay allows screening of antibody responses in single patients. The amount of antibody that is needed does not require pooling of antibodies from different patients.

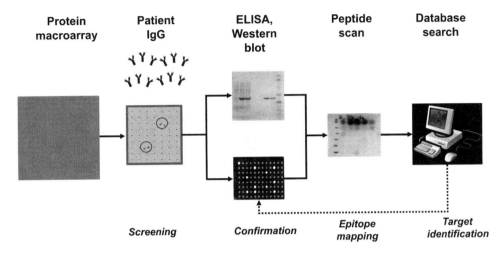

Fig. 2. Experimental setup to define antibody reactivities in CSF by protein macroarrays.

5. Proteins detected on the array are easily produced and purified in large amounts for subsequent experiments, because expression clones for each detected protein are available through the RZPD.
6. This array not only allows detection of CNS proteins but also of artificial mimitopes. Mimitopes can be further characterized by subsequent peptide scan analyses to identify the original, native binding epitopes.
   The disadvantage of the protein array is a result of the expression in *Escherichia coli*. Arrayed proteins lack mammalian glycosylation. Antibodies specific for glycosylated protein domains therefore may not be identified by this technique. Furthermore, the array was constructed from fetal brain and may not contain proteins expressed in the adult brain.

3.3. Identification of Target Antigens of Local Immune Response in MS by Using Protein Macroarrays

The protein array of human fetal brain was applied to search for potential target antigens in MS (*[52]*; **Fig. 2**). CSF of 12 MS patients and five patients with noninflammatory neurological diseases was selected and each probed at the same IgG concentration with a protein array. Binding of IgG was visualized by staining with horseradish peroxidase-conjugated monoclonal anti-human IgG antibodies and subsequent developed by substrate. Expression clones that bound CSF IgG of MS patients but not controls were selected. To identify the target of the antibody binding, corresponding expression clones were ordered through RZPD. Plasmids of the clones were isolated, and the cDNA inserts were sequenced. Subsequently, the sequences were used for a database search. Forty-two unique sequences were defined, which not only made up proteins expressed in the correct reading frame but also sequences that were expressed out of frame, generating polypeptides with sequences that do not occur in nature. Twenty-one expression clones, which showed the strongest reactivity in two or more MS patients were selected (**Table 1**). Ten of these clones expressed genes in the correct frame, and 11 clones expressed out of frame. Protein expression in the clones was induced by isopropyl β-D-thiogalactoside, and the His-tag fusion proteins were purified by nickel chelate chromatography. After confirmation of protein size and specific

Table 1
Expression Clones Which Showed Differential Immune Reactivity in MS Patients and Controls

RZPD clone*	Protein ID	Identity
N04524	B3	Out-of-frame expression
L04523	F3	Out-of-frame expression
E21555	C5	Out-of-frame expression
G17581	D6	Out-of-frame expression
H13574	C6	Out-of-frame expression
E06542	G4	MAZ
H07541	F4	MAZ
M18568	H5	MAZ
L17507	D3	ASPI
A07545	G2	Out-of-frame expression
A03592	H1	TCOF1
C06553	D1	Out-of-frame expression
L02568	G5	RNA-binding motif protein 5
O17528	H3	Out-of-frame expression
M05594	A7	Mitochondrial ribosomal
A11524	A1	Out-of-frame expression
J23564	D5	Out-of-frame expression
A02550	A5	Hemoglobin $\alpha 3$ chain
H15589	E1	Chaperon containing TCRP1 CCT5
F22533	B1	Out-of-frame expression
C10601	B7	RED protein
	GAPDH	GAPDH

*RZPD clone identifiers without prefix 'MPMGp800.'

antibody binding by Western blot (using CSF from the corresponding MS patient), ELISAs were established for all 21 proteins. For some proteins, we observed a significantly higher immunoreactivity in MS patients compared with controls. Whereas 13% of the 132 MS patients had CSF antibodies against the expression product of clone B3, no such antibody responses were detected in any of the 118 controls. In contrast, no reactivity was detected for the control protein glyceraldehyde-3-phosphate dehydrogenase (GAPDH). Immunoreactivities to certain pattern of distinct proteins occurred consistently in different patients. Patients with IgG reactivity to protein B3 usually also had antibodies against four other proteins, suggesting that a similar epitope was present in these expression products that was targeted by the IgG of the patients (pattern I). All five proteins of this pattern represent artificial products that were expressed out of frame. Three other expression clones, which contain different cDNA fragments of Myc-associated zinc (MAZ) finger protein, also showed high and largely overlapping immunoreactivities (pattern II).

Epitopes responsible for the immunoreactivity to pattern I and II were mapped by peptide scan analysis. Subsequent amino acid substitution assays were performed to define the best binding amino acids at each position of both patterns (**Fig. 2**). The Swiss–Prot database was searched with the identified optimal motifs. We found 10 matches with the first motif and 13 with the second motif. Among the matches were some irrelevant proteins, because they were from nonpathogenic species but also epitopes derived from human, bacterial, or viral proteins. Both patterns matched two different proteins of EBV. CSF antibodies of MS patients bound strongest the two peptides derived from EBV. Reactivity to all other proteins, including several self-antigens, was lower. To determine the MS-specific immune reactivity to both EBV proteins, ELISA experiments were performed. Humoral immune responses to these proteins were elevated in the serum and CSF of MS patients compared with controls. The specificity and high affinity of EBV-specific CSF antibodies was confirmed by solution phase assays. Finally, binding to the oligoclonal IgG bands was shown for both proteins, indicating a specific immune reactivity against these identified EBV proteins in the CSF of MS patients. Although other self-antigens were recognized by the antibodies, both EBV antigens showed the best binding. These findings demonstrate that MS patients have an increased and highly specific immune response to EBV in the CNS. These findings also suggest that EBV is involved in the pathogenesis of MS. Possible mechanisms include CNS infection of glia cells or crossreacitivity of EBV antigens with autoantigens in the brain. Further studies will address the specificity of other target proteins that we identified by our macroarray approach.

4. Conclusions

The protein macroarray technology was used for large-scale screening of the target antigens of the focused immune response in the CNS of MS patients. Combined with epitope mapping techniques, we identified two candidate proteins as targets of the local immune response in MS. Both targets were confirmed by additional assays, demonstrating the efficacy of the protein array approach in MS. This approach may not only be suitable to investigate antibody responses in other systemic or focal autoimmune diseases but also may help to investigate antitumor immune responses. However, the success of such an approach is largely dependent on how closely proteins on the array reflect the antigen repertoire in the diseased organ compartment. More advanced arrays containing proper glycosylated proteins of the tissue of relevance are needed to efficiently decrypt antigen responses in human diseases which involve a humoral immune response.

Acknowledgments

This research was supported by the Deutsche Forschungsgemeinschaft projects 2382/5-1 and 7-1 and the Krebshilfe Stiftung.

References

1. Hickey, W. F. (2001) Basic principles of immunological surveillance of the normal central nervous system. *Glia* **36,** 118–124.
2. Cserr, H. F., Knopf, P. M. (1992) Cervical lymphatics, the blood-brain barrier and the immunoreactivity of the brain: a new view. *Immunol. Today* **13,** 507–512.
3. Vandvik, B., Norrby, E., Nordal, H. J., and Degre, M. (1976) Oligoclonal measles virus-specific IgG antibodies isolated from cerebrospinal fluids, brain extracts, and sera from patients with subacute sclerosing panencephalitis and multiple sclerosis. *Scand. J. Immunol.* **5,** 979–992.
4. Smith-Jensen, T., Burgoon, M. P., Anthony, J., Kraus, H., Gilden, D. H., and Owens, G. P. (2000) Comparison of immunoglobulin G heavy-chain sequences in MS and SSPE brains reveals an antigen-driven response. *Neurology* **54,** 1227–1232.

5. Hemmer, B., Archelos, J. J., and Hartung, H. P. (2002) New concepts in the immuno-pathogenesis of multiple sclerosis. *Nat. Rev. Neurosci.* **3**, 291–301.

6. Noseworthy, J. H., Lucchinetti, C., Rodriguez, M., and Weinshenker, B. G. (2000) Multiple sclerosis. *N. Engl. J. Med.* **343**, 938–952.

7. Lucchinetti, C., Bruck, W., and Noseworthy, J. (2001) Multiple sclerosis: recent developments in neuropathology, pathogenesis, magnetic resonance imaging studies and treatment. *Curr. Opin. Neurol.* **14**, 259–269.

8. Brosnan, C. F., and Raine, C. S. (1996) Mechanisms of immune injury in multiple sclerosis. *Brain Pathol.* **6**, 243–257.

9. Babbe, H., Roers, A., Waisman, A., et al. (2000) Clonal expansions of CD8(+) T cells dominate the T cell infiltrate in active multiple sclerosis lesions as shown by micromanipulation and single cell polymerase chain reaction. *J. Exp. Med.* **192**, 393–404.

10. Jacobsen, M., Cepok, S., Quak, E., et al. (2002) Oligoclonal expansion of memory CD8+ T cells in the cerebrospinal fluid from multiple sclerosis patients. *Brain* **125**, 538–550.

11. Skulina, C., Schmidt, S., Dornmair, K., et al. (2004) Multiple sclerosis: brain-infiltrating CD8+ T cells persist as clonal expansions in the cerebrospinal fluid and blood. *Proc. Natl. Acad. Sci. USA* **101**, 2428–2433.

12. Archelos, J. J., Storch, M. K., and Hartung, H. P. (2000) The role of B cells and autoantibodies in multiple sclerosis. *Ann. Neurol.* **47**, 694–706.

13. Walsh, M. J., and Tourtellotte, W. W. (1986) Temporal invariance and clonal uniformity of brain and cerebrospinal IgG, IgA, and IgM in multiple sclerosis. *J. Exp. Med.* **163**, 41–53.

14. Losy, J., Mehta, P. D., and Wisniewski, H. M. (1990) Identification of IgG subclasses' oligoclonal bands in multiple sclerosis CSF. *Acta Neurol. Scand.* **82**, 4–8.

15. Cepok, S., Jacobsen, M., Schock, S., et al. (2001) Patterns of cerebrospinal fluid pathology correlate with disease progression in multiple sclerosis. *Brain* **124**, 2169–2176.

16. Cepok, S., Rosche, B., Grummel, V., et al. (2005) Short-lived plasma blasts are the main B cell effector subset during the course of multiple sclerosis. *Brain* **128**, 1667–1676.

17. Owens, G. P., Burgoon, M. P., Anthony, J., Kleinschmidt-DeMasters, B. K., and Gilden, D. H. (2001) The immunoglobulin G heavy chain repertoire in multiple sclerosis plaques is distinct from the heavy chain repertoire in peripheral blood lymphocytes. *Clin. Immunol.* **98**, 258–263.

18. Baranzini, S. E., Jeong, M. C., Butunoi, C., Murray, R. S., Bernard, C. C., and Oksenberg, J. R. (1999) B cell repertoire diversity and clonal expansion in multiple sclerosis brain lesions. *J. Immunol.* **163**, 5133–5144.

19. Qin, Y., Duquette, P., Zhang, Y., Talbot, P., Poole, R., and Antel, J. (1998) Clonal expansion and somatic hypermutation of V(H) genes of B cells from cerebrospinal fluid in multiple sclerosis. *J. Clin. Investig.* **102**, 1045–1050.

20. Colombo, M., Dono, M., Gazzola, P., et al. (2000) Accumulation of clonally related B lymphocytes in the cerebrospinal fluid of multiple sclerosis patients. *J. Immunol.* **164**, 2782–2789.

21. Remlinger, P. (1928) Les paralysies du traitement antirabique. *Ann. Inst. Pasteur* **55**, 35–68.

22. Rivers, T. M., Sprunt, D. H., and Gerry, B. P. (1933) Observations on attemps to produce acute disseminated encephalomyelitis in monkeys. *J. Exp. Med.* **58**, 39–53.

23. Wekerle, H., Kojima, K., Lannes-Vieira, J., Lassmann, H., and Linington, C. (1994) Animal models. *Ann. Neurol.* **36**, S47–S53.

24. Ota, K., Matsui, M., Milford, E. L., Mackin, G. A., Weiner, H. L., and Hafler, D. A. (1990) T-cell recognition of an immunodominant myelin basic protein epitope in multiple sclerosis. *Nature* **346**, 183–187.

25. Pette, M., Fujita, K., Wilkinson, D., et al. (1990) Myelin autoreactivity in multiple sclerosis: recognition of myelin basic protein in the context of HLA-DR2 products by T lympho-

cytes of multiple-sclerosis patients and healthy donors. *Proc. Natl. Acad. Sci. USA* **87,** 7968–7972.

26. Martin, R., Howell, M. D., Jaraquemada, D., et al. (1991) A myelin basic protein peptide is recognized by cytotoxic T cells in the context of four HLA-DR types associated with multiple sclerosis. *J. Exp. Med.* **173,** 19–24.

27. Wucherpfennig, K. W., and Strominger, J. L. (1995) Molecular mimicry in T cell-mediated autoimmunity: viral peptides activate human T cell clones specific for myelin basic protein. *Cell* **80,** 695–705.

28. Hemmer, B., Fleckenstein, B. T., Vergelli, M., et al. (1997) Identification of high potency microbial and self ligands for a human autoreactive class II-restricted T cell clone. *J. Exp. Med.* **185,** 1651–1659.

29. Benoist, C., and Mathis, D. (2001) Autoimmunity provoked by infection: how good is the case for T cell epitope mimicry? *Nat. Immunol.* **2,** 797–801.

30. Reindl, M., Linington, C., Brehm, U., et al. (1999) Antibodies against the myelin oligodendrocyte glycoprotein and the myelin basic protein in multiple sclerosis and other neurological diseases: a comparative study. *Brain* **122,** 2047–2056.

31. Cross, A. H. (2000) MS: the return of the B cell. *Neurology* **54,** 1214-1215.

32. Lampasona, V., Franciotta, D., Furlan, R., et al. (2004) Similar low frequency of anti-MOG IgG and IgM in MS patients and healthy subjects. *Neurology* **62,** 2092–2094.

33. Berger, T., Rubner, P., Schautzer, F., et al. (2003) Antimyelin antibodies as a predictor of clinically definite multiple sclerosis after a first demyelinating event. *N. Engl. J. Med.* **349,** 139–145.

34. Gilden, D. H. (2002) A search for virus in multiple sclerosis. *Hybrid Hybridomics* **21,** 93–97.

35. Stohlman, S. A., and Hinton, D. R. (2001) Viral induced demyelination. *Brain Pathol.* **11,** 92–106.

36. Hemmer, B., Gran, B., Zhao, Y., et al. (1999) Identification of candidate T-cell epitopes and molecular mimics in chronic Lyme disease. *Nat. Med.* **5,** 1375–1382.

37. Hiemstra, H. S., Duinkerken, G., Benckhuijsen, W. E., et al. (1997) The identification of CD4+ T cell epitopes with dedicated synthetic peptide libraries. *Proc. Natl. Acad. Sci. USA* **94,** 10,313–10,318.

38. Owens, G. P., Burgoon, M. P., Devlin, M. E., and Gilden, D. H. (1996) Strategies to identify sequences or antigens unique to multiple sclerosis. *Mult. Scler.* **2,** 184–194.

39. Scott, J. K., and Smith, G. P. (1990) Searching for peptide ligands with an epitope library. *Science* **249,** 386–390.

40. Cortese, I., Tafi, R., Grimaldi, L. M., Martino, G., Nicosia, A., and Cortese, R. (1996) Identification of peptides specific for cerebrospinal fluid antibodies in multiple sclerosis by using phage libraries. *Proc. Natl. Acad. Sci. USA* **93,** 11,063–11,067.

41. Rand, K. H., Houck, H., Denslow, N. D., and Heilman, K. M. (1998) Molecular approach to find target(s) for oligoclonal bands in multiple sclerosis. *J. Neurol. Neurosurg. Psychiatry* **65,** 48–55.

42. Cortese, I., Capone, S., Luchetti, S., Grimaldi, L. M., Nicosia, A., and Cortese, R. (1998) CSF-enriched antibodies do not share specificities among MS patients. *Mult. Scler.* **4,** 118–123.

43. Gilden, D. H., Burgoon, M. P., Kleinschmidt-DeMasters, B. K., et al. (2001) Molecular immunologic strategies to identify antigens and b-cell responses unique to multiple sclerosis. *Arch. Neurol.* **58,** 43–48.

44. Owens, G. P., Williams, R. A., Burgoon, M. P., Ghausi, O., Burton, D. R., and Gilden, D. H. (2000) Cloning the antibody response in humans with chronic inflammatory disease: immunopanning of subacute sclerosing panencephalitis (SSPE) brain sections with anti-

body phage libraries prepared from SSPE brain enriches for antibody recognizing measles antigen in situ. *J. Virol.* **74,** 1533–1537.

45. Weiner, H., Faupel, T., and Bussow, K. (2004) Protein arrays from cDNA expression libraries. *Methods Mol. Biol.* **264,** 1–13.

46. Hoheisel, J. D., and Lehrach, H. (1993) Use of reference libraries and hybridisation fingerprinting for relational genome analysis. *FEBS Lett.* **325,** 118–122.

47. Bussow, K., Cahill, D., Nietfeld, W., et al. (1998) A method for global protein expression and antibody screening on high-density filters of an arrayed cDNA library. *Nucleic Acids Res.* **26,** 5007–5008.

48. Bussow, K., Nordhoff, E., Lubbert, C., Lehrach, H., and Walter, G. (2000) A human cDNA library for high-throughput protein expression screening. *Genomics* **65,** 1–8.

49. Bussow, K., Quedenau, C., Sievert, V., et al. (2004) A catalog of human cDNA expression clones and its application to structural genomics. *Genome Biol.* **5,** 71.

50. Lee, J., and Bedford, M. T. (2002) PABP1 identified as an arginine methyltransferase substrate using high-density protein arrays. *EMBO Rep.* **3,** 268–273.

51. de Graaf, K., Hekerman, P., Spelten, O., et al. (2004) Characterization of cyclin L2, a novel cyclin with an arginine/serine-rich domain: phosphorylation by DYRK1A and colocalization with splicing factors. *J. Biol. Chem.* **279,** 4612–4624.

52. Cepok, S., Zhou, D., Srivastava, R., et al. (2005) Identification of Epstein-Barr virus proteins as putative targets of the immune response in multiple sclerosis. *J. Clin. Investig.* **115,** 1352–1360.

11

Differential Protein Expression, Protein Profiles of Human Gliomas, and Clinical Implications

Ravi Sirdeshmukh, Vani Santosh, and Anusha Srikanth

Summary

Molecular profiling of tumors at the transcript or protein level has the potential of explaining the biology of tumors as well as classifying them in molecular terms for the purpose of more decisive diagnosis or prognosis, with respect to their aggressive, recurrence, or treatment response. In recent years, these approaches have been used for the study of gliomas, which are the most prevalent and lethal primary brain tumors. The studies on differential protein expression using 2DE/MS-based and other proteomic approaches, which are reviewed here, indicate their usefulness for future studies. We discuss the clinicopathological issues currently encountered with gliomas, the experimental approaches for protein profiling, experimental systems used to find molecular markers for gliomas, followed by a review of recent proteomics studies carried out by us and other investigators. A speculative scheme integrating the observations made in these studies is discussed and the chapter concludes with a perspective of the future developments that can come from constantly improving proteomics technologies.

Key Words: Glioma; glioblastoma; differential expression; proteomics; mass spectrometry; 2DE.

1. Introduction

Gliomas constitute >60% of the primary tumors of the central nervous system (CNS) and encompass a range of neoplasms. Their incidence is low, yet they represent a major source of cancer-related mortality, and the overall outcome of patients with these tumors has not changed much over the past 30 yr. The aim of this chapter is to illustrate recent efforts of molecular analysis of glial tumors and to show the potential of such analyses for understanding these tumors in terms of gene expression profiles, raising the prospects of better clinical prognosis and treatment strategies. Gene expressions can be determined either in terms of the levels of mRNA transcripts or their functional end products, proteins, in a given cellular condition. Their comparative analysis in conditions such as tumor and nontumor tissue would help in identifying the genes differentially expressed and associated with the tumor condition. The two approaches yield complementary information. However, in view of the space limitations, we restrict this chapter to proteomics approaches, differential protein profiles, and their potential clinical utility, but we include a brief account of the application of DNA microarrays and transcript profiling because they have provided important leads to

From: *Bioarrays: From Basics to Diagnostics*
Edited by: K. Appasani © Humana Press Inc., Totowa, NJ

understand tumors by molecular profiling. Also, the focus of our discussion is more on astrocytomas, a histological type that is more prevalent among the gliomas. Other histological types are mentioned in closely relevant contexts. Refer to other recent and specialized reviews on the subject for further details *(1–4)*.

2. Background, Types, Origin of Gliomas, and Clinical Issues

Mature and fully differentiated brain tissue consists of three major types of cells: neurons, astrocytes, and oligodendrocytes. Gliomas are primary tumors involving these cells of the CNS, the most common (>60%) among which are astrocytomas—the tumors involving astrocytes. They are composed of a wide range of neoplasms that differ in their location within the CNS, age and gender distribution, morphological features, growth potential, extent of invasiveness, tendency to progression, and clinical course. The overall classification of gliomas is shown in **Fig. 1A** and putative cellular pathways originating them (discussed subquently) in **Fig. 1B**. There are two distinct categories of these tumors. The first category encompasses at least 75% of the astrocytic tumors and is termed as diffusely infiltrating astrocytomas, which are further subdivided on the basis of their grades. The present grading systems for diffuse astrocytomas is based on histopathological features that are predictive of brain invasion at the tumor–brain interface, high growth potential, and capacity to infiltrate brain for distant spread in the neuraxis. The histological staining of the various types of gliomas is shown in **Fig. 2**. The two currently used grading systems, WHO and St. Anne Mayo, are comparable, and both use the four-tier grading for astrocytoma and the histological criteria considered for grading are nuclear atypia (diffuse astrocytoma/WHO grade II); mitosis (anaplastic astrocytoma; WHO grade III), and microvascular proliferation, necrosis, or both (glioblastoma multiforme; GBM, WHO grade IV and its variants such as giant cell glioblastoma and gliosarcoma). This category inevitably has the more ominous prognosis owing to high tendency of these tumors to undergo malignant transformation, to infiltrate the surrounding brain and to invade leptomeninges. The second category of astrocytic tumors, called circumscribed astrocytomas, clearly exhibits distinct clinicopathological features and consists of Pilocytic astrocytoma (WHO grade I), Pleomorphic xanthoastrocytoma (WHO grade II), and Subependymal giant cell astrocytoma (WHO grade I). In contrast, to the first category, these tumors generally tend to have a more favorable prognosis, because of a reduced capacity for invasive spread and limited potential for growth and anaplastic progression. Therefore, clear recognition of these entities is important. Pilocytic and other grade I tumors are relatively benign and occur in children and young adults. Grade II astrocytomas with a median survival period of 8–10 yr are considered to be low-grade tumors with slow malignant progression, whereas both grade III (median survival period, 2–3 yr) and grade IV (GBMs) are high-grade malignant tumors. GBMs, the most common and the most malignant form of glioma, with a median survival period of <1 yr occur among the older age group and accounts for >60% of the astrocytomas. Even after surgery followed by aggressive chemo- or radiation therapy, they recur with more aggressive behavior. Some low-grade astrocytomas progress to glioblastoma (secondary GBMs), whereas others persist in a dormant state for many years. In contrast, GBMs also can originate *de novo* without any evidence of the low-grade precursor (primary GBMs). Histologically, the two forms are indistinguishable but do associate with different genetic indicators, although they may not reflect much on the prognosis. The other type of glial

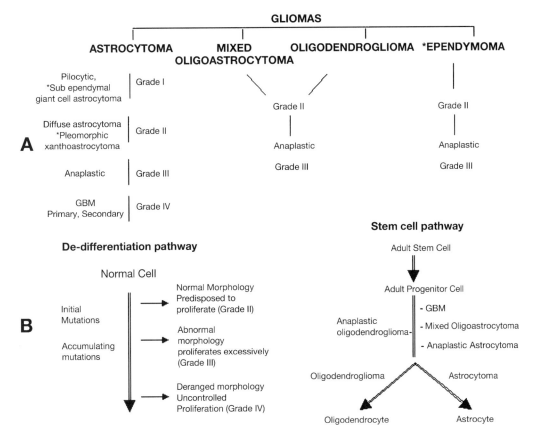

Fig. 1. (A) Classification of gliomas showing histological types and grades. Types marked with asterisks represent tumors of rare occurrence compared with others. **(B)** Diagrammatic representation of the two models suggested for the origin of glial tumors.

tumors, oligodendrogliomas, relates to oligodendrocyte lineage, and there are also mixed gliomas, namely, the oligoastrocytomas. They could be of grade II or III (anaplastic) type. Yet another type, ependymomas (grade II or III), which develop in the vicinity of the ventricular system and the spinal cord, constitute approx 10% of the gliomas (*see* **refs. 5** and **6** for further details).

The cellular origin of gliomas in general is incompletely understood. Tumors of clonal origin are thought to arise from a single normal cell that suffered a genetic alteration resulting in a growth advantage to that cell. Subsequent mutations in cells from this parental clone would result in the outgrowth of a tumor with increasing anaplastic changes and clonal complexity. For the origin of astrocytic tumors, two models can be considered **(Fig. 1B)**. The dedifferentiation hypothesis assumes that the normal cell undergoing mutation is a terminally differentiated astrocyte, oligodendrocyte, or ependymal cell. During gliomagenesis, the differentiated mutant normal cell triggers its growth, resulting in tumors of low grade (grade II). Further mutations would cause the tumor cells to acquire a less differentiated and more malignant morphological features (grade III), which progress into further dedifferentiated state with little resemblance to the cell(s) of origin, accompanied by loss of cell-specific markers (grade IV).

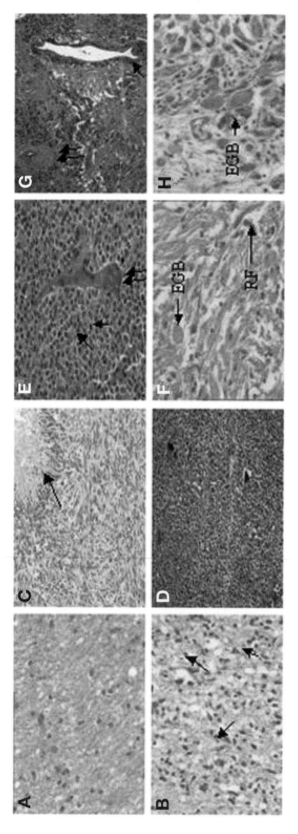

Fig. 2. Images of H&E-stained formalin-fixed sections of representative types of gliomas tissues. (**A**) Diffuse astrcytoma, grade II with mild nuclear atypia. (**B**) Anaplastic astrocytoma, grade III, tumor exhibits nuclear atypia and mitosis (arrow). (**C**) Glioblastoma multiforme, grade IV with mitosis and necrotic areas (arrow). (**D**) Oligodendroglioma grade II, tumor shows the characteristic "honeycomb" pattern of cells with "chicken wire" blood vessels. (**E**) Anaplastic oligodendroglioma grade III, tumor cells are pleomorphic, show mitotic activity (arrow), and microvascular proliferation (double arrow). (**F**) Pilocytic astrocytoma, tumor shows characteristic rosenthal fibers (RF) and eosinophil granular bodies (EGB). (**G**) Ependymoma with ependymal canal (arrow) and perivascular rosettes (double arrow). (**H**) Pleomorphic Xanthoastrocytoma, tumor makes up pleomorphic astrocytes and several EGBs.

Evidence for this progression model comes from patients who clinically present with a low-grade tumor and who subsequently develop a tumor recurrence of higher grade and eventually form GBM.

The alternative misdifferentiation hypothesis is based on the concept that glial cells originate from self-renewing stem cells differentiating into progenitor cells that eventually develop into three cell types of the mature glia. The pool of multipotential adult stem/progenitor cells that are induced to undergo proliferation may be particularly vulnerable to neoplastic transformation. During development and in adult stem cells, occurrence of asymmetric cell division marks the generation of daughter cells of variant proliferative and differentiation status *(7)*. And it is an aberration in the normal mechanisms that results in generation of tumor precursors and the onset of cancer. Mutations in the progenitor cells that are capable of differentiating into both oligodendrocytes and astrocytes could give rise to the heterogenous GBMs, which have either oligodendroglial, astrocytic, or mixed oligoastrocytic features. Mutations in the progenitor cells further along the differentiation pathway may give rise to either anaplastic oligodendrogliomas or anaplastic astrocytomas. Finally, mutations in the cells furthest along the differentiation pathway would result in low-grade oligodendrogliomas or astrocytomas having histomorphological features resembling astrocytes or oligodendrocytes. Certain hematological malignancies are known to originate from transformation of normal stem cells, and the existence of stem cells in restricted zones of adult brain raises the possibility of stem cell origin of gliomas *(8)*. It was recently shown that a small subset of cells carrying a surface marker CD133 (CD133+ cells) in human brain cancer could drive tumorigenesis. A very small number of them when injected into mice could form tumors unlike the CD133– cancer cells *(9)*. This finding adds strong experimental support to the stem cell model for gliomas. Furthermore, cellular heterogeneity is common in cancers. For glioma, it is a hallmark. The progenitor cells that can differentiate along several pathways may undergo aberrant differentiation and may unfold its potential generate greater heterogeneity encountered in gliomas, compared with the more restricted precursors

Grade I tumors, such as pilocytic astrocytomas, are essentially benign tumors and have significantly long median survival period. Grade II diffuse astrocytomas are slow-growing tumors, but they can progress into aggressive grade III or grade IV GBMs with dismal prognosis. Many grade IV tumors behave like grade III in terms of their clinical outcome, and grade III like grade IV. Their chemo- or radiation response is also relatively poor, posing a great challenge for their management. Whereas pure astrocytomas project such a complex picture, the complexity is further increased when one has to distinguish them from oligodendrogliomas and mixed oligoastrocyomas, which are different in terms of histology, treatment response, and overall prognosis. With this kind of variation, accurate typing and grading of these tumors is crucial for better clinical management. Also, glial tumors may develop through one of the two pathways described above, are highly heterogenous, and the therapeutic strategies applied to them may be different. Accurate typing and grading of these tumors is thus crucial for better clinical management. However, the high level of heterogeneity of gliomas both with respect to the cell type as well as progression, results in a high degree of interobserver variations in their histopathology analysis and grading *(10)*. Thus incorporating additional parameters, such as their distinction in terms of molecular markers, that allow us to discriminate between their different types and grades would be very important.

3. Molecular Genetic Changes and Molecular Markers and Importance of Molecular Profiles

Several genetic aberrations and gene expression changes associated with oncogenesis and malignant progression of the astrocytic tumors have been identified . Amplification of epidermal growth factor receptor (EGFR) and other genes, such as CDK4 and MDM2, and deletions or inactivating mutations for P53, Rb, and other regulatory factors such as phosphatase and tensin homolog and platelet-derived growth factor constitute the genetic pathways implicated for astrocytomas *(3)*. All of these negatively regulate specific activities in normal glial cells. Genomic microarrays were used to investigate amplification of oncogenes throughout the genome of GBMs *(11)*. High-level amplifications were identified with respect to oncogenes including CDK4, MYC, MDM2, and several others. These genes are suggested to be involved in the GBM tumorigenesis. Recently, Hayashi et al. *(12)* have shown that glioblastoma can be classified into distinct genetic subsets on the basis of presence of p53 mutation, EGFR amplification, or absence of both.

These molecular differences and others have helped in the use of immunohistochemistry approaches to complement histopathological analysis of the tumors. For example, immunohistochemistry to assess proliferative activity in the routine neuropathological evaluation of diffuse astrocytic tumors provides a more direct and objective index of the growth potential. Two cell cycle-associated nuclear proteins that help in estimation of growth potential and that have been widely exploited for immunohistochemical techniques are Ki-67 antigen and proliferating cell nuclear antigen. Both can be detected by immunohistochemical techniques. The monoclonal antibody MIB-1 against the Ki-67 antigen recognizes a nuclear nonhistone protein, Ki-67 protein, which is expressed during the rest of the phases of the cell cycle, except the G_0 phase *(13,14)*. Multiple studies with astrocytic tumors indicate that Ki-67 immunoreactivity is a valuable tool to estimate the growth potential. There seems to be a general correlation between histopathological grade and MIB-1–labeling indices. Glial fibrillary acidic protein (GFAP) is an intermediate filament protein (IFP) specifically expressed by astrocytic cells. Immunoreactivity for GFAP is considered indicative of the glial phenotype *(15–17)*. In diffuse astrocytomas, GFAP is differentially expressed and its expression in anaplastic astrocytomas and GBMs is highly variable. It tends to decrease during glioma progression, but it is not used as a prognostic factor. Expression of vimentin, the major IFP present in the fetal stages, is common in GBMs and is also used as a marker. The two subtypes of GBMs follow different genetic pathways *(6)*. Primary GBMs are characterized by EGFR amplification or mutation consistent with its overexpression, whereas the P53 mutations are associated with secondary GBMs as they are with their low-grade precursors. Immunohistochemistry for these two antigens is used to support the clinical and histopathological analysis to distinguish these two types. Despite these advances, distinctions based on single-molecule differences between the histological types and grades are not precise and adequate.

Gene expression profiling of cancer tissues would be useful to identify signature gene sets that could distinguish different types or grades of tumors and their outcome. A comprehensive characterization of the gene expression that may correlate with the clinical behavior of gliomas would add precision in their molecular classification and help to predict the prognosis and treatment response of patients *(2,18)*. DNA microarray analysis has given rise to significant molecular expression data on glial tumors, which has initiated their characterization and molecular taxonomy. However, it is not always possible to translate the gene expression data at the mRNA level into proteins that are the end determinants of the gene function. Moreover, the majority of the drug targets and diagnostic tests are protein based. Protein profiles and their posttranslational modifications may offer useful and perhaps better tools that can define tumor types, their chemo- or radiosensitivity as well as distinguish aggressive tumors from less aggressive tumors *(19)*. When integrated with clinical information such as patient data on age and sex, symptoms, imaging, other clinical conditions, histopathological analysis, other clinical conditions, and so on, gene expression profiling by microarray or proteomics approaches would be highly useful to define tumor groups in terms of transcript or proteomics profiles and to subsequently correlate them for biological relevance or histological or survival-related distinctions.

4. Experimental System: Glioma Cell Lines Vs Primary Tumors

For any molecular analysis to identify disease lesions, the first important consideration is the choice of an appropriate experimental system. Tumor-derived cell lines constitute a useful experimental system to study molecular aspects relevant to the tumors and are an indispensable tool for cancer research. The creation and characterization of permanent cell lines derived from primary human gliomas permitted access to unlimited, renewable material for studying development of the brain tumors. However, it is possible that cell lines of the same pathological group may exhibit heterogeneity in gene expression, gene mutation, and cellular response to various treatments *(20)*.Several previous reports describe study of expression of mRNA or protein associated with glioma cell lines, and cell line-associated RNA or protein expression was examined as a diagnostic or functional (e.g., multiple drug resistance and invasiveness) marker for glioblastoma tumors *(21)*. They also have established and characterized five glioma cell lines derived from GBM or anaplastic astrocytoma and demonstrated that the pattern of P53 and EGFR expression as observed in the original tumor is retained in the cell lines making them useful for analysis of tumor growth and progression *(22)*. Vogel et al. *(23)* have studied protein patterns between GBMs and five different glioma cell lines and identified differential proteins whose expression may be driven by in vitro culturing of cells. Such characterization of protein expression to distinguish solid tumors from the cell lines should bring greater value for the use of cells (and tissues) to study molecular analysis in relation to the biology of these tumors. Also, new cell lines are being established that may advance the studies *(24)*. However, these cell lines are already tumorigenic and selected for growth in culture, limiting the amount of information that could be gathered about the events that led to the formation of their tumors of origin. Several groups have demonstrated the existence of the stem cells in human brain tumors and their existence seems to correlate with the heterogeneity seen in tumor cells with respect to proliferation and differentiation *(25)*, and the thinking that gliomas and other brain tumors may originate from stem cells is gaining increasing support *(8)*. Two cell lines, HNGC-1 and HNGC-2, derived from high-grade malignant gliomas have been shown to represent low-grade and high-grade malignancy, respectively *(26)*. The slower proliferating HNGC-1 cells progressively transform into the rapidly proliferating HNGC-2 cell line; thus, they exhibit stem cell-like patterns and the ability to sequentially mimic the process of tumorigenesis. These cells thus carry a rich potential as a model system to study molecular events associated with gliomagenesis and progression.

The use of primary tumors to understand lesions important in tumor formation suffers from two drawbacks. First, the availability of adequate amount of material for molecular analysis is often limiting, particularly in gliomas. Second, the extent of variability and heterogeneity in specimens brings in further limitations for clinical correlations to pass statistical acceptability. Direct comparison of molecular analysis between the tumor and the nontumor tissue from the same patient is also difficult in astrocytomas in which the tumor boundaries are very diffuse, forcing the use of normal tissues from other surgeries on other patients. Surgical resections from temporal lobe epilepsy surgeries are therefore taken and used as experimental controls. Because such tissues are taken from other individuals, many of them are to be used to arrive at a normalized protein expression pattern for the purpose of comparison. Despite these limitations, primary tumors have been used with reasonable success, because the present methods of molecular analysis are highly sensitive and allow handling of very small amounts of samples and are yielding significant information.

5. DNA Microarray Analysis and Transcript Profiles

Gene expression profiling with either oligonucleotide- or cDNA-based microarrays has been used for understanding glial tumor biology and behavior (reviewed in **refs. 2,4,** and **27**). Details of this topic and the methodology are described elsewhere in other chapters. Either of these types of arrays can access virtually the entire genome, and the challenge would be to analyze several data sets to pick the most reliable, differentially expressed genes. Then, these genes are subjected to hierarchical clustering to identify groups of genes or tumors with global expression profiles. Several reports describe using DNA microarrays to pinpoint differentially expressed genes between normal brain and diffuse astrocytomas and between astrocytomas of different grades *(28–30)*. To understand the molecular basis of astrocytoma progression, transcriptional profile of approx 7000 genes in primary grade II gliomas and corresponding recurrent high-grade (grade III or IV) gliomas were compared using oligonucleotide-based microarray analysis, and the mRNAs associated with these groups of tumors were studied by van den Boom et al. *(31)*. These studies allowed distinction of primary grade II and progressive grade III and IV tumors in terms of transcript sets. Other investigators also have reported differential gene expression to distinguish subtypes of glioblastomas *(32,33)*. Oligodendrogliomas, a subtype of gliomas, have different histology, exhibit different chemosensitivity, and have different prognosis. Microarray-based studies also distinguished genes whose expression correlated with this type of tumors compared with the other histotypes *(34,35)*. Another type, the cystic, slowly proliferating pilocytic astrocytoma, is a distinct type of glioma and could be reliably differentiated from the common and rapidly proliferating, diffuse astrocytomas by using cDNA microarray analysis. Of the expression of 7073 genes, apolipoprotein D, was overexpressed almost 10-fold in pilocytic astrocytomas and emerged as a potential pylocytic tumor marker *(36)*. Necrosis-associated gene expression in GBMs was studied by Raza et al. *(37)*. The DNA microarray-based expression analysis is also found to be helpful to uncover previously unrecognized patient subsets that differ in their survival and thus translate microarray-based expression results into prognostic assays for clinical use *(38)*.

6. Proteomics Approaches and Protein Profiles

6.1. Methodology

Expression profiling at the protein level is relatively more complex for the following reasons: 1) The number of proteins for each gene can be as many as six to eight times higher if we include protein variants and essential posttranslational modifications. 2) The protein complement is dynamic and subject to changes driven by multiple factors; thus, protein abundance can vary by 6 to 8 orders of magnitude adding further level of complexity. 3) Expression is not measured against preannotated proteins, but protein annotation is indeed part of the experiment and a challenge by itself. And 4), to access various proteins for identification, they need to be separated, and the success of the separation will depend upon the resolving power of the technique. The sensitivity of the techniques and their dynamic range are such that only small sets of proteins can be accessed in a given experimental condition. The complications are further enhanced as examples emerge to suggest that the protein networks are divergent in differentiated cells of the same organism. Therefore, an integrated view of the mutual, interdependent functioning of the living systems is important.

6.1.1. Protein and Peptide Separations and Mass Spectrometry (MS)

The biggest challenge in protein analysis for proteomics studies is dealing with not only an enormous number of molecules in a high-throughput manner but also molecules that are similar but still contain a gradient of chemical heterogeneity. During the last several decades, biochemistry was dominated by development of protein separation principles and methods. Many primary separation approaches and their variations are already available and encompass electrophoresis, liquid chromatography as well as affinity principles. Rapid developments over the last decade in the application of these methods of protein separation, their miniaturization, evolution of MS, methods for proteins, and new bioinformatics tools have enabled evolution of powerful experimental approaches for proteomic explorations. A general scheme, as shown in **Fig. 3**, is to separate proteins or peptides, subject them to one more kinds of mass spectrometric analysis, and use this combined information for data base queries using various algorithms and bioinformatics tool (reviewed in **ref. 39**). MS is based on ionization of the analyte molecules in the gas phase and provides a way for accurate determination of masses of the molecules from mass by charge (m/z) ratios of the molecular ions. The advent of soft ionization techniques such as matrix-assisted laser desorption ionization (MALDI) and electrospray ionization have helped the application of MS to biological macromolecules such as proteins. However, the mass differences among large molecules such as proteins are not discriminatory enough to allow identification of proteins. A protein after its separation using biochemical principles is therefore subjected to proteolytic digestion by using enzymes that cleave at specific sites in the protein chain. The peptide fragments thus generated are a function of the location and number of the target amino acid residue in the protein chain and yield a mass fingerprint by MALDI. Such a fingerprint can be used for its identification from the database searches. MALDI-MS is therefore routinely used for high throughput protein identifications with reasonable degree of statistical success. Often, peptide mass fingerprint data are not adequate for acceptable protein identifications, and small sequence information is required for use in conjunction with the fingerprint data.

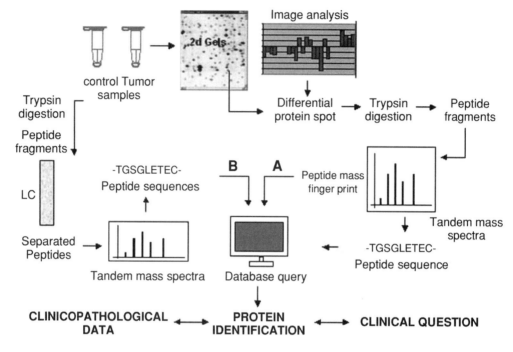

Fig. 3. General scheme of the workflow in MS-based proteomics.

Protein sequencing by chemical method is common, but MS offers an alternative to the chemical method. The chemistry and the labiality of the peptide bonds linking amino acid residues in proteins allow fragmentation of proteins along its backbone. In the tandem mass spectrometers with two mass analyzers or in ion trap instruments, a parent ion of selected mass can be fragmented in a controlled manner by using gas-induced collisions, and a mass ladder corresponding to peptide fragments differing by one amino acid can be generated. The mass differences between two contiguous fragments can be used to assign amino acids. Sequences up to 10–15 residues can be generated by this method and used independently or in combination with other MS data for protein identifications. In addition to their sequence and abundance, a property of proteins that may be interesting from a functional point of view is the state of protein modifications. Diverse posttranslational modifications such as proteolytic processing, phosphorylations, and glycosylations introduce important chemical variations in the protein and change the mass of the protein or peptides. Mass spectrometry can be used for detecting such changes and the sites at which they occur. Space constraints do not allow us any detailed discussion on this aspect.

One of the most commonly used and established work flows in mass spectrometry-based proteomics studies is two-dimensional gel electrophoresis (2DE) combined with mass spectrometry (2DE/MS) (**Fig. 3**). The 2DE/MS method has very high resolving power and is sensitive enough to facilitate the analysis of small amounts of clinical samples. The proteins from tumor and nontumor tissues are first separated on 2DE gels and detected as single protein spots by specific stating. By comparative analysis of gel images, differentially expressed proteins displayed on the gels are selected and the differential protein spots are subjected to in-gel tryptic digestion and processed for peptide mass fingerprinting (MALDI-MS) followed by sequencing of some selected peptides by tandem mass spectrometry (MS/MS). The combined information for each protein is then used for protein identifications. In another method *(40)*, a shotgun approach, proteins are subjected to en masse tryptic digestion without any preseparation, and thousands of resulting peptides in the total tryptic digest are then separated by multidimensional HPLC and protein identifications are done using MS/MS-based sequencing of a large number of peptides. In this latter approach, several peptides originating from the same protein may fractionate at different places on HPLC and identify or match with the protein from the database. Therefore, success of a protein identification will be determined by the number of peptide hits for the same protein along the chromatographic run. In a simplified version, proteins are separated on one-dimensional sodium dodecyl sulfate-polyacrylamide gel electrophoresis, each protein band, which might contain few to tens of protein, is digested with trypsin, the resulting peptides separated on reverse phase-HPLC and sequenced by MS/MS method. Protein identifications are then made on the basis of sequence information as above.

In a different experimental approach, protein mass patterns, especially for low-mass-value proteins, are generated by picking up a set of low-mass-value proteins or peptides from cells or tissues by using an affinity principle. Protein masses are then determined by surface enhanced laser desorption ionization-MS method *(41)*. This approach does not permit identification of proteins but images of protein mass profiles can be rapidly obtained that can be used to represent the tissue condition. Intact protein mass profiles of tissue sections also can be obtained using MALDI-MS *(42)*. On average, 500–1000 individual peptides or proteins in a much broader range of molecular weights can be accessed to generate tissue images representing cellular phenotypes. All the aforementioned methods and approaches have been used for molecular analysis of the glial tumors (*see* **Subheading 6.2.**).

6.2. Experimental Findings

One broad aim of all the studies on differential gene or protein expression in gliomas is to identify molecules that can help in more precise tumor grouping, classification for therapeutic and prognostic assessments. Hidden among the tumor-associated proteins, there may be information on biological functions and aberrations underlying tumorigenesis and progression that reveal important molecules with potential therapeutic targets. Broadly, the clinical questions that have been addressed in the proteomic profiling studies so far involve molecular correlation with histological grading of gliomas; molecular distinction of primary and secondary GBMs; proteins associated with varying survival rates; and aberrated protein functions relevant to tumor growth, progression, and invasiveness.

6.2.1. Grade-Related Protein Expression and Potential Biomarkers

Previous studies by several investigators represent the study of differential expression of some selected protein candidates between tumor and normal tissues by using Western blotting, immunohistochemistry, or both and their correlation with tumor type. In some of these studies, protein analysis was complemented with differential expression studied at the transcript level. Differentially expressed proteins revealed from some of these studies are as follows: expression of interleukin-6 in glioblastomas but not in low-grade astrocytomas or oligodendrogliomas; increased expression of insulin-like growth factor-binding protein 2 in relation to advancing tumor grade *(43,44)*. A distinct set of proteins identified in several of the previous studies are the matrix metalloproteases-2 and -9 and their possible association with tumor invasiveness (reviewed in **ref. 45**). A recent study reported differential expression of YKL-40 and identified it as a potential tumor marker for gliomas *(46)*.

In one of the first attempts to understand global protein expression in gliomas, Hanash and co-workers *(47)* analyzed several high- and low-grade gliomas by 2DE and identified 22 differential protein spots specific to high-grade gliomas, many of which were identified using a 2DE gel database and serum proteins, cytoskeleton proteins, and enzymes (**Table 1**). In another study, Hiratsuka et al. *(48)*, using 2DE/MS and peptide mass fingerprinting, compared proteins from gliomas of different grades with nontumor tissues. Serum proteins, albumin, transthyretin, hemopexin, and apolipoprotein A-1 were found to be upregulated in high-grade tumors. Another differentially expressed protein, PEA-15, is a regulator of extracellular signal-regulated kinase (ERK) mitogen-activated protein kinase (MAPK) and is a modulator of cell proliferation and cell death. We have studied and reported differential protein expression between normal tissue and high-grade astrocytomas and identified 29 differentially expressed proteins *(49)*. Differentially expressed proteins revealed in this study were secretory, heat-shock proteins (HSPs), cytoskeletal, mitochondrial, neuronal cell proteins, and other proteins or enzymes that have regulatory roles in the cells. Higher levels of serum proteins observed in our studies and previous studies may be because of increased vascularity and presence of higher blood volume. Similarly lower levels of neuronal cell-specific proteins such as β-synuclein, synaptosomal-associated protein, and *N*-ethylmaleimide sensitive factor-attachment protein (in grade III tumors) may be a result of increased astrocytic cellularity in the tumor areas. Some of the differentially expressed proteins with regulatory functions observed in these studies include proteins such as the HSPs, tumor suppressor protein prohibitin, and Rho-GDP dissociation inhibitor or Rho -G –D inhibitor or Rho-guanine nucleotide dissociation inhibitor (GDI) implicated in cell signaling. In both these studies, changes associated with the cytoskeleton or cytoskeleton-related proteins also were observed. Hiratsuka et al. *(48)* observed differential regulation of a group of cytoskeleton-related proteins, prominent among them being downregulation of Sirtuin homolog 2 (SIRT2), suggesting the involvement of cytoskeleton modulation in glioma pathogenesis. We observed *(49)* destabilization of the intermediate filament protein (IFP) GFAP in tumors. Intermediate filament proteins constitute structural scaffolding in cells as well as forming part of the regulatory dynamics of the cells by associating with several proteins. The 49-kDa form (GFAPα) is the main form of the protein reported for the normal glial filaments. Smaller proteolytic fragments of the protein also exist representing soluble fraction of the filaments *(50)*. We observed, in our studies, that the 49-kDa species increased in the grade III tumors but at the same time smaller proteolytic fragments also increased. This increase may be consistent with a steady-state condition representing GFAP expression and its concomitant destabilization in proliferating astrocytes.

In the aforementioned studies, grade-related differential protein expression has been the major objective to identify possible genes involved in glioma tumorigenesis and to identify proteins that might serve as potential molecular markers. SIRT2 is suggested to be a potential glioma marker by Hiratsuka et al. *(48)*. GFAP is already used as marker

Table 1
Differentially Expressed Proteins Observed in Gliomas Analyzed by 2DE/MS-Based Proteomics Approaches

Name of protein	Biological function	Cellular location	Chromosomal location	Ref
Albumin	Metabolic transporter	Secretory protein	4q11-q13	48
				49
Transthyretin	Thyroid hormone-binding protein	Secretory protein	18q12.1	48
				47
Hemopexin	Regulates heme function	Secretory protein	11p15.5-p15.4	48
Apolipoprotein A-1	Lipid transport	Secretory protein	11q23-q24	48
				47
Chaperonin (HSP60)	Molecular chaperone	Mitochondrial protein	2q33.1	49
α-Crystallin, -B	Molecular chaperone	Cytoskeleton/ cytoplasmic	11q23.1	57
				49
Hsp70	Molecular chaperone	Endoplasmic reticulum lumen	9q33-q34.1	49
Heat shock protein 27 kDa(HSP27)	Molecular chaperone	Cytoplasmic in interphase cells	7q11.23	57
				47
Protein disulfide isomerase A3	Molecular chaperone	ER lumen	15q15.3	57
Prohibitin	Antiproliferative	Mitochondrial; nuclear	17q21	48
	Molecular chaperone			49
Oncogene DJ1	Protein folding	Cytoplasmic, ribosome	1p36.33-p36.12	49
Protein kinase Cγ	Signal transduction			57
MAPK/ ERK kinase 1	Signal transduction	Cytoplasm	15q22.1-q22.33	57
Oncoprotein 18	Signal transduction	Cytoplasmic	1p36.1-p35	47
Ezrin (p81; ctyovillin)	Signal transduction	Membrane–associated	6q25.2-q26	57
Tyrosine tryptophan monooxygenase activation protein	Signal transduction		20q13.1	49

(continued)

161

Table 1 (*Continued*)

Name of protein	Biological function	Cellular location	Chromosomal location	Ref
PEA-15	Signaling protein, Regulates ERK/MAPK	Cytoplasmic	1q21.1	48
RAB3A	GTP binding protein	Cytoplasmic	19p13.11	57
Ras related protein Ral-A	GTP binding protein	Cytoplasmic	7p14.1	57
Transforming protein Rho A	GTP binding protein	membrane-associated	3p21.31	57
Rac 1	GTP-binding protein	Inner surface of plasma membrane	7p22.1	57
NDPK A	ATP-binding, kinase activity	Nucleus	17q21.3	47
ATPase	Metabolic enzyme	Cytoplasmic/transmembrane	3q13.2-q13.31	49
Phosphoglycerate mutase 1	Metabolic enzyme	Cytoplasmic	10q24.1	57
Glutathione *S*-transferase M	Metabolic enzyme	Cytoplasmic	1p13.3	57
Glutathione *S*-transferase P	Metabolic enzyme	Cytoplasmic	11q13.2	57
Glutamate dehydrogenase 1	Metabolic enzyme	Mitochondrial matrix	10q23.2	57
Phosphopyruvate hydratase	Metabolic enzyme	Cytoplasmic	1p36.23	57
CRMP-4	Nucleic acid metabolism	Cytoplasmic	5q32	48
Plasminogen activator inhibitor-1	Protease	Cytoplasmic	7q21.3-q22	57
Cathespin D	Protease/apoptosis induction	Lysosomal	11p15.5	57
Proteasome β subunit	Threonine endopeptidase activity	Nuclear and cytoplasmic	16q22.1	47
SIRT2	NAD-dependent deacetylase. Cell cycle, cytoskeleton regulation	Cytoplasmic	19q13.2	48
Profilin 2	Actin-binding protein	Cytoplasmic	3q25.1-q25.2	48
tropomodulin 2	Formation of actin profilament	Cytoplasmic	15q21.1-q21.2	49
Tropomyosin 4	Actin binding protein	Cytoplasmic	19p13.1	49
Neurocalcin delta	Regulator of rhodopsin, binds to actin	Cytoplasmic	8q22-q23	48
Glial fibrillary acidic protein	Major intermediate filament protein Astrocyte marker	Cytoplasmic	17q21	49
Vimentin	Intermediate filament protein	Cytoplasmic	10p13	49

Protein	Function	Location	Chromosome	Ref.
Moesin	Activates erzin	Cytoplasmic	Xq11.2-q12	*47*
Synaptonemal associated protein 25	Regulation of neurotransmitter Release Protein	Vesicle membrane, Presynaptic plasma membrane	20p12-p11.2	*49*
N-ethylmaleimide-sensitive factor attachment protein, gamma	Vesicular transport Between the ER and Golgi	Cytoplasmic membrane protein	18p11.22	*49*
β-synuclein	Protects neurons from 6ODHA Caspase activation	Cytoplasmic	5q35	*49*
Rho GDP dissociation Inhibitor α	Regulates the GDP/GTP exchange Reaction of the Rho proteins	Cytoplasmic	17q25.3	*49*
Ferritin, light polypeptide	Iron homeostasis	Mitochondrial	19q13.3-q13.4	*49*
Sorcin isoform a or A	Calcium homeostiasis	Cytoplasmic	7q21.1	*49*
Chain a, human mitochondrial Adenylate dehydrogenase	Mitochondria	Mitochondria		*49*
Ubiquitin carboxy-terminal hydrolase L1	Ubiquitin-protein hydrolase	Cytoplasmic	4p14	*49*
Adenosine deaminase	Deaminase function, RNA editing	Nuclear	20q12-q13.11	*49*
ATP synthase	ATP synthesis	Mitochondrial matrix	12q13.13	*49*
78 kDa glucose-regulated protein	ER lumen	Nuclear and cytoplasmic	9q33-q34.1	*57*
Cyclin-dependent kinase inhibitor	Cell cycle regulation	Nucleus	6p21.31	*57*
Eukaryotic initiation factor 4A	Translation regulation		17p13.1	*57*
Annexin ii (lipocortin ii)	Phospolipid-binding protein	Lamina beneath plasma membrane	15q22.2	*57*
Annexin v (lipocortin v)	Anticoagulant	Lamina beneath the plasma membrane	4q28-q32	*49* *47*

All proteins listed in the original reports discussed here have been included. Proteins from ref. *47* were identified on the basis of 2D gel databases. Other studies used peptide mass fingerprinting (*48*), peptide sequencing by MS/MS, or both (*49,57*). In the reports in which differential expression was indicated collectively for high-grade tumors, the same trend of regulation was assumed both for grade III and IV and is indicated accordingly in the table.

163

for astrocytes *(13)*. The proteolytic fragments of GFAP observed by us are distinct and their proportion in relation to the intact GFAP may vary with grades of the tumor. The progressive proteolytic fragmentation patterns may be additionally useful to define grades III and IV of astrocytomas. Low levels of Rho-GDI are implicated in high-grade tumors of the breast and metastasis *(51)* and its altered expression is also shown in drug-induced apoptosis in non-small cell lung carcinoma *(52)*. Expectedly, differential expression of SIRT2, prohibitin, Rho-GDI as well as destabilization of GFAP and appearance of its fragments will receive special attention for further investigations as potential markers for malignant gliomas.

Oligodendocytes represent a distinct histological type of glioma and differ from astrocytomas in chemoresponse. There are no reliable molecular markers to distinguish these tumors. Preliminary analysis of Luider et al. *(53)* indicated the possibility of differential appearance of GFAP and its fragments in these tumors compared with astrocytomas. Recently, Mokhtari et al. *(54)* analyzed many glial tumors made up of both astrocytomas and oligodendrogliomas, and they examined expression and level of oligo2, GFAP, p53, and loss of chromosome 1p. They observed oligodendrogliomas to be Oligo+/GFAP–. However, on the basis of the expression of these two proteins, their results suggest two additional subgroupings corresponding to Oligo–/GFAP+ and Oligo+/GFAP– status, among astrocytomas and oligoastrocytomas. Thus, by patterns of the expression of these two proteins, the tumors could be classified into "oligodendrogliomas" and "astrocytomas and oligoastrocytomas." The aforementioned findings, however, need further experimental support for any application in the clinical environment.

6.2.2. Protein Profiles and GBM Subtypes

As discussed under **Subheading 2**, GBMs can be subdivided into two sub types: primary tumors, which arise *de novo* and secondary GBMs, which progress from lower-grade gliomas. The two types are distinct disease entities, have different genetic pathways, and may differ in prognosis. Primary GBMs are characterized by overexpression of EGFR, phosphatase, and tensin homolog mutations. In contrast, secondary GBMs carry P53 mutations. Thus, clinical and genetic findings to distinguish these two types are available and are exploited in immunohistochemical analysis by several workers. In a blinded study, Furuta et al. *(55)* recently analyzed pure populations of cells, obtained from these two types of tumors by microdissection, for protein profiles by 2DE followed by their identification by MS/MS and arrived at a set of 11 distinct proteins that were expressed in one or the other type, but not in both. These proteins included 1) glycolytic enzyme such as enolase; 2) extracellular matrix proteins involved in cell migration and in embryogenesis and tumor invasion; 3) molecules related to cell–cell adhesion; 4) proteins implicated in nuclear functions, and so on. Similarly, Rb2/p130 expression and their inverse correlation with the degree of malignancy, i.e., high expression in low-grade gliomas and low expression in high-grade gliomas, was observed by Li et al. *(56)*, suggesting Rb2/p130 expression as potential prognosis factor.

6.2.3. Survival-Related Protein Profiles

In a comprehensive and recent study, Iwadate et al. *(57)* examined the proteomes of many gliomas of defined histological grades (grade II, III, and IV) against normal tissue

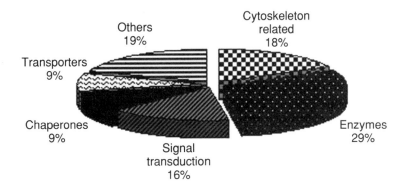

Fig. 4. Functional grouping of differentially expressed proteins observed in gliomas and included in Table 1. Functional categories were assigned basically as defined and indicated in the original studies. Proteins with unknown functions or in small numbers under several functional are shown as others.

samples by using 2DE. They correlated the protein patters with patient survival periods over a 5-yr period. Proteome-based clustering could distinguish normal vs tumor samples as well as more aggressive tumors from less aggressive tumors even within a given grade. Using peptide mass fingerprinting, they identified 37 tumor-associated proteins from these clusters, which may provide a novel, clinically more relevant molecular classification to define tumors on the basis of their aggressiveness independent of the conventional histological classification. Many of the overexpressed proteins in the high-grade tumors could be designated as signal transduction proteins, including small G proteins, whereas others included molecular chaperons, transcription and translation regulators, extracellular matirix-related proteins, and cell adhesion molecules, and they could be developed into markers for prognostication.

6.2.4. Differential Protein Expression and Understanding Malignancy

Amplification or mutations in genes such as EGFR, P53, and others involved in the signaling pathway form the basis for the onset of uncontrolled cell proliferation. However, there are several downstream biochemical events, particularly at the level of protein functions, that contribute to the maintenance of this state as well as alter the physiology of the differentiated cells. Irrespective of the primary objective of the study, the aforementioned reports have identified several differentially expressed proteins. The known functions of at least some of these proteins permit some speculative discussion on the biological effects underlying these tumors and their malignant state and suggest their therapeutic implications. **Table 1** is list of differentially expressed proteins identified in these studies, from different grades of gliomas. Their functional grouping is shown in **Fig. 4**.

One of the distinct groups of functionally related proteins revealed in these studies includes many small G proteins of the Rho family *(57)*. Rho proteins that have similarities with RAS oncogenes were initially thought to regulate actin cytoskeleton. However, they are now implicated in signal transduction pathways, regulate gene expression and cell proliferation, and Rho function is required in RAS transformation. Rho proteins cycle between the membrane and the cytoplasm as GTP-bound or GDP-bound forms, respectively *(58)*. GTP-bound forms promoted by Rho-guanine nucleotide exchange factors and Rho-GTPase activating proteins bring about GTP hydrolysis. Another related factor, Rho-GDI, which regulates the Rho GDP–GTP equilibrium and consequent GTPase activity by sequestering Rho-GDP in the cytoplasm, was found to be underexpressed in the majority of the malignant (grade III and IV) tumor specimens studied *(49)*. Its downregulation along with higher levels of G proteins implies promotion of Rho-GTPase activities required for cell growth signaling. Thus, Rho- and Rho-related protein functions may play a role in gliomagenesis, although RAS mutations have not been implicated in gliomas.

Another putative regulatory protein found to be downregulated in malignant (particularly grade III) astrocytomas is prohibitin *(49)*. The cellular function of this protein is not fully understood. It is implicated in the nuclear as well as mitochodrial functions *(59,60)*. Its role in cell cycle arrest may be facilitated by its involvement in the Rb pathway and interaction with the downstream transcriptional factors to inhibit cell proliferation. Mutations in prohibiton genes are known to be associated with sporadic breast cancer *(61)*, and thus it has been discussed in the literature as a tumor suppressor protein. Some studies have reported upregulation of prohibitin in cancers *(48,62)*; however, no clear functional explanation for this effect was discussed. Its relative levels in different grades of astrocytomas, in our studies, suggest overall (~60% of the tumors) downregulation of this protein in grade III tumors. In grade IV, this underexpression was comparatively less pronounced, if at all. In a few grade III tumors, we observed marginal upregulation of this protein. In the light of its growth suppressor function, its down regulation as observed by us may have implications in the proliferation state of the gliomas. Rb pathway has been implicated in glioma and our results suggest possible involvement of prohibitin in this pathway.

HSPs are overexpressed in response to various stress stimuli, including cancer, inflammation, and atherosclerosis *(63)*. Major biological function of HSPs is the chaperone activity. In astrocytes, they also are involved in maintaining the integrity of the cytoskeleton *(64)*. HSPs have been reported to be upregulated in many cancers. In gliomas, the situation seems to be mixed. Increased levels of crystalline alpha B, HSP70 and -27 were observed in malignant tumors by some workers *(47,57)*. In contrast, we observed some increase in grade IV tumors but in grade III high-grade tumors, their levels (particularly crystalline alpha B, HSP60) were found to be lower than the controls *(49)*. We have tried to explain this variant finding in terms of low induction of HSP promoters in differentiated astrocytes and transport of HSPs out of the cells; however, they need to be experimentally proven. Although the levels of HSPs were found to be higher in grade IV than the control, these levels may not be still adequate to support higher chaperone requirement in these rapidly growing tumors. Nevertheless, lower levels of HSPs may imply lowering of their chaperoning and cytoskeleton-stabilizing function in cells. If true, destabilization of GFAP in these tumors observed by us may even be a result of this alteration. Through their posttranslational modifications, interactions with associated proteins, IFPs including GFAP form dynamic, regulatory networks related to structural rearrangements during various cellular processes, including cell division *(65)*. The dynamics between the soluble and insoluble forms seems to be controlled by phosphorylation of GFAP, and work in progress (unpublished data) in

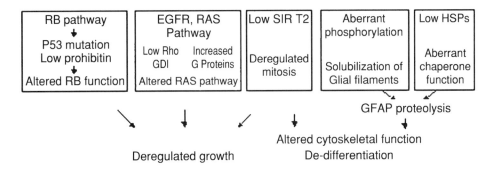

Fig. 5. Schematic representation of biochemical effects speculated on the basis of the known functions of differentially expressed proteins and their combined contribution to the tumor state.

our laboratory suggests changes in phosphorylation status of this protein, presumably promoting its proteolysis. A steady-state situation that may result would be continued GFAP expression in increasing astrocytic population in the growing tumor associated with its simultaneous destabilization and proteolysis. Grade IV tumors with greater dedifferentiated characteristics may lose the ability to make GFAP but continue its proteolysis, resulting in overall underexpression of the protein. It is interesting to note that even other cytoskeleton-related proteins are altered in gliomas *(48)*. SIRT2 was significantly decreased in all grades of gliomas. Sirtuin homolog 2 is thought to regulate microtubule network and is involved in the control of mitosis. However, this may be regulated at the transcriptional level. Some other cytoskeleton-related proteins whose expression was altered are CRMP-2 (involved in microtubule dynamics) and profilin 2 (actin binding). Together, any destability of the IFP-like GFAP or alterations in other cytoskeleton components imply change in the cytoskeleton functions and may have cascading effects in the differentiated state of the tumor cells.

Malignancy of brain tumors is different from other types of cancers in that the tumor cells do not migrate to other organ sites in the body but infiltrate into surrounding brain tissue, thus spreading the tumor locally and leading to tumor recurrence. Investigations of many groups on protease profiling indicate increase in a number of proteases notably, serine protease (uPA), cystein protease (cathepsin), and matrix metalloproteases-2 and -9 and implicate their role in modulation of the extracellular matrix and glioma invasion. Inhibition strategies involving each of these proteases have shown to significantly reduce tumor invasion and growth (*see* **ref. 45**).

Thus, downregulation of prohibitin and Rho-GDI, overexpression of Rho GTPases may promote cell proliferation. SIR T2 levels may also be responsible for the deregulated mitosis. Concomitant proteolysis of GFAP presumably as a combined effect of low HSPs and tumor associated GFAP phosphorylation may cause destabilization of the cytoskeleton. All these effects together may contribute to the overall cell proliferating and dedifferentiation pathways constituting the tumor phenotype, as diagramatically shown in **Fig. 5**.

Experimental observation made so far with grade III astrocytomas seem to be consistent with the scheme. It is possible that the same effects hold true for progressive secondary GBMs. In contrast, we are not sure whether the scheme could be extended to primary GBMs, which may have more independent genetic and biochemical pathway. Chromosome (chr) 19 deletions have been already implicated in gliomas. The molecular profiling findings reviewed in this chapter include at least four differentially expressed proteins mapping on this chromosome. Other chromosomes carrying five or more differentially expressed genes are chr 1, 11, 17, and 20. In particular, alterations on chr 20 map virtually on the same loci. chr 11 has been implicated in other types of malignancies such as leukemias, whereas differential proteins such as Rho-GDI identified on chr 17 are discussed as a potential marker in breast and lung carcinomas. chr 1 and 20 may be investigated further in relation to gliomas.

7. Future Perspective

Gliomas represent a unique group of cancers when it comes to resistance to chemo- and radiation treatments and variation in prognosis. Despite the increasing number of tumor-associated genes and proteins that have been identified during the past several years, enhancing our understanding of the tumors considerably, the progress over clinical outcome of gliomas has been dismal. Guidelines or criteria for histological assessments of tumor types and grades are in practice but inadequate to deal with the great degree of variability observed with real cases. Genetic alterations such as EGFR amplification or P53 deletion for GBMs and their use and the use of other antigens such Ki-67, GFAP, and vimentin provide immunohistochemical tools and help in histopathological analysis. Still, definition of tumors in terms of global molecular changes would be highly useful for their precise and more reliable classification, and most of the efforts of molecular analysis are directed toward this aim. DNA microarray and proteomics allow simultaneous analysis of expression of large number of genes and proteins. Initial attempts using these methods of molecular analysis support the feasibility of developing a new basis for tumor taxonomy. Transcript profiling has been explored to distinguish tumor histotypes such as astrocytomas from oligodendrogliomas and tumor grades as well as to identify patient subsets accordingly to their survival outcomes.

Protein are the ultimate determinants of the gene function, and protein profiles offer a stronger basis for the biological aberrations and may be more robust for applications in the clinical settings. Most of the studies carried out for proteomics analysis of gliomas have relied upon 2D gel electrophoretic separation of proteins followed by their identification by microsequencing or MS-based sequencing. This approach has so far been largely applied for the study of astrocytomas and has yielded useful information to identify altered protein expressions in tumors of different grades and different subtypes and to find survival correlations. Although this method generally displays 1000–1500 well-resolved proteins, it has limitations accessing low-abundant proteins or membrane proteins, many of which may be crucial in relation to the regulation of the tumor growth and progression. In addition, the method is time-consuming and may not be suitable for large-scale application. An alternative, liquid chromatography–MS/MS approach is more sensitive *(39)* and has the potential to cover greater dynamic range of proteins in the analysis of tumor vs normal tissues. New methods for enrichment of membrane proteins and use of liquid chromatography–MS/MS approach to identify them have been developed and have been used for membrane proteins from normal brain as well as glioma cell line *(66)*. These developments, when applied to study primary tumors, are sure to yield valuable information on membrane-associated and other low-abundant proteins and further unravel the basis of tumor heterogeneities. Comparison of proteins and protein patterns between primary tumors and cell lines and identification of cell line-associated proteins that may be driven by culturing in vitro *(23)* strengthens the prospects of successfully using cell lines as an experimental system to understand tumors. Similarly, new glioma cell lines such as the lines developed by Shiras et al. *(26)* may allow study of molecular profiles to understand transition to rapid proliferation state of the cells and can prove to be a useful model system to identify new genes and proteins associated with tumorigenesis and progression.

Finally, discoveries from molecular analysis of tumors have to be validated and translated into feasible tests in the clinical environment, which is going to be an enormous effort and has to be initiated globally. Large-scale screening methods will have to be developed. Direct tissue imaging based on protein mass profiles determined by surface enhanced laser desorption ionization- or MALDI-MS is under rapid development and has potential for large-scale applications *(67)*. However, these methods do not permit identification of profiled proteins. Strategies to combine or adapt these methods to validate or screen for tumor-specific proteins identified by other approaches have to be developed. Possibility of the identification and accurate mass characterization of unique proteotypic peptides from the protein digests is being explored, and databases of such peptides may be developed in the near future. Thus, possibility of direct MS-based screening of tissue protein digests for proteotypic peptides and identification of proteins by accurate mass measurements is not very distant *(68)*. Body fluids are an important source for identifying biomarkers associated with a disease *(69,70)*. For postsurgery monitoring of glioma patients, imaging or tissue analysis may not be the obvious and preferred choice, and biomarkers in the body fluids will be useful. Methods for direct analysis and exploration of fluids such as blood plasma, serum, or cerebrospinal fluid by MS are still in their infancy. Existence of blood-brain barrier also may render some limitations in the case of CNS tumors. But alternative experimental strategies cannot be ruled out and can be investigated. Identification of SOX6 as a glioma marker, by screening testis cDNA expression library by antisera from glioma patients *(71)*, is an example of the usefulness of the alternative strategies. Thus, multiple experimental strategies and experimental systems are being used to study gliomas, and there should be little doubt that these efforts will further unravel the processes underlying these tumors and help in development of more definitive therapies or new biomarkers for the clinical management of gliomas.

Acknowdgments

Cancer proteomics efforts in the Sirdeshmukh laboratory are supported under the NMITLI Program of the Council of Scientific and Industrial Research (CSIR), Government of India. A.S. is a student associate under CSIR Program on Youth for Leadership in Science. We thank all the members of the Sirdeshmukh proteomics laboratory for efforts on glioma proteomics cited in this chapter and A. N. Suhasini for help in the preparation of this manuscript.

References

1. Zhu, Y., and Parada, L. F. (2002) The molecular and genetic basis of neurological tumours. *Nat. Rev. Cancer* **2**, 616–626.
2. Mischel, P. S., Cloughesy, T. F., and Nelson, S. F. (2004) DNA-microarray analysis of brain cancer: molecular classification for therapy. *Nat. Rev. Neurosci.* **5**, 782–792.
3. Rich, J. N., and Bigner, D. D. (2004) Development of novel targeted therapies in the treatment of malignant glioma. *Nat. Rev. Drug Discov.* **3**, 430–446.
4. Sanson, M., Thillet, J., and Hoang-Xuan, K. (2004) Molecular changes in gliomas. *Curr. Opin. Oncol.* **16**, 607–613.
5. Collins, V. P. (2004) Brain tumours: classification and genes. *J. Neurol. Neurosurg. Psychiatry* **75**, 2–11.
6. Keihues, P., and Cavenee, W. K. (2000) WHO classification of tumors. In: *Pathology and Genetics of Tumors of Nervous System*, IARC Press, Lyon, France.
7. Berger, F., Gay, E., Pelletier, L., Tropel, P., and Wion D. (2004) Development of gliomas: potential role of asymmetrical cell division of neural stem cells. *Lancet Oncol.* **5**,511–514.
8. Singh, S. K., Clarke, I. D., Hide, T., and Dirks, P. B. (2004) Cancer stem cells in nervous system tumors. *Oncogene* **23**, 7267–7273.
9. Singh, S. K., Hawkins, C., Clarke, I. D., et al. (2004) Identification of human brain tumour initiating cells. *Nature* **432**, 396–401.
10. Daumas-Duport, C. (2000) The future of neuropathology. *Clin. Neurosurg.* **47**, 112–120.
11. Hui, A. B., Lo, K. W., Yin, X. L., Poon, W. S.. and Ng, H. K. (2001) Detection of multiple gene amplifications in glioblastoma multiforme using array-based comparative genomic hybridization. *Lab Invest.* **81**, 717–723.
12. Hayashi, Y., Yamashita, J., and Watanabe, T. (2004) Molecular genetic analysis of deep-seated glioblastomas. *Cancer Genet. Cytogenet.* **153**, 64–68.
13. Morrison, C. D., and Prayson, R. A. (2000) Immunohistochemistry in the diagnosis of neoplasms of the central nervous system. *Semin. Diagn. Pathol.* **17**, 204–215.
14. Neder, L., Colli, B. O., Machado, H. R., Carlotti, C. G., Jr., Santos, A. C., and Chimelli, L. (2004) MIB-1 labeling index in astrocytic tumors—a clinicopathologic study. *Clin. Neuropathol.* **23**, 262–270.
15. Reeves, S. A., Helman, L. J., Allison, A., and Israel, M. A. (1989) Molecular cloning and primary structure of human glial fibrillary acidic protein. *Proc. Natl. Acad. Sci. USA* **86**, 5178–5182.
16. Rutka, J. T., Murakami, M., Dirks, P. B., et al. (1997) Role of glial filaments in cells and tumors of glial origin: a review. *J. Neurosurg.* **87**, 420–430.
17. Eng, L. F., Ghirnikar, R. S., and Lee, Y. L. (2000) Glial fibrillary acidic protein : GFAP-thirty-one years (1969-2000). *Neurochem. Res.* **25**, 1439–451.
18. Kim, S., Dougherty, E. R., Shmulevich, I., et al. (2002) Identification of combination gene sets for glioma classification. *Mol. Cancer Ther.* **13**, 1229–1236.
19. Wang, H., and Hanash, S. M. (2002) Contributions of proteome profiling to the molecular analysis of cancer. *Technol. Cancer Res. Treat.* **1**, 237–246.
20. Rhee, C. H., Hess, K., Jabbur, J., et al. (1999) cDNA expression array reveals heterogeneous gene expression profiles in three glioblastoma cell lines. *Oncogene* **18**, 2711–2717.
21. Zhang, W. (2002) Identification of combination gene sets for glioma classification. *Mol. Cancer Ther.* **1**, 1229–1236.
22. Zhang, R., Tremblay, T. L., McDermid, A., Thibault, P., and Stanimirovic, D. (2003) Identification of differentially expressed proteins in human glioblastoma cell lines and tumors. *Glia* **2**, 194–208.
23. Vogel, T. W., Zhuang, Z., Li, J., et al. (2005) Proteins and protein pattern differences between glioma cell lines and glioblastoma multiforme. *Clin Cancer Res.* **11**, 3624–3632.

24. Machado, C. M., Schenka, A., Vassallo, J., et al. (2005) Morphological characterization of a human glioma cell line. *Cancer Cell Int.* **1**, 13.
25. Reya, T., Morrison, S. J., Clarke, M. F., and Weissman, I. L. (2001) Stem cells, cancer, and cancer stem cells. *Nature* **414**, 105–111.
26. Shiras, A., Bhosale, A., Shepal, V., et al. (2003) A unique model system for tumor progression in GBM comprising two developed human neuro-epithelial cell lines with differential transforming potential and coexpressing neuronal and glial markers. *Neoplasia* **5**, 520–532.
27. Boudreau, C. R., and Liau, L. M. (2004) Molecular characterization of brain tumors. *Clin. Neurosurg.* **51**, 81–90.
28. Huang, H., Colella, S., Kurrer, M., Yonekawa, Y., Kleihues, P., and Ohgaki, H. (2000) Gene expression profiling of low-grade diffuse astrocytomas by cDNA arrays. *Cancer Res.* **60**, 6868–6874.
29. Mischel, P. S., Shai, R., Shi, T., et al. (2003) Identification of molecular subtypes of glioblastoma by gene expression profiling. *Oncogene* **22**, 2361–2373.
30. Shai, R., Shi, T., Kremen, T. J., et al. (2003) Gene expression profiling identifies molecular subtypes of gliomas. *Oncogene* **22**, 4918–4923.
31. van den Boom, J., Wolter, M., Kuick, R., et al. (2003) Characterization of gene expression profiles associated with glioma progression using oligonucleotide-based microarray analysis and real-time reverse transcription-polymerase chain reaction. *Am. J. Pathol.* **163**, 1033–1043.
32. Palotie, A., Liau, L. M., Cloughesy, T. F., and Nelson, S. F. 2003. Identification of molecular subtypes of glioblastoma by gene expression profiling. *Oncogene* **22**, 2361–2373.
33. Somasundaram, K., Sreekanth Reddy, P., Vinnakota, K., et al. (2005) Upregulation of ASCL1 and inhibition of Notch signaling pathway characterize progressive astrocytoma. *Oncogene* **24**, 7073–7083.
34. Watson, M. A., Perry, A., Budhjara, V., Hicks, C., Shannon, W. D. and Rich, K. M. (2001) Gene expression profiling with oligonucleotide microarrays distinguishes World Health Organization grade of oligodendrogliomas. *Cancer Res.* **61**, 1825–1829.
35. Huang, H., Okamoto, Y., Yokoo, H., et al. (2004)Gene expression profiling and subgroup identification of oligodendrogliomas.*Oncogene* **23**, 6012–6022.
36. Hunter, S., Young, A., Olson, J., et al. (2002) Differential expression between pilocytic and anaplastic astrocytomas: identification of apolipoprotein D as a marker for low-grade, non-infiltrating primary CNS neoplasms. *J. Neuropathol. Exp. Neurol.* **61**, 275–281.
37. Raza, S. M, Fuller, G. N, Rhee, C. H, et al. (2004) Identification of necrosis-associated genes in glioblastoma by cDNA microarray analysis. *Clin Cancer Res.* **10**, 212–221.
38. Freije, W. A., Castro-Vargas, F. E., Fang, Z., et al. (2004). Gene expression profiling of gliomas strongly predicts survival. *Cancer Res.* **64**, 6503–6510.
39. Aebersold, R., and Mann, M. (2003) Mass spectrometry-based proteomics. *Nature* **422**, 198–207.
40. Washburn, M. P., Wolters, D., and Yates, J. R., 3rd. (2001) Large-scale analysis of the yeast proteome by multidimensional protein identification technology. *Nat. Biotechnol.* **19**, 242–247.
41. Wright, G. L., Jr. (2002) SELDI proteinchip MS: a platform for biomarker discovery and cancer diagnosis. *Expert Rev. Mol. Diagn.* **2**, 549–563.
42. Chaurand, P., Schwartz, S. A., and Capriolo, R. M. (2004) Profiling and imaging proteins in tissue sections by MS. *Anal. Chem.* **76**, 87A–93A.
43. Elmlinger, M. W., Deininger, M. H., Schuett, B. S., et al. (2001) In vivo expression of insulin-like growth factor-binding protein-2 in human gliomas increases with the tumor grade. *Endocrinology* **142**, 1652–1658.
44. Sasaki, A., Ishiuchi, S., Kanda, T., Hasegawa, M., and Nakazato, Y. (2001) Analysis of interleukin-6 gene expression in primary human gliomas, glioblastoma xenografts, and glioblastoma cell lines. *Brain Tumor Pathol.* **18**, 13–21.

45. Rao, J. S. (2003) Molecular mechanisms of glioma invasiveness: the role of proteases. *Nat. Rev. Cancer* **3,** 489–501.

46. Nutt, C. L., Betensky, R. A., Brower, M. A., Batchelor, T. T., Louis, D. N., and Stemmer-Rachamimov, A. O. (2005) YKL-40 is a differential diagnostic marker for histologic subtypes of high-grade gliomas. *Clin. Cancer Res.* **11,** 2258–2264.

47. Hanash, S. M., Bobek, M. P., Rickman, D. S., et al. (2002) Integrating cancer genomics and proteomics in the post-genome era. *Proteomics* **2,** 69–75.

48. Hiratsuka, M., Inoue, T., Toda, T., et al. (2003) Proteomics-based identification of differentially expressed genes in human gliomas: down-regulation of SIRT2 gene. *Biochem. Biophys. Res. Commun.* **26,** 558–566.

49. Chumbalkar, V. C., Subhashini, C., Dhople, V. M., et al. (.2005) Differential protein expression in human gliomas and molecular insights. *Proteomics* **5,** 2702.

50. Bigbee, J. W., Bigner, D. D., Pegram, C., and Eng, L. F. (1983) Study of glial fibrillary acidic protein in a human glioma cell line grown in culture and as a solid tumor. *J. Neurochem.* **40,** 460–467.

51. Jiang, W. G., Watkins, G., Lane, J., et al. (2003) Prognostic value of rho GTPases and rho guanine nucleotide dissociation inhibitors in human breast cancers. *Clin. Cancer Res.* **9,** 6432–6440.

52. MacKeigan, J. P., Clements, C. M., Lich, J. D., Pope, R. M., Hod, Y., and Ting, J. P. (2003) Proteomic profiling drug-induced apoptosis in non-small cell lung carcinoma: identification of RS/DJ-1 and RhoGDI alpha. *Cancer Res.* **63,** 6928–6934.

53. Luider, T. M, Kros, J. M, Sillevis, Smitt P. A., van den Bent, M. J, and Vecht, C. J. (1999) Glial fibrillary acidic protein and its fragments discriminate astrocytoma from oligodendroglioma. *Electrophoresis* **20,** 1087–1091.

54. Mokhtari, K, Paris, S., Aguirre-Cruz, L., et al. Olig2 expression, GFAP, p53 and 1p loss analysis contribute to glioma subclassification. *Neuropathol. Appl. Neurobiol.* **31,** 62–69.

55. Furuta, M., Weil, R. J., Vortmeyer, A. O., et al. (2004) Protein patterns and proteins that identify subtypes of glioblastoma multiforme. *Oncogene* **40,** 6806–6814.

56. Li, Q., Sakurai, Y., Ryu, T., et al. (2004) Expression of Rb2/p130 protein correlates with the degree of malignancy in gliomas. *Brain Tumor Pathol.* **21,** 121–125.

57. Iwadate, Y., Sakaida, T., Hiwasa, T., et al. (2004). Molecular classification and survival prediction in human gliomas based on proteome analysis. *Cancer Res.* **64,** 2496–2501.

58. Sahai, E., and Marshall, C. J. (2002) RHO-GTPases and cancer. *Nat. Rev. Cancer* **2,** 133–142.

59. Wang, S., Nath, N., Adlam, M., and Chellappan, S. (1999) Prohibitin, a potential tumor suppressor, interacts with RB and regulates E2F function. *Oncogene* **18,** 3501–3510.

60. Kirchman, P. A., Miceli, M. V., West, R. L., Jiang, J. C., Kim, S., and Jazwinski, S. M. (2003) Prohibitins and Ras2 protein cooperate in the maintenance of mitochondrial function during yeast aging. *Acta Biochim. Pol.* **50,** 1039–1056.

61. Sato, T., Saito, H., Swensen, J., and Olifant, A. (1992) The human prohibitin gene located on chromosome 17q21 is mutated in sporadic breast cancer. *Cancer Res.* **52,** 1643–1646.

62. Byrjalsen, I., Mose Larsen, P., Fey, S. J., Nilas, L., Larsen, M. R., and Christiansen, C. (1999) Two-dimensional gel analysis of human endometrial proteins: characterization of proteins with increased expression in hyperplasia and adenocarcinoma. *Mol. Hum. Reprod.* **5,** 748–756.

63. Mosser, D. D., and Morimoto, R. I. (2004) Molecular chaperones and the stress of oncogenesis. *Oncogene* **16,** 2907–2918.

64. Helfand, B. T., Chang, L., and Goldman, R. D. (2004) Intermediate filaments are dynamic and motile elements of cellular architecture. *J Cell Sci.* **117,** 133–141.

65. Omary, M. B., Coulombe, P. A., and McLean, W. H. (2004) Intermediate filament proteins and their associated diseases. *N. Engl. J. Med.* **351,** 2087–2100.

66. Nielsen, P. A., Olsen, J. V., Podtelejnikov, A. V., Andersen, J. R., Mann, M., and Wisniewski, J. R. (2005) Proteomic mapping of brain plasma membrane proteins. *Mol Cell Proteomics* **4,** 402–408.
67. Schwartz, S. A., Weil, R. J., Johnson, M. D., Toms, S. A., and Caprioli, R. M. (2004) Protein profiling in brain tumors using mass spectrometry: feasibility of a new technique for the analysis of protein expression in tissue sections. *Clin. Cancer Res.* **10,** 981–987.
68. Pan, S., Zhang, H., Rush, J., et al. (2005) High throughput proteome screening for biomarker detection. *Mol Cell Proteomics* **4,** 182–190.
69. Anderson, N. L., and Anderson, N. G. (2002) The human plasma proteome: history, character, and diagnostic prospects. *Mol. Cell Proteomics* **1,** 845–867.
70. Zheng, P. P., Luider, T. M., Pieters, R., et al. (2003) Identification of tumor-related proteins by proteomic analysis of cerebrospinal fluid from patients with primary brain tumors. *J. Neuropathol. Exp. Neurol.* **62,** 855–862.
71. Ueda, R., Iizuka, Y., Yoshida, K., Kawase, T., Kawakami, Y., and Toda, M. (2004) Identification of a human glioma antigen, SOX6, recognized by patients' sera. *Oncogene* 23, 1420–1427.

12

Antibody-Based Microarrays
From Focused Assays to Proteome-Scale Analysis

Christer Wingren and Carl A. K. Borrebaeck

Summary

Our research in immunotechnology focuses on the cells and molecules of the immune system for various biomedical applications. Applied research as well as projects addressing fundamental issues is being pursued within four main project areas: 1) antibody technology, 2) proteomics, 3) cancer, and 4) allergy. Within the proteomics project, we have taken advantage of the strong background in antibody engineering to design antibody-based micro- and nanoarrays for applications ranging from focused assays to proteome-scale analysis. Antibody-based microarray is a novel technology that holds great promise in proteomics. The microarray can be printed with thousands of recombinant antibodies carrying the desired specificities, the biological sample added (e.g., an entire proteome), and virtually any specifically bound analytes detected. The microarray patterns generated can then be converted into proteomic maps, or molecular fingerprints, revealing the composition of the proteome. Global proteome analysis and protein expression profiling, by using this tool, will provide new opportunities for biomarker discovery, drug target identification, and disease diagnostics and will give insight into disease biology. Ultimately, we apply this novel technology platform within our cancer and allergy projects to perform high-throughput disease proteomics.

Key Words: Antibody microarrays; biomarker discovery recombinant antibody library; disease diagnostics; protein expression profiling; proteome analysis; proteomics.

1. Background

Entering the postgenomic era, proteomics, or the large-scale analysis of proteins, has become a key discipline for identifying, characterizing, and screening all proteins encoded by the genome (*1–3*). The human proteome is thought to be composed of >300,000 different proteins, distributed among approx 200 cell types. Because aberrant expression and function of any of these proteins in the proteome can result in disease, the impact of proteome analysis in medicine will be substantial (*1–3*). In traditional proteomics, various separation methods, such as two-dimensional (2D) gels, coupled with mass spectrometry (MS), have evolved into a versatile tool commonly used (*2–7*). However, because the number of proteomic projects has increased during the past few years, it has become clear that 1) highly multiplex, high-throughput proteomic approaches displaying high specificity-selectivity and sensitivity will be needed, and 2) that integration of different data sets generated by multiple strategies

From: *Bioarrays: From Basics to Diagnostics*
Edited by: K. Appasani © Humana Press Inc., Totowa, NJ

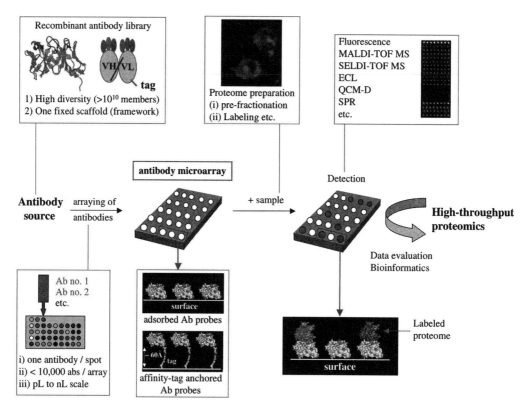

Fig. 1. Schematic illustration of the antibody microarray concept.

and technology platforms will be required *(2,3)*. Antibody-based microarrays are among the novel class of rapidly emerging proteomic technologies that will offer new opportunities for proteome-scale analysis *(8)*. This chapter discusses our approach to develop the antibody microarray technology into a proteomic research tool and our present and future applications thereof.

2. Antibody Microarrays: A Short Introduction

Protein microarrays have the potential to mirror the breakthroughs in genomics enabled by high-density DNA microarrays developed during the past years *(9)*. To this end, affinity protein microarrays in which specific binders, such as antibodies, are used to detect and quantitate protein analytes (quantitative proteomics) **(Fig. 1)** has raised high expectations *(8)*. The technology will undoubtedly play a significant role within proteomics, supplementing current technologies, such as 2D gels and MS *(8,10–15)*. The array patterns generated can be converted into detailed proteomic maps revealing the composition of the proteome. The technology will thus provide high-throughput means to perform comparative proteome analyses of, e.g., healthy vs diseased samples. This approach will allow scientists to address, e.g., signaling and metabolic pathways; to identify disease-specific proteins; to examine protein–protein interactions of functional networks; and to perform differential protein expression profiling, disease diagnostics, and biomarker discovery *(8,10,12–15)*. Furthermore, the potential to study disease development and progression and to assess response to treatment will have a major impact on how we analyze and examine several diseases, such as human cancer.

Currently, major efforts are under way to explore and exploit the antibody microarray technology within (disease) proteomics. In the past few years, the first applications in which antibody arrays have been used for protein expression profiling and limited proteomic profiling have been published *(16–19)*. In these examples, low- to medium-density arrays were used for multiplex analysis of complex samples (e.g., cell lysates and serum) targeting mainly water-soluble analytes *(16–19)*. Recently, also the possibility to target membrane proteins by using antibody arrays was outlined *(20,21)*. Despite great progress, the full potential of the technology will be taken advantage of only when the remaining key issues, such as scaling up the arrays, has been solved. Still, antibody-based microarrays belong to the novel class of rapidly evolving proteomic technologies that in the near future will display a true high-throughput format *(8)*.

3. Quest for Developing Antibody Microarrys: Our Approach

Our long-term interest and previous achievements within antibody engineering and applications thereof have provided a strong basis for our approach *(22)*. Our specific aim was to develop an antibody-based microarray technology platform for high-throughput (disease) proteomics. Our long-term goals were to study human diseases, such as cancer, and to perform disease diagnostics, biomarker discovery, and drug target identification by using our antibody-based microarrays. Ultimately, these efforts would then be combined with our similar and parallel activities already being pursued at the genomic level within the areas of cancer and allergy *(22)*.

To accomplish our goals, we have taken on a broad approach and identified five critical subareas that had to be addressed: 1) content, 2) array, 3) sample, 4) assay, and 5) data processing (**Fig. 2**). This chapter focuses on the progress made within these five areas that have resulted in high-performing affinity microarrays based on human recombinant antibody fragments.

3.1. Content

3.1.1. Choice of Content

The choice of probes is a key issue in the process of designing affinity microarrays for proteome analysis *(10,11,13,16)*. The rational behind selecting antibodies over various antibody mimics as content have been reviewed recently *(8)* and are not be discussed here. Among the different antibody formats available, we strongly favor recombinant antibody libraries as probe source *(8,16)* over monoclonal reagents *(10,23,24)*. The key reasons for this choice are briefly outlined below (reviewed in **ref.** *8*):

Availability:
- Large (>10^{10} members) recombinant antibody libraries that will provide access to probes with virtually any desired specificities are available *(25–27)*
- The number of available monoclonal antibodies is small, creating major difficulties in finding antibodies with the desired specificities

Scaling up:
- Not a problem (see availability)
- The task of producing thousands of monoclonal antibodies will be overwhelming, making the process of fabricating high-density arrays extremely demanding

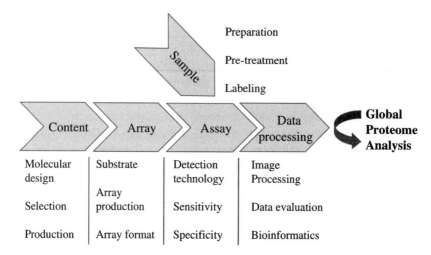

Fig 2. Overall scheme of our proteomics projects, outlining the five critical subareas addressed, aiming to design antibody-based affinity microarrays for high-throughput disease proteomics.

Properties:
- in libraries based on one fixed scaffold (framework), all members will display very similar molecular properties (e.g., on-chip stability)
- monoclonal antibodies, based on many different scaffolds, will display a wide range of properties, making them less suited as probes.

Redesign:
- upon optimization of the selected scaffold, the best molecular properties can readily be redesigned into all members in the library in a one-step procedure
- the design of monoclonal antibodies would have to be optimized individually, making such large-scale improvements virtually impossible.

3.1.2. Design of Content

We have designed a human recombinant single-chain Fv (scFv) antibody library genetically constructed around one framework, the n-CoDeR library containing 2×10^{10} clones[10] *(25)* as probe source for our microarray applications *(16)*. The scaffold, composed of the variable-region frameworks VH3-23 and VL1-47 was carefully selected with the purpose to generate high-performing antibody fragments *(25)*. The results showed that highly functional, well-expressing antibody fragments specific for several different ligands, including peptides, proteins (of human and nonhuman origin), carbohydrates, and haptens, could be generated *(25)*. Indeed, the library design may even provide an antibody diversity exploring a specificity space that supersedes the evolutionary process provided naturally *(27)*. Furthermore, these antibody fragments commonly exhibited dissociation constants in the subnanomolar range *(25)*. Most importantly, we have recently demonstrated that these single framework antibody fragments (SinFabs) were an excellent source of probes for microarray applications *(16,28–34)*. Thus, we have already bypassed one of the major bottlenecks identified within many protein array projects, the availability of almost limitless number of probes displaying high functionality, specificity, and sensitivity.

The structural and functional on-chip stability of arrayed probes is often identified as a key issue. Commonly, proteins tend to unfold and lose their activity when printed onto a solid support and allowed to dry out. For example, early work showed that <20% of arrayed monoclonal antibodies displayed adequate reactivity *(23)*. To illustrate the importance of the precise choice of framework when designing antibody libraries for array applications, we have recently compared the biophysical properties (functional on-chip stability in particular) of scFv antibodies based on four different scaffolds *(28)*. The results showed that the on-chip half-life (how long a fabricated array could be stored in a dried out state and still display 50% reactivity) varied significantly, from only 7 d (VH5-51/VL2-23), 39 d (VH3-30/VKIIIa), and 42 d (VH5-51/VKIIIb) to >180 d (VH3-23/VL1-47, i.e. our library design) *(28)*. Furthermore, structural analysis indicated that the two key molecular parameters of buried surface area and the number of interdomain van der Waals interactions correlated well with the observed on-chip stability *(28)*. In recent work, using even further optimized array setups, we have shown that our scFv antibodies displayed functional on-chip stability (80–100% retained antigen-binding activity) in room temperature exceeding 240 d *(34a)*. To date, these are the highest on-chip stabilities reported for recombinant antibodies in microarray applications.

Furthermore, our library has been designed so that each antibody carries two C-terminal affinity-tags (a flag- and His-, or myc-tag) *(25)* that may be used for purification, anchoring of the probes onto the substrates, or both (**Fig. 1**) *(29,31)*. Other tag systems also may be used, but many available affinity-tags display binding affinities that are too low, thereby reducing or impairing the efficiency of coupling. However, we have recently shown that one possible way of addressing this problem is to redesign the single His-tag to introduce a double His-tag *(34a)*.

3.1.3. Selection and Production of Content

We have an in-house capacity to select and produce antibodies in small scale by using *Escherichia coli* or *Pichia pastoris*. To meet the logistic challenges regarding large-scale selection and production, we have established a long-term collaboration with Bioinvent International AB (Lund, Sweden) *(35)*, where we have access to a fully automated facility for large-scale screening, selection, and production of recombinant antibody fragments *(36)*. Their system can handle target antigens in various formats, including peptides, soluble proteins as well as cell-bound targets at a high throughput. Depending on the array design (*see* **Subheading 3.2.**), the number of clones required will vary significantly, from a few clones to several thousand. However, the process of selecting numerous, well-characterized probes is a challenge even for specialized companies. Selecting binders for an entire proteome is further complicated by no proteome currently being available as individually purified proteins. This problem can, to some extent, be addressed by adopting novel array designs (*see* **Subheading 3.2.**) *(32)*.

3.2. Array

3.2.1. Substrate Design

To succeed in the efforts of developing high-performing antibody-based microarrays, or protein arrays in general, the design of the substrate is also essential *(8,30,31,37–39)*. Four key properties of the solid supports are as follows: 1) high biocompatability, 2) high and selective probe binding capacity, 3) ability to bind the probes in a favorable orientation, and 4) low nonspecific binding (background) (reviewed in **refs. *8,30,37–39***).

A wide range of substrates are commercially available, but the choice is not obvious, and recent data do not suggest one support to be the optimal substrate for any given protein microarray application *(8,30,37–39)*. The precise choice will depend on several factors, such as the probe source, the coupling chemistry needed, the sample complexity, and the sensitivity required, and so on. To date, our preferred commercial substrates are black polymer Maxisorp microarray slides (plastic/adsorption) (low background, high biocompatibility, high sensitivity) *(40)*, protein-binding glass slides (polymer-modified glass; adsorption) (low background, high biocompatibility) *(40)*, FAST-slides (nitrocellulose-coated glass, adsorption) (high biocompatibility, high probe binding capacity) *(41)*, and Xenoslide N glass slides (Ni^2-chelate derivatized/affinity binding) *(42)*.

To increase the selection of potential substrates, we also have setup two cross-disciplinary projects to design our own in-house substrates suitable for antibody microarray applications *(30,31)*. In collaboration with Prof. Höök and others (Lund Institute of Technology), patterns of DNA-labeled and scFv-antibody–carrying lipid vesicles directed by material-specific immobilization of DNA and supported lipid bilayer formation on an Au/SiO$_2$ template have been generated *(31)*. Apart from providing very inert substrates with low nonspecific binding, this is one of the first designs toward self-addressable protein arrays. In addition, the setup also may be used for fabricating membrane protein arrays.

Together with Prof. Laurell and others (Lund Institute of Technology), a variety of silicon-based supports, including planar silicon, micro- and macroporous silicon as well as nitrocellulose-coated variants thereof, were recently designed and evaluated *(30)*. The model surfaces were scored based on biocompatibility and probe binding capacity as judged by spot morphology, signal intensities, signal-to-noise ratios, dynamic range, sensitivity, and reproducibility. A set of five commercially available substrates were used as reference surfaces. The results showed that several well-performing silicon-based supports readily could be designed, where in particular a nitrocellulose-coated macroporous variant, MAP3-NC7 (**Fig. 3**), received the highest scores *(30)*. In comparison, MAP3-NC7 displayed properties equal or better to those of the reference substrates. Together, designed surfaces based on silicon can undoubtedly meet the requirements of the next generation of solid supports for antibody microarrays, and they were also recently used in microarrays with a dual readout system based on fluorescence and MS *(51)*.

3.2.2. Array Production and Handling

We have established an in-house antibody microarray facility for fabrication, handling and analysis of the chips. To fabricate the arrays, we have a BioChipArrayer1 *(43)*, a noncontact spotter, based on piezzo technology, that deposits the probes in the picoliter (pL) scale. The choice of substrate determines whether the probes will have to be prepurified before the dispensing. When fabricated, the chips are fed into a Protein Array Workstation for fully automatic handling (blocking, washing, addition of sample, and so on) of several chips at the same time *(43)*. Finally, a ScanArrayExpress system (confocal fluorescence scanner) *(43)* is used for detection and quantification of the arrays.

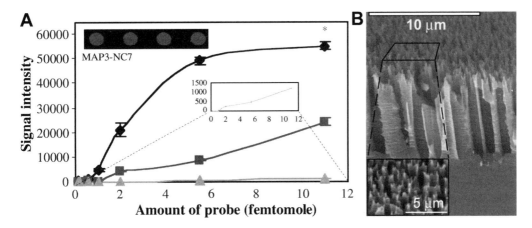

Fig. 3. Properties of the designed macroporous silicon-based nitrocellulose coated substrated MAP3-NC7. **(A)** Signal intensities and spot morphologies. Three identical 8 × 8 arrays, composed of an anti-choleratoxin SinFab molecule arrayed in seven serial dilutions ranging from 11 fmol/spot to 111 atmol/spot, were incubated with Cy5-labeled analyte, choleratoxin subunit B, at a concentration of 80 nM (blue diamonds), 8 nM (red squares), or 0.8 nM (cyan triangles) nM. A nonspecific scFv (clone FITC-8) was used as negative control. Signal intensified after subtracting local background and negative control are shown. A close-up of the data obtained for the 0.8 nM analyte is shown. **(B)** Scanning electron micrograph images. The cross-section and inserted top view of the substrate are shown. Data adapted from **ref. *30*.**

3.2.3. Array Format

The choice of array design depends on the application, ranging from small focused assays targeting a few analytes to proteome-scale analysis addressing thousands of analytes. To date, we have fabricated mainly low- to medium-density microarrays, but we have in-house capacity to also produce dense microarrays (i.e., a few thousand probes per array). Today, this is the "top-of-the-line" design that most array laboratories are capable of generating, providing that the availability of probes is not the limiting factor.

However, to be able to perform proteome-scale analysis, microarray designs allowing >> 10,000 analytes to be addressed simultaneously must be developed. Recently, we outlined our view of the array format of tomorrow—megadense nanoarrays (>100,000 probes) *(32)*. Here, we have proposed that nanotechnology will provide us with the tools required to design and fabricate such nanoarrays *(32)*. At this point, the term "probe specificity" is no longer relevant. This technology would clearly facilitate the probe selection step , because we would no longer need to know the fine specificity of each probe at forehand. Instead, these highly dense arrays would be analysed based on pattern recognition, by using, for example, artificial neural networks. In differential proteome analysis, patterns that differ between healthy vs diseased sample would first be analyzed and identified using the full arrays. Based on these observations, smaller and more focused microarrays could then be rapidly designed, based on perhaps <250 antibodies, to analyze and characterize the observed differences in more detail. Ultimately, the combination of megadense antibody nanoarrays with self-adressable array designs (instead of having to dispense the probes one by one) and label-free analytical sensing principles (instead of having to label complex proteomes) may prove to be the array design for high-throughput (disease) proteomics.

3.3. Sample

All samples generated in a soluble format can be analyzed by antibody-based microarrays *(8)*. The reduced sample consumption in the microarray format is essential, because only minute volumes of precious samples are often available. We have shown that volumes in the picoliter scale may be sufficient if also the sample is arrayed using conventional spotters *(29)*.

The sample complexity is a key feature that may impair the analysis by 1) making it difficult to label the samples in a representative manner, by 2) causing high nonspecific binding and thereby significantly reducing the assay sensitivity, or a combination of 1 and 2. In traditional proteomics, various prefractionation strategies, removal of high-abundant proteins, or both have been successfully applied in preparation for 2D gel analysis, which significantly improved the detection of low-abundant proteins *(4–6)*. However, in antibody array applications, little attention has so far been placed upon optimizing the sample format. We have recently shown that a simple one-step fractionation (based on size) of complex proteomes considerably enhanced the detection of low molecular weight (<50 kDa) and low-abundant analytes (subpicogram per milliliter range) *(52)*. Recently, we also have shown that nonfractionated proteomes, such as human serum, could be directly screened for low-abundant analytes (sub picogram per milliliter) by optimizing the microarray design (choice of substrate, blocking reagent, sample buffer, labeling reagent, and so on) *(53)*. This optimization owes to the inherent power of selectivity (specificity) displayed by affinity microarrays and is considered to be a major advantage compared with MS-based approaches.

3.4. Assay

3.4.1. Specificity

Using readily available "on-the-shelf" antibody reagents for designing antibody arrays may cause problems, because these reagents have not been designed and selected for microarray applications *(3,16)*. This reason is probably as why several recent studies have raised serious concerns as to whether antibodies are specific enough to act as content *(3,10,44–46)*. In our case, we have used an antibody probe source designed for microarray applications that also was found to display outstanding properties with respect to specificity *(8,16,25,29)*. The lack of any detectable cross-reactivities is a prominent feature of our antibody microarray technology platform, irrespective of analyte format (peptide, hapten, protein, intact cells, and so on) and degree of sample purity ranging from pure analytes to complex samples, such as human serum or crude cell lysates applied (**Fig. 4**) *(8,16,29)*.

3.4.2. Analytical Principles and Limit of Detection

We have adopted fluorescence as the main analytical principles for our antibody microarray technology platform. We have recently shown that a limit of detection (LOD) in the subpicomolar to femtomolar range was regularly obtained, whether the antigen was a large protein, a peptide or a small hapten, even when applied as complex

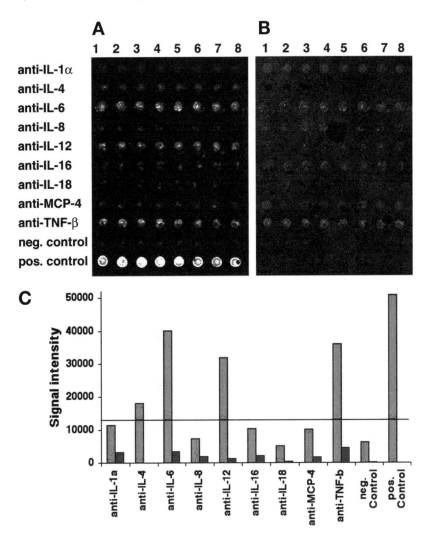

Fig. 4. Comparative proteome analysis of nonstimulated vs stimulated dendritic cells, by using a nine-probe anti-cytokine antibody microarray. The sample, in the formate of crude cell lysate, was directly labeled in a two-color approach. (**A**) Cy5-scanned microarray image of activated sample. (**B**) Cy3-scanned microarray image of nonactivated sample. (**C**) Comparison of cytokine levels in activated vs nonactivated sample. Four cytokines were found to be upregulated in the activated sample. Data adopted from **ref. *16***.

mixtures *(29)*. Moreover, a limit of detection (LOD) corresponding to only 300 zeptomoles (~5000 molecules) has been achieved without using any signal amplification *(29)*. These LODs clearly demonstrated the power of our antibody microarrays (reviewed in **ref. *8***). Furthermore, by using our optimized SinFab microarrays designs (with respect to buffer systems, blocking reagents, choice of label, and so on), we have recently shown that an LOD in the subpicogram per milliliter range could indeed be accomplished also for directly labeled serum samples, without having to prefractionate the serum *(53)*. To perform adequately, it has been suggested that antibody-based microarrays must have an LOD in the picogram (attomolar) range *(47)*. Thus, our current antibody microarray technology platform is already within the suggested range of LODs to be able to perform adequate proteome analysis. Currently, we are exploring the possibilities of performing signal amplifications to increase the assay sensitivity even further.

We also have pursued alternative avenues for interfacing label-free analytical principles to circumvent the inherent problems associated with label-dependent readout systems. In these efforts, we have observed LODs in the attomolar range (subnanomolar) (matrix-assisted laser desorption ionization/time of flight MS) *(33)*, femtomolar range (surface-enhanced laser desorption ionization MS) (Wingren and Borrebaeck, unpublished observations), and in the picomolar range (quarts crystal microbalance with dissipation monitoring) *(34)*. These approaches are promising, but they have so far only been used for small prospective arrays. Ultimately, analytical principles, such as MS and tandem MS, may allow the user to both detect and identify the bound analyte(s) in a one-step procedure directly on the chip.

The range of analytical principles evaluated within protein microarrays as whole and the LODs observed have been reviewed previously *(8)* and are therefore not discussed further here. Briefly, a sandwich setup should be a preferred design to improve the specificity and sensitivity of the microarray assay *(47–50)*. The sandwich approach works fine as long as small focused arrays are constructed. However, as soon as the arrays need to be scaled up, this approach is no longer a viable approach, because the process of generating high-quality sandwich antibody pairs against thousands of analytes will be overwhelming. In addition, a threshold of approx 50 probes per sandwich array has been proposed to still maintain adequate assay features (e.g., specificity) *(15)*.

3.5. Data Processing

We have implemented currently available software and know-how to successfully quantitate, analyzes, and evaluate our array data. Still, the handling of protein array data is in general terms in its infancy compared with the rigid procedures adopted within the established DNA microarray technology. Significant progress will occur within this area during the next years, when commercial software are adapted to antibody microarrays, allowing data to be presented as supervised or nonsupervised hierarchial clusters, heat maps, and so on.

4. Antibody Microarray Applications: Current and Future

The first printed antibody microarrays that opened perspectives for rapid large-scale protein analysis were reported only a few years ago *(23,24)*, and the major efforts have since focused on developing the basic technology platform *(10,11,13,14)*. The number of applications is still small, but it is anticipated to increase significantly within the next years *(8)*. To date, applications have been developed mainly within diagnostics, small-scale screening, and focused protein expression profiling by using low- to medium-density arrays (<500) based predominantly on intact monoclonal antibodies (reviewed in **refs.** *8,10,11,13,15*).

We have shown that our antibody microarray technology platform based on optimized recombinant SinFabs successfully could be used to screen for haptens, peptides, and intact proteins (water-soluble proteins as well as membrane proteins) in a format

Probes   Replicates

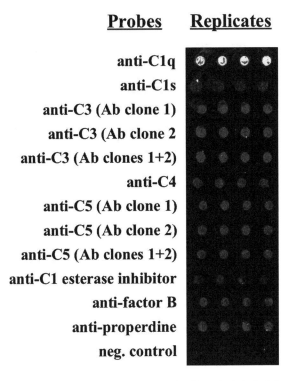

anti-C1q
anti-C1s
anti-C3 (Ab clone 1)
anti-C3 (Ab clone 2)
anti-C3 (Ab clones 1+2)
anti-C4
anti-C5 (Ab clone 1)
anti-C5 (Ab clone 2)
anti-C5 (Ab clones 1+2)
anti-C1 esterase inhibitor
anti-factor B
anti-properdine
neg. control

Fig. 5. Screening of human complement proteins in directly labeled human serum by using a focused anticomplement protein sinFab microarray. All eight complement proteins targeted could be detected.

ranging from pure analytes to complex proteomes, such as sera or crude cell lysates *(8,16,29–34,53,44)*. The potential of this platform was demonstrated by serum proteome screening of high-abundant (nanomolar range, submicrogram per milliliter) and low-abundant (subpicogram per milliliter) analytes *(29,53,54)*. In the former case, a focused sinFab array composed of 10 probes directed against eight complement proteins were used for evaluation (**Fig. 5**). The results showed that directly labeled human serum readily could be screened for content of high-abundant analytes (**29**; Ingvarsson et al., unpublished data). Work is currently in progress to evaluate this platform for screening human patient serum samples for complement protein deficiencies *(53;* Ingvarsson et al., unpublished data). For low-abundant analytes, a focused anticytokine chip was used to screen human serum samples. The results showed that analytes present in the subpicogram per milliliter range could be detected in directly labeled complex proteomes *(53)*. Considering the complexity of human serum *(5–7)*, these data demonstrates proof-of-concept for our array platform, illustrating the high specificity displayed by our content *(16,29)*.

In recent experiments, differential cytokine expression profiling was successfully performed on dendritic cells challenged with a proinflammatory cytokine cocktail, by using recombinant scFv antibodies *(16,52;* Ingvarsson et al., unpublished data). In these studies, directly labeled crude cell lysates, cell supernatants, or both were analyzed on arrays based on 10 to 84 SinFabs. The data showed that specific upregulation of several cytokines could be detected after 24 h (**Fig. 4**). Similarly, the kinetics (0, 4, 8, 16, 24, and 48 h) of the cytokine expression also was successfully studied using our antibody arrays. Most importantly, these data were further corroborated on both gene and protein levels by matching DNA microarray analysis and ELISA experiments (*16*; Ingvarsson et al., unpublished data).

To generate correct and complete maps of the entire proteome, both water-soluble proteins as well as membrane proteins must be addressable. Membrane proteins, which constitute an extremely important group of proteins being one of the most common targets for disease diagnostic, biomarker discovery and therapeutic antibodies, are, however, often considered as a difficult group of proteins to analyze. Recently, Belov and co-workers reported on a successful application of antibody-based microarrays for immunophenotyping of leukemias by targeting membrane-bound cell surface proteins on intact cells (*20,21*). We are currently also developing an array technology platform based on recombinant SinFabs for membrane protein profiling (Dexlin et al., unpublished data).

5. Summary and Conclusions

We have successfully developed the first generation of high-performing antibody microarrays based on recombinant antibody fragments for complex sample analysis, demonstrating great potential within disease proteomics. A highly specific and sensitive assay could be designed targeting a wide range of analytes. We also have outlined the array format of tomorrow, indicating the progress desired before the technology truly will evolve into the high-throughput proteomic research tool needed by the research community. The final product will then allow us to perform high-throughput, comparative proteomics with a resolution that no technology is even close to today. Ultimately, this approach will open up new avenues for disease diagnostics, biomarker discovery and drug target identification with important implications for disease proteomics.

Acknowledgments

This study was supported by grants from SWEGENE Protein Array Development Program, the Swedish National Research Council (VR-NT), Bioinvent International AB, the Åke Wiberg Foundation, the Gunnar Nilsson Cancer Foundation, the Magnus Bergwall Foundation, and the Swedish Society of Medicine.

References

1. Phizicky, E., Bastiaens, P. I. H., Zhu, H., Snyder, M. and Fields, S. (2004) Protein analysis on a proteomic scale. *Nature* **422**, 208–215.
2. Hanash, S. (2003) Disease proteomics. *Nature* **13**, 226–232.
3. Zhu, H., Bilgin, M., and Snyder, M. (2003) Proteomics. *Annu. Rev. Biochem.* **72**, 783–812.
4. Tirumalai, R. S., Chan, K. C., Prieto, D. A., Issaq, H. J., Conrads, T. P. and Veenstra, T. D. (2003) Characterisation of the low molecular weight human serum proteome. *Mol. Cell Proteomics* **2**, 1096–1103.
5. Adkins, J. N., Varnum, S. M., Auberry, K. J, et al. (2002) Toward a human blood serum proteome. *Mol. Cell. Proteomics* **1**, 947–955.

6. Pieper, R., Gatlin, C., Makusky, A. J., et al. (2003) The human serum proteome: Display of nearly 3700 chromatographically separated protein spots on two-dimensional electrophoresis gels and identification of 325 distinct proteins. *Proteomics* **3,** 1345–1364.

7. Anderson, N. L., Polanski, M., Pieper, R., et al. (2004) The human plasma proteome: a non-redundant list developed by combination of four separate sources. *Mol. Cell. Proteomics* **3,** 311–326.

8. Wingren, C., and Borrebaeck, C. A. K. (2004) High-throughput proteomics using antibody microarrays. *Expert Rev. Proteomics* **1,** 355–364.

9. Staudt, L. M. (2002) Gene expression profiling of lymphoid malignacies. *Annu. Rev. Med.* **53,** 303–318.

10. MacBeath, G. (2002) Protein microarrays and proteomics. *Nat. Genet.* **32,** 526–532.

11. Pavlickova, P., Schneider, E. M. and Hug, H. (2004) Advances in recombinant antibody microarrays. *Clin. Chim. Acta* **343,** 17–35.

12. Borrebaeck, C. A. K. (2000) Antibodies in diagnostics - from immunoassays to protein chips. *Immunol. Today* **21,** 379–382.

13. Zhu, H., and Snyder, M. (2003) Protein chip technology. *Curr. Opin. Chem. Biol.* **7,** 55–63.

14. Wilson, D. S., and Nock, S. (2003) Recent developments in protein microarray technology. *Ang. Chem.* **42,** 494–500.

15. Haab, B. B. (2003) Methods and applications of antibody microarrays in cancer research. *Proteomics* **3,** 2116–2122.

16. Wingren, C., Ingvarsson, J., Lindstedt, M., and Borrebaeck, C. A. K. (2003) Recombinant antibody microarrays–a viable option? *Nat. Biotechnol.* **21,** 223.

17. Miller, J. C., Zhou, H., Kwekel, J., et al. (2003) Antibody microarray profiling of human prostate cancer sera: antibody screening and identification of potential biomarkers. *Proteomics* **3,** 56–63.

18. Knezevic, V., Leethanakul, C., Bichsel, V. E., et al. (2001) Proteomic profiling of the cancer microenvironment by antibody arrays. *Proteomics* **1,** 1271–1278.

19. Sreekumar, A., Nyati, M. K., Varambally, S., et al. (2001) Profiling of cancer cells using protein microarrays: discovery of novel radiation-regulated proteins. *Cancer Res.* **61,** 7585–7593.

20. Belov, L., de la Vega, O., dos Remedios, C. G., Mulligan, S. P., and Christopherson, R. I. (2001) Immunophenotyping of leukemias using a cluster of differentiation antibody microarray. *Cancer Res.* **61,** 4483–4489.

21. Belov, L., Huang, P., Barber, N., Mulligan, S. P., and Christopherson, R. I. (2003) Identification of repertoires of surface antigens on leukemias using an antibody microarray. *Proteomics* **3,** 2147–2154.

22. http://www.immun.lth.se. Accessed September 11, 2006.

23. Haab, B. B., Dunham, M. J., and Brown, P. O. (2001) Protein microarrays for highly parallel detection and quantification of specific proteins and antibodies in complex solutions. *Genome Biol.* **2,** 1–22.

24. MacBeath, G., and Schreiber, S. L. (2000) Printing proteins as microarrays for high-throughput function determination. *Science* **289,** 1760–1763.

25. Söderlind, E., Strandberg, L., Jirholt, P., et al. (2000) Recombining germline-derived CDR sequences for creating diverse single-framework antibody libraries. *Nat. Biotechnol.* **18,** 852–856.

26. Knappik, A., Ge, L., Honegger, A., et al. (2000) Fully synthetic human combinatorial antibody libraries (HuCAL) based on modular concensus frameworks and CDRs randomized with trinucleotides. *J. Mol. Biol.* **296,** 57–86.

27. Borrebaeck, C. A. K., and Ohlin, M. (2002) Antibody evolution beyond Nature. *Nat. Biotechnol.* **20,** 1189–1190.

28. Steinhauer, C., Wingren, C., Malmborg-Hager, A., and Borrebaeck, C. A. K. (2002) Single framework recombinant antibody fragments designed for protein chip applications. *Biotechniques* **33**, 38–45.
29. Wingren, C., Steinhauer, C., Ingvarsson, J., Persson, E., Larsson, K., and Borrebaeck, C. A. K. (2005) Microarrays based on affinity-tagged scFv antibodies: sensitive detection of analyte in complex proteomes. *Proteomics* **5**, 1281–1291.
30. Steinhauer, C., Ressine, A., Marko-Varga, G., Laurell, T., Borrebaeck, C. A. K., and Wingren, C. (2004) Biocompatability of surfaces for antibody microarrays: design of macroporous silicon substrates. *Anal. Biochem.* **341**, 204–213.
31. Svedhem, S., Pfeiffer, I., Larsson, C., Wingren, C., Borrebaeck, C. A. K., and Höök, F. (2003) Patterns of DNA-labelled and protein/scFv-carrying lipid vesicles directed by preferential protein adsorption and supported lipid bilayer formation on an Au/SiO^2 template. *Chem. Biol. Chem.* **4**, 339–343.
32. Wingren, C., Montelius, L., and Borrebaeck, C. A. K. (2004) Mega-dense nanoarrays–the challenge of novel antibody array formats. In: *Protein Microarrays* (M. Schena and S. Weaver, eds.), Jones and Bartlett Publishers, Sudbury, MA, pp. 339–352.
33. Borrebaeck, C. A. K., Ekström, S., Malmborg-Hager, A. C., Nilsson, J., Laurell, T., and Marko-Varga, G. (2001) Protein chips based on recombinant antibody fragments: a highly sensitive approach as detected by mass spectrometry. *Biotechniques* **30**, 1126–1132.
34. Larsson, C., Bramfeldt, H., Wingren, C., Borrebaeck, C., and Hook, F. (2005) Gravimetric antigen detection utilizing antibody-modified lipid bilayers. *Anal. Biochem.* **345**, 72–80.
34a. Steinhauer, C., Wingren, C., Khan, F., He, M., Taussig, M. J., Borrebaeck, C. A. (2006) Improved affinity coupling for antibody microarrays: engineering of double-(His)6-tagged single framework recombinant antibody fragments. *Proteomics* **6**, 4227–4234.
35. http://www.bioinvent.se. Accessed September 11, 2006.
36. Hallborn, J., and Carlsson, R. (2002) Automated screening procedure for high-throughput generation of antibody fragments. *Biotechniques* **33**, 30-37.
37. Kusnezow, W., and Hoheisel, J. D. (2003) Solid supports for microarray immunoassays. *J. Mol. Recognit.* **16**, 165–176.
38. Angenendt, P., Glokler, J., Sobek, J., Lehrach, H. and Cahill, D. J. (2003) Next generation of protein microarray support materials: evaluation for protein and antibody microarray applications. *J. Chromatogr. A* **1009**, 97–104.
39. Angenendt, P., Glokler, J., Murphy, D., Lehrach, H., and Cahill, D. J. (2002) Toward optimized antibody microarrays: a comparison of current microarray support materials. *Anal. Biochem.* **309**, 253–260.
40. http://www.nuncbrand.com. Accessed September 11, 2006.
41. http://www.whatman.com. Accessed September 11, 2006.
42. http://www.xenopore.com. Accessed March 10, 2005.
43. http://www.perkinelmer.com. Accessed September 11, 2006.
44. Kingsmore, S. F., and Patel, D. D. (2003) Multiplexed protein profiling in antibody-based microarrays by rolling circle amplification. *Curr. Opin. Biotechnol.* **14**, 74–81.
45. Mitchell, P. (2002) A perspective on protein microarrays. *Nat. Biotechnol.* **20**, 225–229.
46. Service, R. F. (2001) Searching for recipes for protein chips. *Science* **294**, 2080–2082.
47. Kusnezow, W., and Hoheisel, J. D. (2002) Antibody microarrays: promises and problems. *Biotechniques* **33**, 14–23.
48. Beator, J. (2002) From protein microarrays to cytokine detection chips. *Biotech. Int.* December 20–22.
49. Pawlak, M., Schick, E., Bopp, M. A., Schneider, M. J., Oroszlan, P., and Ehrat, M. (2003) Zeptosen's protein microarrays: a novel high performance microarray platform for low abundance protein analysis. *Proteomics* **2**, 383–393.

50. Schweitzer, B., Wiltshire, S., Lambert, J., et al. (2000) Immunoassays with rolling circle DNA amplification: a versatile platform for ultrasensitive antigen detection. *Proc. Natl. Acad. Sci. USA* **97,** 10,113–10,119.

51. Finnskog, D., Ressine, A., Laurell, T., and Marko-Varga, G. (2004) Integrated protein microchip assay with dual fluorescent- and MALDI read-out. *J. Proteome Res.* **3,** 988–994.

52. Ingvarsson, J., Lindstedt, M., Borrebaeck, C. A., Wingren, C. (2006) One-step fractionation of complex proteomes enables detection of low abundant analytes using antibody-based microarrays. *J. Proteome. Res.* **5,** 170–176.

53. Wingren, C., and Borrebaeck, C. A. (2006) Antibody microarrays: current status and key technological advances. *OMICS* **10,** 411–427.

54. Ellmark, P., Ingvarsson, J., Carlsson, A., Lundin, B. S., Wingren, C., and Borrebaeck, C, A. (2006) Identification of protein expression signatures associated with *Helicobacter pylori* infection and gastric adenocarcinoma using recombinant antibody microarrays. *Mol. Cell Proteomics* **5,** 1638–1646.

13

Glycoprofiling by DNA Sequencer-Aided Fluorophore-Assisted Carbohydrate Electrophoresis

New Opportunities in Diagnosing and Following Disease

Wouter Laroy and Roland Contreras

Summary

In recent years, the importance of glycosylation has been realized. Carbohydrates not only serve as decoration for proteins and lipids but also form an integral part of the glycoconjugate, adding to its functional and structural properties. The specific type of glycosylation may change with altering physiological conditions, such as disease. Likewise, altering glycosylation may determine disease or disease properties. Studying glycosylation has long been a major problem, mainly because of the lack of proper analytical technology. Only minute quantities of the analytes are available from natural sources, and they cannot be amplified like DNA. The possibility of branching, different isomeric linkages, and the use of building blocks with the same mass lead to isomeric and isobaric structures. Lately, technology has become available that is able to differentiate these structures and that comforts enough sensibility and throughput to analyze samples from natural sources. DNA sequencer-aided fluorophore-assisted carbohydrate electrophoresis (DSA-FACE) is one of these new techniques with great potential. Here, we briefly introduce the method and discuss some of its applications in diagnosis of disease.

Key Words: Congenital disorders of glycosylation; diagnosis; disease; DNA sequencer-aided fluorophore-assisted carbohydrate electrophoresis; DSA-FACE; glycosylation; liver.

1. Introduction

Glycosylation is the enzymatic addition of sugars or oligosaccharides to macromolecules such as proteins (glycoproteins) and lipids (glycolipids). Generally, glycosyla-tion occurs in the secretory pathway, i.e., in the endoplasmic reticulum and the Golgi apparatus. Consequently, >90% of all secreted and membrane-bound proteins are glycosylated. Together with other post- and cotranslational modifications, of which phosphorylation is probably the most studied, glycosylation helps in diversifying gene products. Indeed, extensive genomic studies *(1)* have shown that the functioning of a complex organism requires more than the gene products directly encoded in the genome *(2,3)*. In contrast to phosphorylation, most forms of glycosylation are related to extracellular functions *(4,5)*.

In glycobiology, the glycan structures and their influence on the macromolecule or the environment are studied. It has been shown that glycosylation is not a random process. Depending on protein structure, both during and after folding, specific sites are modified with carbohydrates both during and after folding. To add to this complex-

From: *Bioarrays: From Basics to Diagnostics*
Edited by: K. Appasani © Humana Press Inc., Totowa, NJ

ity, glycosylation and its presence or absence can depend on the physiological state of the cell and on its environment *(6,7)*. There is no direct code in the genome that determines the state of glycosylation. Although all cells in an organism have the same genomic information, the physiological state of a specific cell and its environment determine which part of the glycosylation machinery is active. We can state that a specific cell under specific conditions glycosylates specific sites on specific proteins in a specific way. Consequently, a particular glycoprotein should not be considered a single entity but rather as a group of different glycoforms *(8)*. Different glycoforms of the same protein may have different functional, kinetic, or physical properties.

The so-called paradox of glycosylation states that no specific function can be associated with a specific carbohydrate structure *(9)*. This statement may seem to contradict the aforementioned statement that glycosylation has specific functions, but this is not necessarily so. These functions depend on the specific protein to which the carbohydrate is linked and on the environment in which it is expressed. When considering glycans as a structural feature of proteins, just like amino acids, this role can be easily understood. A classical example is the effect of glycosylation of the heavy chain of immunoglobulin G on the structure of the whole Fc moiety, whereby the effector functions associated with the Fc portion of IgGs are tuned by its glycosylation *(10)*. Regardless, the exact role of a carbohydrate depends on its interactions with the protein backbone and with other proteins, carbohydrates, or other types of molecules. This view allows us to understand that a specific carbohydrate structure can serve many functions. Some general functions can be assigned to specific structures, but they usually do not cover the whole story. Terminal sialic acid molecules on glycoproteins are known to protect them from clearance from the blood by liver receptors. However, as part of a larger terminal carbohydrate structure, they may influence many other functions or properties of the specific protein in a specific environment, illustrating that for full comprehension of complex life, glycomics is as essential as proteomics and genomics. Because the condition of the cell largely controls which part of the glycosylation machinery is active at a certain time, changes in these conditions result in altered glycosylation. Alternatively, the types of glycoconjugates produced by a cell, tissue, or organism reflect their current physiological state. In pathology, this knowledge can be used in two ways. First, these changes can indicate a certain disease, just like a change in protein concentration. Second, the altered glycosylation can be studied as a function of the pathology. In the former use, diagnosis is the focus *(11)*, whereas in the latter use the main goal is therapy *(12,13)*. In what follows, we discuss the use of glycosylation changes in diagnosis, making use of newly developed technology.

2. Glycosylation in Diagnosis: Current Use and Limitations

Until now, the direct use of glycosylation in diagnosis of diseases or physiological conditions is limited. However, some of the current applications are based on the occurrence of different isoforms of a protein; different oligosaccharides present on a particular protein backbone result in different glycoforms of that protein. When a charged sugar is involved, the protein isoforms have different isoelectric points, which facilitates their separation and gives them predictive value. For example, specific changes in isoforms of human transferrin are directly related to alcohol abuse *(14–16)*. Therefore, carbohydrate-deficient transferrin is currently used for the follow-up of alcoholic patients under treatment. The same approach is taken for the detection of abuse of exogenous erythropoietin (EPO) in sports *(17)*; the test is based on differences in EPO isoforms between human endogenous and recombinantly expressed EPO.

Specific antibodies are commonly used in diagnosis to detect changes in protein concentrations. Many antibodies recognize specific sugars as such or as part of a combined sugar–peptide epitope. Alternatively, lectins can be used for the detection of specific types of glycosylation. Currently, these carbohydrate recognition proteins are used in some diagnostic tests, most commonly in the detection of specific carbohydrate cancer markers *(7)*. However, it is not the intention of this chapter to go into further detail about these applications.

One of the major current obstacles for the use of carbohydrates in diagnosis has been the lack of appropriate analytic techniques. Carbohydrate-recognizing proteins have proven useful as diagnostic tools, but they also have their limitations. They do not allow us to get a complete picture of the different oligosaccharides present because they only recognize a specific part of the molecule. Techniques that do allow further characterization are nuclear magnetic resonance, X-ray crystallography, and mass spectrometry (MS). Nuclear magnetic resonance is currently the best for obtaining full structural data. However, because large amounts of pure product are needed, this method cannot be used for diagnostic purposes. Crystallization of oligosaccharides and even more so glycoproteins remains very difficult and impractical for diagnostic applications. MS has been extensively used in the analysis of oligosaccharides *(18)*, but it has some disadvantages, too. A major disadvantage is associated with the occurrence of isomeric and isobaric branched structures in natural oligosaccharides, many of which cannot be differentiated by classical MS. More advanced multiple MS techniques *(19)* do allow differentiation of these structures, but they lack the advantages of high throughput and sensitivity.

Because the amount of biological sample available is frequently limited, and because of the characteristics of oligosaccharides, the perfect technique for glycodiagnosis should combine high-sensitivity, high-resolving power (including the isomeric and isobaric structures), high throughput, and low costs (material and personnel). In the remaining part of this chapter, we focus on a new carbohydrate analysis technique that approaches these criteria and may make glycodiagnostics a reality.

3. Glycosylation Analysis: DNA Sequencer-Aided Fluorophore-Assisted Carbohydrate Electrophoresis (DSA-FACE)

Carbohydrates, like DNA and proteins, can be separated electrophoretically on polyacrylamide gels *(19)*. To allow detection, and in some cases to add charge, the oligosaccharides are fluorescently labeled. Classical fluorophore-assisted carbohydrate electrophoresis (FACE) allows separation and detection of oligosaccharides, although with relatively low sensitivity and poor resolving power. A major breakthrough came when slab gel DNA sequencing equipment was successfully used for the profiling and analysis of oligosaccharides *(20,21)*. When an appropriate fluorescent dye is used, usually 8-aminopyrene-1,3,6-trisulfonic acid, DSA-FACE (**Fig. 1**) allows separation of many isomeric and isobaric structures with high resolution and sensitivity. Moreover, the high sensitivity permits the use of minute amounts of glycoproteins for the analysis of all the attached glycans. Now that this technique is available, high-throughput analysis of biological samples becomes possible, and its use in glycodiagnosis is emerging.

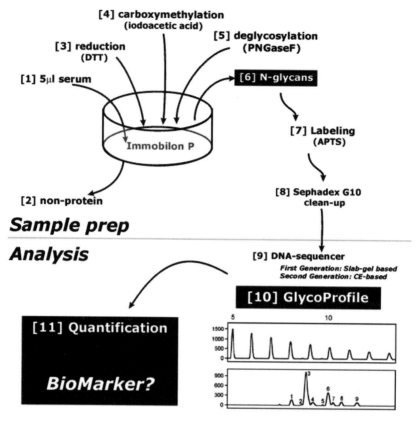

Fig. 1. DSA-FACE for glycoprofiling of serum and diagnosis of disease. During sample preparation, glycoproteins are immobilized on Immobilon P in a 96-well filtration plate (**1**). After reduction and carboxymethylation, efficient deglycosylation results in the *N*-glycan pool (**3–6**). After fluorescent labeling and clean-up (**7,8**), electrophoretic separation on a DNA sequencer platform leads to the serum GlycoProfile. Analysis of this profile may be informative for diseases or other physiological conditions.

4. DSA-FACE in Glycodiagnosis

4.1. N-Glycosylation Changes in Liver Disease

To exploit a disease marker in practice, it is not sufficient to have a good analytical method, getting a pool of molecules including the marker must be easy and preferentially cheap. Because blood sampling is a common and harmless practice, serological markers are the most common. Usually, a change in the relative amount of a specific protein is measured.

Chronic liver pathology is a complex disease. It evolves through a continuous process that has been divided into stages (*22*). Administration of hepatotoxic agents (alcohol and viruses) leads to inflammatory necrosis of liver tissue, which activates hepatic stellate cells to secrete extracellular matrix components as part of the healing process. Because of the chronic nature of the necrosis, however, this process results in excessive deposition of these components, and the ensuing fibrosis disturbs liver architecture. Different degrees of fibrosis are distinguished, and different scoring systems have been used for staging it. The Metavir score, based on histological examination of liver biopsy material, is commonly used. Metavir score F0 reflects the healthy liver. Cirrhosis, the most severe form of fibrosis, has a Metavir score of F4. This stage is characterized by clusters of abnormally replicating hepatocytes, named regenerative nodules, which constitute a major risk factor for the development of hepatocellular carcinoma, one of the most aggressive cancers known. Staging diseases in a limited number of classes is a classical practice in medicine; however, it is a very limited view: a disease evolves continuously from one stage to another.

The invasive liver biopsy procedure leads not only to major discomfort to the patient but also to social and economic consequences. Moreover, a substantial portion of liver biopsies leads to severe complications and even death *(23)*. Now that treatments for liver disease are emerging (but still largely experimental), other means for follow-up of liver conditions are needed because repetitive biopsies are not advisable. Some significant changes in serological values have been shown in liver disease. Total bilirubin, hyaluronic acid *(24)*, and α2-macroglobulin *(25)* are just some examples (reviewed in **ref.** *26*). However, adequate predictive value is obtained only when they are combined in a mathematical model *(27,28)*.

In a recent study on liver disease, the assumption was made that different active glycosylation machineries are active in healthy and diseased hepatocytes *(29)*. Because the majority of glycoproteins in the serum are synthesized by these cells, the total *N*-glycome of these proteins may reflect liver condition. Using DSA-FACE, total serum *N*-glycoprofiles were obtained. A new serological marker based on specific changes the serum glycome was tested on a cohort of 60 blood donors and 188 liver disease patients. Clear changes in the *N*-glycan profiles were observed in cirrhosis (**Fig. 2**): more agalacto forms, more structures containing bisecting *N*-acetylglucosamine, and less branching. Based on these profiles, the so-called GlycoCirrhoTest was defined (**Fig. 3**), allowing differentiation of cirrhosis from noncirrhotic chronic hepatitis with a sensitivity of 92%, a specificity of 93% and an efficiency of 85–90%. Interestingly, when combined with FibroTest (a commercially available liver test based on several known serological markers), compensated cirrhosis was detected with 100% specificity, 75% sensitivity and an overall classification efficiency of 93%. Decompensated cirrhosis was detected with 100% sensitivity, but this may have less clinical relevance.

A more detailed study of the glycan profiles revealed delicate changes in the different fibrosis stadia (**Fig. 3**). These changes are of considerable clinical importance, because biopsy is still the only practical starting point for grading fibrosis. A marker is of even more importance in the follow-up of the staged patients, because new therapies are becoming available. In these cases, repeated biopsies are not advisable and are most commonly refused by the patient. Moreover, repeated biopsy would make an already expensive treatment a huge burden on social security or on the patient. Thus, the aforementioned and other ways for the serological diagnosis of liver disease are closely followed by the hepatology field.

Although staging of fibrosis, e.g., by Metavir score, remains difficult, we can conclude that follow-up may become possible using the serum glycome tests alone, or in combination with any of the other markers or algorithms. Although specific differentially glycosylated proteins have been used previously, GlycoCirrhoTest would be the first marker based exclusively on total glycome profiles. Finally, researchers and clini-

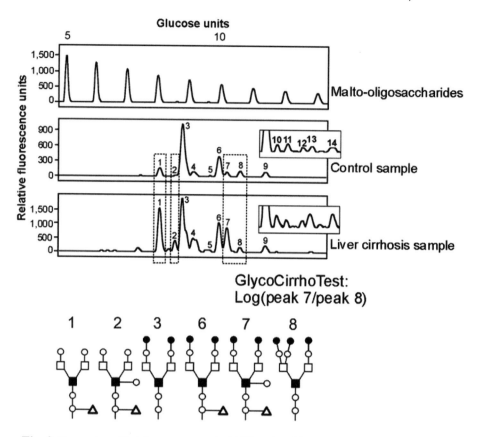

Fig. 2. Examples of total serum protein *N*-GlycoProfiles. (**Top**) Maltooligosaccharide reference. (**Middle**) Typical electropherogram of desialylated *N*-glycans derived from proteins in control serum sample. (**Bottom**) Representative electropherogram obtained from a cirrhosis patient. Structures of *N*-glycans of relevance to this study are shown below the panels; peaks that are important for fibrosis and cirrhosis markers are boxed. (Figure reproduced from Callewaert et al. (2004). Copyright permission was granted by the Nature Publishing Group.)

cians have to question whether categorical staging (biopsy) or continuous staging (serological markers) are the most appropriate. The former is still the gold standard in diagnosing liver disease *(30)*. The latter surely has advantages, but its acceptance requires a different way of looking at the progressing disease.

4.2. Glycosylation Changes in Congenital Disorders of Glycosylation (CDG)

Congenital disorders of glycosylation (CDG) is a booming area in pediatrics *(31)*. Since the first report of this disease in 1980, 20 different types of CDG have been described. CDGs are genetic diseases caused by defects in the glycosylation machinery. Two types are distinguished. CDGs type I have defects that affect the synthesis of the *N*-glycan lipid-linked precursor, or the transfer of this precursor oligosaccharide to specific sites on the forming polypeptide. Thus, fewer sites on a protein are occupied, resulting in abnormal glycoforms. The original test for CDG was based on transferrin isoelectrofocusing to detect different sialylation degrees. When defects in the processing of the transferred precursor glycan occur, the disease is classified as CDG type II.

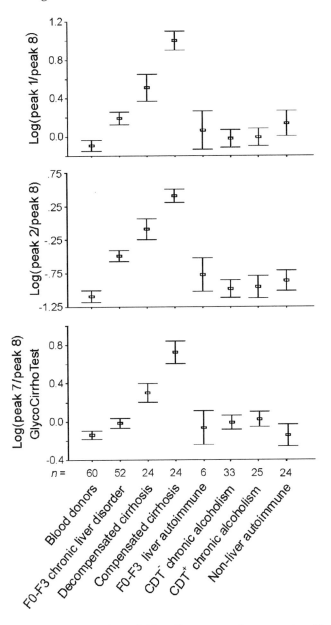

Fig. 3. Trends in derived diagnostic variables. Serum samples were classified in eight clini-cally relevant groups. Three diagnostic variables were derived from the profiles in **Fig. 2**. Ordi-nate scale is logarithmic. Error bars represent 95% confidence intervals for the means.(Figure reproduced from Callewaert et al. (2004). Copyright permission was granted by the Nature Publishing Group.)

Transferrin eukaryotic translation initiation factor IEF is still the most frequently used diagnostic assay: it is cheap, rapid, relatively reliable, and suitable for screening *(32,33)*. So far, its major clinical application has been for screening. However, there are some disadvantages, too. It does not detect defects in *O*-glycan biosynthesis, and not even all defects in *N*-glycan biosynthesis. Indeed, normal transferrin can only bear a limited amount of specific *N*-glycan structures. Thus, only defects in the synthesis of those will be detected. We can conclude that abnormal transferrin IEF indicates CDG, but normal transferrin IEF does not exclude it.

Better technology is now available for the detection of abnormalities, at least for the profiling of *N*-glycans, in the form of DSA-FACE. This technology permits analysis of the glycome not only of one protein but of all proteins in the serum. As shown previously, altered *N*-glycan profiles are observed in different subtypes of CDG type I *(34)*. The observed changes seem to be a milder manifestation of some of the changes observed in adult liver cirrhosis patients, which agrees with the reported steatosis and fibrosis in CDG type I patients. These changes include increased core fucosylation of the biantennary glycan and reduced abundance of the trigalacto triantennary structure. Other changes associated with liver cirrhosis, such as strong undergalactosylation and an increased presence of structures with a bisecting *N*-acetylglucosamine, are less frequently observed in the CDG type I serum *N*-glycome or not at all. At least for CDG-Ia, this result correlates with the aberration observed in the transferrin IEF pattern.

Several CDGs are caused by or associated with defects in *O*-glycosylation *(35)*. *O*-Mannosylation has been shown to be affected in a range of congenital muscular dystrophies *(36)*. Defects in proteoglycan synthesis have been associated with Ehler–Danlos sclera *(37)* and hereditary multiple exostoses *(38,39)*. So far, these kinds of changes cannot be detected by DSA-FACE; thus, other assays need to be developed.

5. Conclusions

So far, total glycome profiles have not been used for the diagnosis of diseases, although this link has been reported in the literature. The main reason is that methodology for glycome analysis in a clinical setting has not been available. DSA-FACE combines some strong features that may allow its application in clinical practice. Its sensitivity allows analysis of tiny amount of samples, a major breakthrough in the analysis of materials from biological sources. The examples in this chapter may not be problematic in this aspect, but consider be helpful in analyzing subsets of proteins in the serum or glycans from other sources where less material is available, e.g., mucus and cerebrospinal fluid. One big advantage of DSA-FACE over other sensitive methods is its ability to separate many isobaric and isomeric structures. This ability is of huge importance in the analysis of sugars. Increasing the resolution improves the diagnostic value of a profile. Finally, DSA-FACE also has a throughput that allows screening of many samples. Because conditions of separation are based only on the carbohydrates and not on their source, these conditions can be the same for any analysis. Thus, one DNA sequencer, whether it is gel based or capillary electrophoresis based, can analyze all diseases to which *N*-glycosylation changes are associated. Indeed, the aforementioned two examples are those that have been studied so far, but they are not the only possible applications. The detection of several forms of cancer, inflammation, and so on are other applications. A new way of diagnosis, which we tend to call GlycoDiagnosis may evolve from this strategy.

References

1. Consortium, I. H. G. S. (2001) Initial sequencing and analysis of the human genome. *Nature* **409**, 860–921.

2. Dennis, J. W., Granovsky, M., and Warren, C. E. (1999) Protein glycosylation in development and disease. *Bioessays* **21,** 412–421.

3. Rudd, P. M., Elliott, T., Cresswell, P., Wilson, I. A., and Dwek, R. A. (2001) Glycosylation and the immune system. *Science* **291,** 2370–2376.

4. Crocker, P. R., and Feizi, T. (1996) Carbohydrate recognition systems: functional triads in cell-cell interactions. *Curr. Opin. Struct. Biol.* **6,** 679–691.

5. Solis, D., Jimenez-Barbero, J., Kaltner, H., et al. (2001) Towards defining the role of glycans as hardware in information storage and transfer: basic principles, experimental approaches and recent progress. *Cells Tissues Organs* **168,** 5–23.

6. Haltiwanger, R. S., and Lowe, J. B. (2004) Role of glycosylation in development. *Annu. Rev. Biochem.* **73,** 491–537.

7. Sell, S. (1990) Cancer-associated carbohydrates identified by monoclonal antibodies. *Hum. Pathol.* **21,** 1003–1019.

8. Opdenakker, G., Rudd, P. M., Ponting, C. P., and Dwek, R. A. (1993) Concepts and principles of glycobiology. *FASEB J.* **7,** 1330–1337.

9. Varki, A. (1993) Biological roles of oligosaccharides: all of the theories are correct. *Glycobiology* **3,** 97–130.

10. Krapp, S., Mimura, Y., Jefferis, R., Huber, R., and Sondermann, P. (2003) Structural analysis of human IgG-Fc glycoforms reveals a correlation between glycosylation and structural integrity. *J. Mol. Biol.* **325,** 979–989.

11. Durand, G., and Seta, N. (2000) Protein glycosylation and diseases: blood and urinary oligosaccharides as markers for diagnosis and therapeutic monitoring. *Clin. Chem.* **46,** 795–805.

12. Dove, A. (2001) The bittersweet promise of glycobiology. *Nat. Biotechnol.* **19,** 913–917.

13. Dwek, M. V., Ross, H. A., and Leathem, A. J. (2001) Proteome and glycosylation mapping identifies post-translational modifications associated with aggressive breast cancer. *Proteomics* **1,** 756–762.

14. Stibler, H. (1991) Carbohydrate-deficient transferrin in serum: a new marker of potentially harmful alcohol consumption reviewed. *Clin. Chem.* **37,** 2029–2037.

15. Wuyts, B., and Delanghe, J. R. (2003) The analysis of carbohydrate-deficient transferrin, marker of chronic alcoholism, using capillary electrophoresis. *Clin. Chem. Lab. Med.* **41,** 739–746.

16. Wuyts, B., Delanghe, J. R., Kasvosve, I., Gordeuk, V. R., Gangaidzo, I. T., and Gomo, Z. A. (2001) Carbohydrate-deficient transferrin and chronic alcohol ingestion in subjects with transferrin CD-variants. *Clin. Chem. Lab. Med.* **39,** 937–943.

17. Lasne, F., Martin, L., Crepin, N., and de Ceaurriz, J. (2002) Detection of isoelectric profiles of erythropoietin in urine: differentiation of natural and administered recombinant hormones. *Anal. Biochem.* **311,** 119–126.

18. Zaia, J. (2004) Mass spectrometry of oligosaccharides. *Mass Spectr. Rev.* **23,** 161–227.

19. Sheeley, D. M., and Reinhold, V. N. (1998) Structural characterization of carbohydrate sequence, linkage, and branching in a quadrupole Ion trap mass spectrometer: neutral oligosaccharides and N-linked glycans. *Anal Chem.* **70,** 3053–3059.

20. Callewaert, N., Geysens, S., Molemans, F., and Contreras, R. (2001) Ultrasensitive profiling and sequencing of N-linked oligosaccharides using standard DNA-sequencing equipment. *Glycobiology* **11,** 275–281.

21. Callewaert, N., Vervecken, W., Van Hecke, A., and Contreras, R. (2002) Use of a meltable polyacrylamide matrix for sodium dodecyl sulfate-polyacrylamide gel electrophoresis in a procedure for N-glycan analysis on picomole amounts of glycoproteins. *Anal. Biochem.* **303,** 93–95.

22. Ishak, K.G. (2002) Inherited metabolic diseases of the liver. *Clin. Liver Dis.* **6,** 455–479.

23. Menon, K. V., and Kamath, P. S. (2000) Managing the complications of cirrhosis. *Mayo Clin. Proc.* **75,** 501–509.
24. Engstrom-Laurent, A., Loof, L., Nyberg, A., and Schroder, T. (1985) Increased serum levels of hyaluronate in liver disease. *Hepatology* **5,** 638–642.
25. Naveau, S., Poynard, T., Benattar, C., Bedossa, P., and Chaput, J. C. (1994) Alpha-2-macroglobulin and hepatic fibrosis. Diagnostic interest. *Dig. Dis. Sci.* **39,** 2426–2432.
26. Afdhal, N. H., and Nunes, D. (2004) Evaluation of liver fibrosis: a concise review. *Am. J. Gastroenterol.* **99,** 1160–1174.
27. Imbert-Bismut, F., Ratziu, V., Pieroni, L., Charlotte, F., Benhamou, Y., and Poynard, T. (2001) Biochemical markers of liver fibrosis in patients with hepatitis C virus infection: a prospective study. *Lancet* **357,** 1069–1075.
28. Poynard, T., Imbert-Bismut, F., Ratziu, V., et al. (2002) Biochemical markers of liver fibrosis in patients infected by hepatitis C virus: longitudinal validation in a randomized trial. *J. Viral Hepat.* **9,** 128–133.
29. Callewaert, N., Van Vlierberghe, H., Van Hecke, A., Laroy, W., Delanghe, J., and Contreras, R. (2004) Noninvasive diagnosis of liver cirrhosis using DNA sequencer-based total serum protein glycomics. *Nat. Med.* **10,** 429–434.
30. Cadranel, J. F., Rufat, P., and Degos, F. (2000) Practices of liver biopsy in France: results of a prospective nationwide survey. For the Group of Epidemiology of the French Association for the Study of the Liver (AFEF). *Hepatology* **32,** 477–481.
31. Jaeken, J., and Carchon, H. (2004) Congenital disorders of glycosylation: a booming chapter of pediatrics. *Curr. Opin. Pediatr.* **16,** 434–439.
32. Jaeken, J., van Eijk, H.G., van der Heul, C., Corbeel, L., Eeckels, R., and Eggermont, E. (1984) Sialic acid-deficient serum and cerebrospinal fluid transferrin in a newly recognized genetic syndrome. *Clin. Chim. Acta* **144,** 245–247.
33. Carchon, H. A., Chevigne, R., Falmagne, J. B., and Jaeken, J. (2004) Diagnosis of congenital disorders of glycosylation by capillary zone electrophoresis of serum transferrin. *Clin. Chem.* **50,** 101–111.
34. Callewaert, N., Schollen, E., Vanhecke, A., Jaeken, J., Matthijs, G., and Contreras, R. (2003) Increased fucosylation and reduced branching of serum glycoprotein N-glycans in all known subtypes of congenital disorder of glycosylation I. *Glycobiology* **13,** 367–375.
35. Marquardt, T., and Denecke, J. (2003) Congenital disorders of glycosylation: review of their molecular bases, clinical presentations and specific therapies. *Eur. J. Pediatr.* **162,** 359–379.
36. Yoshida, A., Kobayashi, K., Manya, H., et al. (2001) Muscular dystrophy and neuronal migration disorder caused by mutations in a glycosyltransferase, POMGnT1. *Dev Cell.* **1,** 717–724.
37. Quentin, E., Gladen, A., Roden, L., and Kresse, H. (1990) A genetic defect in the biosynthesis of dermatan sulfate proteoglycan: galactosyltransferase I deficiency in fibroblasts from a patient with a progeroid syndrome. *Proc. Natl. Acad. Sci. USA* **87,** 1342–1346.
38. Lind, T., Tufaro, F., McCormick, C., Lindahl, U., and Lidholt, K. (1998) The putative tumor suppressors EXT1 and EXT2 are glycosyltransferases required for the biosynthesis of heparan sulfate. *J. Biol. Chem.* **273,** 26,265–26,268.
39. Cheung, P. K., McCormick, C., Crawford, B. E., Esko, J. D., Tufaro, F., and Duncan, G. (2001) Etiological point mutations in the hereditary multiple exostoses gene EXT1: a functional analysis of heparan sulfate polymerase activity. *Am. J. Hum. Genet.* **69,** 55–66.

14

High-Throughput Carbohydrate Microarray Technology

Denong Wang, Ruobing Wang, Dhaval Shah,
Shaoyi Liu, Aili Wang, Xiaoyuan Xu, Ke Liu,
Brian J. Trummer, Chao Deng, and Rong Cheng

Summary

One of our long-term interests is to explore the immunogenic sugar moieties that are important for "self-" and "nonself" discrimination and host immune responses. We have established a high-throughput platform of carbohydrate microarrays to facilitate these investigations. Using this technology, carbohydrate-containing macromolecules of distinct structural configurations, including polysaccharides, natural glycoconjugates, and mono- and oligosaccharides coupled to lipid, polyacrylamide, and protein carriers, have been tested for microarray construction without further chemical modification. Here, we discuss issues related to the establishment of this technology and areas that are highly promising for its application. We also provide an example to illustrate that the carbohydrate microarray is a discovery tool; it is particularly useful for identifying immunological sugar moieties, including differentially expressed complex carbohydrates of cancer cells and stem cells as well as sugar signatures of previously unrecognized microbial pathogens.

Key Words: Antibodies; antigens; carbohydrates; glycans; glycoconjugates; microarrays; microspotting; nitrocellulose; polysaccharides; severe acute respiratory syndrome-associated coronavirus; SARS-CoV.

1. Introduction

Our group has focused on development of a carbohydrate-based microarray technology to facilitate investigation of carbohydrate-mediated molecular recognition and anticarbohydrate immune responses (1–3). Like nucleic acids and proteins, carbohydrates are another class of the essential biological molecules. Because of their unique physicochemical properties, carbohydrates are capable of generating structural diversity, and so they are prominent in display on the surfaces of cell membranes or on the exposed regions of macromolecules (4–6). As a result, carbohydrate moieties are suitable for storing biological signals in the forms that are identifiable by other biological systems. In this chapter, we discuss 1) our theoretical consideration for developing high-throughput carbohydrate microarrays, 2) a unique approach we took to establish carbohydrate microarrays, 3) a practical carbohydrate microarray platform that is currently in use by our laboratory, and 4) an example that illustrates a highly promising area for carbohydrate microarray technology to explore.

From: *Bioarrays: From Basics to Diagnostics*
Edited by: K. Appasani © Humana Press Inc., Totowa, NJ

2. Theoretical Considerations in Developing Carbohydrate Microarrays

The Genome Project has led to the discovery that only approx 30,000 genes in the human genome must account for all the complexity of the human organism. This discovery raised an important question about the roles of protein processing and structural modification in modulating the biological activities of proteins and cellular functions. In higher eukaryotic species, most secretory and membrane-bound proteins are decorated with sugar moieties, which is achieved by a critically important posttranslational protein modification, called glycosylation. In many physiological and pathophysiological conditions, expression of cellular glycans, in the form of either glycoproteins or glycolipids, is differentially regulated. There are documented examples that show that cell display of precise complex carbohydrates is characteristically associated with the stages or steps of embryonic development, cell differentiation, and transformation of normal cell to abnormally differentiated tumor or cancer cells *(7–10)*. Sugar moieties are also abundantly expressed on the outer surfaces of the mass majority of viral, bacterial, protozoan, and fungal pathogens. Many sugar structures are pathogen specific, which makes them important molecular targets for pathogen recognition, diagnosis of infectious diseases, and vaccine development *(4,11–15)*.

In spite of the biological magnitude of carbohydrate molecules, the characterization of carbohydrate structures and the exposition of their function have lagged compared with other major classes of biological molecules, such as nucleic acids and proteins. For example, whereas the microarray-based high-throughput technologies for nucleic acids *(16,17)* and proteins *(18,19)* were developed years ago, the first carbohydrate microarray research was published in 2002 *(1,20–24)*.

Our endeavor has focused on establishment of a high-throughput carbohydrate microarray platform that is technically equivalent to the state-of-the-art technologies of the cDNA microarray. Carbohydrates are strikingly different from nucleic acids in structure, physicochemical properties, and cellular function. Thus, the fundamental principles that are at the basis of establishment of a carbohydrate-based assay differ from the basic principles of the nucleic acid-based assays, such as the cDNA microarray and oligonucleotide biochips. For the nucleic acid-based biochips, the detection specificity is determined by the A::T and C::G base pairing, and there is no need to preserve the three-dimensional (3D) structures of the nucleic acid molecules. By contrast, carbohydrate microarrays require preservation of the 3D conformations and topological configurations of sugar moieties on a chip to permit a targeted molecular recognition by the corresponding cellular receptors to take place *(4,5)*.

Therefore, several technical difficulties must conquer to establish a high-throughput carbohydrate microarray technology. These difficulties take into account whether carbohydrate macromolecules of hydrophilic character can be immobilized on a chip surface by methods that are suitable for high-throughput production of microarrays; whether immobilized carbohydrate-containing macromolecules preserve their immunological properties, such as expression of carbohydrate-epitopes or antigenic determinants and their solvent accessibility; whether the carbohydrate microarray system reaches the sensitivity, specificity, and capacity to detect a broad range of antibody specificities in clinical specimens; and eventually whether this technology can be applied to investigate the carbohydrate-mediated molecular recognition on a titanic scale that was previously impossible.

3. Experimental Approach to Establishment
of High-Throughput Carbohydrate Microarrays

Our intent was to introduce immunological specificity to microarray technology to establish a microarray-based broad-range immunosensor for the exploration of immunological diversity of carbohydrates and the immune responses to carbohydrate antigens. In experimental design, we applied a well-studied model system of carbohydrate–anticarbohydrate interaction, α(1,6)dextran and anti-α(1,6)dextran antibodies *(25–27)*, for our initial investigation *(1)*. To address whether carbohydrate macromolecules of hydrophilic character can be immobilized on a chip surface, we applied the fluorescein isothiocyanate (FITC)-conjugated α(1,6)dextrans as probes to screen available chip substrates that were produced for printing cDNA microarrays for their potential use in carbohydrate microarrays. This investigation guided us to the discovery that the nitrocellulose-coated glass slides are suitable for immobilization of carbohydrate-containing macromolecules.

To test whether the size and molecular weight of polysaccharides influence their surface immobilization, FITC-α(1,6)dextran preparations of different molecular weights and of similar molar ratios of FITC/glucose were applied. A structurally distinct polysaccharide, inulin, was chosen as a control to see whether surface immobilization of polysaccharides is restricted to a specific carbohydrate structure. This investigation demonstrated that dextran preparations of different molecular weights, ranging from 20 to 2000 kDa, and inulin of 3.3 kDa could be printed and immobilized on the nitrocellulose-coated slide without chemical conjugation. The linear range of the material transferred and surface immobilized, however, differs significantly among dextran preparations of different molecular weights *(1)*.

To investigate whether immobilized carbohydrate antigens preserve their antigenic determinants, dextran preparations of different linkage compositions and with different ratios of terminal to internal epitopes were printed on nitrocellulose-coated glass slides. These preparations included N279, displaying both internal linear and terminal nonreducing end epitopes, B1299S, heavily branched and expressing predominantly terminal epitopes, and LD7, a synthetic dextran composed of 100% α(1,6)-linked internal linear chain structure. The dextran microarrays were incubated with monoclonal antibodies of defined specificities, either a groove-type anti-α(1,6)dextran 4.3F1 (IgG3) *(28)* or a cavity-type anti-α(1,6)dextran 16.4.12E (IgA) *(29)*. The former recognizes the internal linear chain of α(1,6)dextrans, whereas the latter is specific for the terminal nonreducing end structure of the polysaccharide.

The groove-type monoclonal antibody (mAb), 4.3F1, bound well to the dextran preparations with predominantly linear chain structures, N279 and LD7, but bound poorly to the heavily branched α(1.6)dextran, B1299S. By contrast, when the cavity-type mAb 16.4.12E was applied, it bound to the immobilized dextran preparations with branches (N279 and B1299S) but not to those with only internal linear chain structure (LD7). These patterns of antigen–antibody reactivities are typically identical to those recognized by an ELISA binding assay for either the groove type or cavity type of anti-dextran mAbs. Therefore, the immunological properties of dextran molecules are well preserved when immobilized on a nitrocellulose-coated glass slide. Their nonreducing end structure, recognized by the cavity-type anti-α(1,6)dextrans as well as the internal linear chain epitopes bound by the groove-type anti-α(1,6)dextrans are displayed on the surface after immobilization and are accessible to antibodies in an aqueous solution. This approach was then extended to test a large panel of carbohydrate-containing macromolecules to assess their immobilization on chip and to evaluate expression of antigenic structures for antibody detection. We demonstrated that polysaccharides and glycoconjugates of distinct structural configurations and of diverse sugar chain contents were applicable for this biochip platform, i.e., a method of nitrocellulose-based noncovalent immobilization for high-throughput construction of carbohydrate microarrays *(1)*.

Nitrocellulose polymer is a fully nitrated derivative of cellulose in which free hydroxyl groups are substituted by nitro groups, and it is thus hydrophobic in character. Documented investigations have suggested that immobilization of proteins on a nitrocellulose membrane requires revelation of their hydrophobic surfaces to the membrane *(30,31)*. The molecular forces for the carbohydrate–nitrocellulose interaction remain to be characterized. Perhaps the 3D microporous configuration of the nitrocellulose coating on the slides, the macropolymer characteristics of polysaccharides and the polyamphypathic properties of many carbohydrate-containing macromolecules are key factors for the stable immobilization of carbohydrate antigens on the slide. Given the structural diversity of carbohydrate antigens, we highly recommend that each preparation must be tested on this chip substrate.

4. Practical Platform of Carbohydrate Microarrays

The aforementioned experimental investigations have guided our research to the establishment of a high-throughput platform of carbohydrate microarrays. As illustrated in the **Fig. 1**, this approach applies to existing cDNA microarray systems, including spotter and scanner, for carbohydrate array production and applications. A key technical element of this array platform is the introduction of nitrocellulose-coated microglass slides to immobilize unmodified carbohydrate antigens on the chip surface.

A high-precision robot designed to produce cDNA microarrays was used to spot carbohydrate antigens onto a chemically modified glass slide. The microspotting capacity of this system is approx 20,000 spots per chip. The antibody-stained slides were then scanned for fluorescent signals with a Biochip scanner that was developed for cDNA microarrays.

For microspotting, antigens and antibodies were printed using PIXSYS 5500C (Cartesian Technologies, Irvine, CA). Supporting substrate was FAST Slides (Whatman Schleicher and Schuell, Keene, NH). For immunofluorescence staining, the staining procedure used is essentially identical to regular immunofluorescent staining of tissue sections. For microarray scanning, a ScanArray 5000 Standard Biochip Scanning System and its QuantArray software (Perkin-Elmer, Boston, MA) were used for scanning and data capture.

4.1. An Eight-Chamber Subarray System for Customized Arrays

We have designed an eight-chamber subarray system to create customized carbohydrate microarrays for defined purposes. As illustrated in the **Fig. 2**, each microglass slide contains eight separated subarrays. The microarray capacity is approx 600 microspots per subarray. A single slide is designed to enable eight microarray analyses. A similar design with array capacity of approx 100 microspots is also commer-

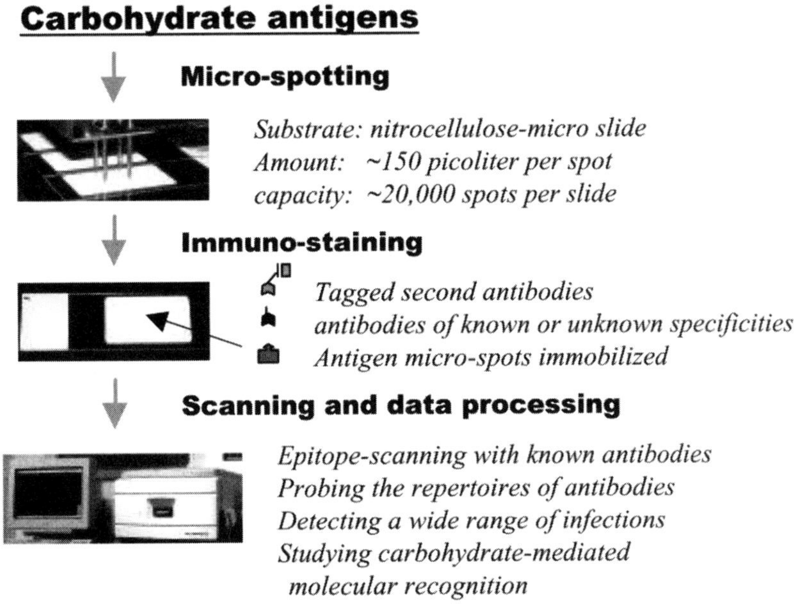

Carbohydrate antigens

↓

Micro-spotting

Substrate: nitrocellulose-micro slide
Amount: ~150 picoliter per spot
capacity: ~20,000 spots per slide

↓

Immuno-staining

Tagged second antibodies
antibodies of known or unknown specificities
Antigen micro-spots immobilized

↓

Scanning and data processing

Epitope-scanning with known antibodies
Probing the repertoires of antibodies
Detecting a wide range of infections
Studying carbohydrate-mediated
molecular recognition

Fig. 1. A high-throughput biochip platform for constructing carbohydrate-based microarrays.

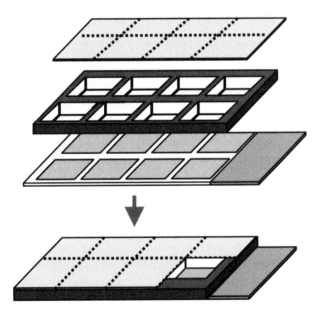

Fig. 2. Graphical presentation of the eight-chamber subarrays.

cially available (Whatman Schleicher and Schuell). For eight-chamber subarrays. Each microglass slide contains eight subarrays of identical content. There is chip space for 600 microspots per subarray, with spot sizes of approx 200- and 300-μm intervals, center to center. A single slide is, therefore, designed to permit eight detections. For repeats and dilutions, each antigen will be printed at 0.5–1.0 mg/mL and also at a 1:10 dilution of the initial concentration. A given concentration of each preparation will be repeated at least three times to allow statistical analysis of detection of identical preparation at given antigen concentration. For antibody isotype standard curves, antibodies of IgG, IgA, and IgM isotype of corresponding species serve as standard curves for antibody detection and normalization.

4.2. Examination of the Presence of Antigens and Antibodies on the Array

To verify that we have successfully "printed" proteins, synthetic peptides, and carbohydrates, we incubate microarrays with antibodies, receptors, or lectins known to react with the printed substance. The reaction is detected either directly by conjugating with a fluorochrome to the "detector" or by a second-step staining procedure.

4.3. Staining and Scanning of Carbohydrate Microarrays

Immediately before use, the printed microarrays are rinsed with phosphate-buffered saline (PBS), pH 7.4, with 0.05% Tween 20 and then blocked by incubating the slides in 1% bovine serum albumin in PBS containing 0.05% NaN_3 at 37⌡C for 30 min. They are then incubated at room temperature with serum specimens at given dilutions in 1% bovine serum albumin PBS containing 0.05% NaN_3 and 0.05% Tween-20. Next, anti-human (or other species) IgG, IgM, or IgA antibodies with distinct specific fluorescent tags (Cy3, Cy5, or FITC) are applied to reveal the bound antibodies according to their Ig heavy chain isotypes. The stained slides are rinsed five times with PBS with 0.05% Tween-20 after each staining step. ScanArray 5000A is used to scan the microarray. This instrument is a standard biochip scanning system (Perkin-Elmer) equipped with multiple lasers, emission filters and ScanArray acquisition software.

4.4. Analysis of Microarray Data

Fluorescence intensity values for each array spot and its background are calculated using QuantArray software analysis packages. Data for at least three replicates for each substance analyzed are collected on each chip ("triple spotting").

4.5. Validation and Further Investigation of Microarray Observations

It is always astute to verify microarray data by other experimental approaches. We check our carbohydrate microararry findings by doing conventional immunoassays, e.g., ELISA, dot blot, Western blot, flow cytometry, and immunohistology. Examples of such investigations were described in our recent publications *(1,3)*.

5. Promising Areas for Exploring Carbohydrate Microarray Technology

As described above, carbohydrates of multiple molecular configurations and of diverse sugar chain structures can be stably immobilized on a nitrocellulose-coated glass slide without chemical conjugation *(1,3)*. A direct application of this technology is in exploring the repertoires of human antibodies with anticarbohydrate activities. When a large collection of microbial carbohydrate antigens is arrayed on a sugar chip, such array would allow a simultaneous detection and characterization of a wide range of antibody reactivities and provide specific diagnostic information of infectious diseases. In combination with the use of semisynthetic glycoconjugates, which display unique oligosaccharide chains, a biochip-based characterization of the epitope-binding specificity of a single antibody or lectin and determination of the dominant antigenic responses elicited by a natural infection or a vaccination would be practically achievable. Similarly, this technology can be extended to monitor autoantibodies and tumor-specific or -associated anticarbohydrate activities.

This novel approach can be applied to other biological systems in which a carbohydrate–carbohydrate or carbohydrate–protein interaction plays a significant role. For example, a sugar chip can be applied in screening for the carbohydrate-based cellular receptors of a microorganism. Experimentally, a candidate protein or the whole cell of a microbe can be placed on a sugar chip to probe the carbohydrate structures that a microbial pathogen may bind and selectively colonize in certain type of host cells or tissue environments. Such investigation is noteworthy, because it may lead to a better understanding of the host–microbe biological relationship as well as the pathogenesis mechanism of human pathogens. A reverse type of application is to print a large panel of structurally uncharacterized polysaccharides or glycoconjugates, such as those isolated from mixtures of natural herbs of traditional eastern medicines, to react with antibodies or lectins of known carbohydrate-binding specificities. In this way, the sugar chain epitope profile of these printed preparations can be rapidly recognized, providing important clues for drug discovery. In addition, such sugar chain epitope mapping strategy is technically straightforward and is suitable to serve as a scanning method to verify and control the quality of a complex formula of herbal medicine or nutrition additives, which contain significant quantities of glycan components or lectins with carbohydrate-binding reactivities.

Recently, we have focused on the establishment of a glycomics strategy to facilitate identification of the sugar moieties of biological significance. We proposed to take advantage of the highly evolved immune systems of mammals to recognize the immunogenic sugar moieties of microbes. Specifically, we use carbohydrate microarrays to capture specific antibodies elicited by a microorganism (see **Fig. 3** for a schematic illustration of this approach). To critically evaluate this strategy, we have chosen a previous unrecognized viral pathogen, severe acute respiratory syndrome-associated coronavirus (SARS-CoV) *(32–34)*, as a model for our investigation. This task was difficult and challenging, because information regarding the sugar moieties of this virus was entirely unavailable. This information is, however, very important for our consideration of vaccination strategy against SARS-CoV as well as investigation of pathogenic mechanisms of SARS. Therefore, we constructed glycan arrays to display carbohydrate antigens of defined structures and subsequently applied these tools to detect carbohydrate-specific antibody "fingerprints" that were elicited by a SARS vaccine. Our rational was that if SARS-CoV expressed antigenic carbohydrate structures, then immunizing animals by using the whole virus-based vaccines would have the possibility to elicit antibodies specific for these structures. In addition, if SARS-CoV displayed a carbohydrate structure that mimics host cellular glycans, then vaccinated animals may develop antibodies with autoimmune reactivity to their corresponding cellular glycans.

By characterizing the SARS-CoV neutralizing antibodies elicited by an inactivated SARS-CoV vaccine, we detected autoantibody reactivity specific for the carbohydrate moieties of an abundant human serum glycoprotein asialo-orosomucoid (ASOR) **(Fig. 3B)**.

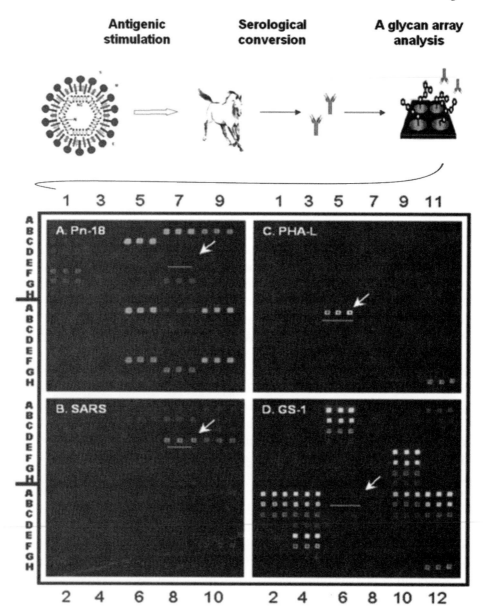

Fig. 3. Glycan arrays are used to characterize the antibody profiles of vaccinated animals (glycan array I) and to scan for ASOR-specific immunological probe (glycan array II). Antigen preparations spotted on each glycan array and their array locations are summarized in supplemental Tables S1 and S2 of **ref. 3** (available at the Physiological Genomics website at http://physiolgenomics.physiology.org/cgi/content/full/00102.2004/DC1).

For array I (**Fig. 3A,B**), a glycan array that contains 51 antigens (0.5 ng/microspot) was constructed and applied to scan horse anti-Pn 18 serum (**Fig. 3A**) as well as anti-SARS neutralizing antibodies (**Fig. 3B**).

For array II (**Fig. 3C,D**), a glycan array that displays 24 antigens, many of preparations of Gal-containing carbohydrate antigens, was stained by lectin PHA-L (**Fig. 3C**), which is specific for Galβ1,4-*N*-acetylglucosamine-linked units, and by lectin GS1-B4 (**Fig. 3D**), which is considered to be specific for Galα1-3Gal.

This surveillance provides important clues for the selection of specific immunological probes to further examine whether SARS-CoV expresses antigenic structures that imitate the host glycan. We found that lectin PHA-L (for *Phaseolus vulgaris* L.) is specific for a defined complex carbohydrate of ASOR (**Fig. 3C,D**). Using this reagent as a probe, we confirmed that only the SARS-CoV–infected cells express a PHA-L–reactive antigenic structure (*see* **ref.** *3* for the details). We obtained, therefore, immunologic evidences that a carbohydrate structure of SARS-CoV shares antigenic similarity with host glycan complex carbohydrates. This viral component is probably responsible for the stimulation of the autoantibodies directed at a cellular glycan complex carbohydrate.

These observations raise important questions about whether autoimmune responses are indeed elicited by SARS-CoV infection and whether such autoimmunogenicity contributes to SARS pathogenesis. ASOR is an abundant human serum glycoprotein and the ASOR-type complex carbohydrates are also expressed by other host glycoproteins *(35,36)*. Thus, the human immune system is generally nonresponsive to these self-carbohydrate structures. However, when similar sugar moieties were expressed by a viral glycoprotein, their cluster configuration could differ significantly from those displayed by a cellular glycan, and in this manner, generate a novel nonself antigenic structure. A documented example of such antigenic structure is a broad-range HIV-1 neutralization epitope recognized by a monoclonal antibody 2G12. This antibody is specific for a unique cluster of sugar chains displayed by the gp120 glycoprotein of HIV-1 *(37)*. It is, hence, important to examine whether naturally occurring SARS-CoV expresses the ASOR-type autoimmune reactive sugar moieties. During a SARS epidemic spread, the viruses replicate in human cells. Their sugar chain expression may differ from the monkey cell-produced viral particles. Scanning of the serum antibodies of SARS patients by using glycan arrays or other specific immunological tools may endow with information to shed light on this question.

In synopsis, recent establishment of carbohydrate-based microarrays, and especially the availability of different technological platforms to meet the multiple needs of carbohydrate research, marks an important developmental stage of postgenomic research *(1,20–24,38)*. Our laboratory has established a simple, precise, and highly efficient experimental approach for the construction of carbohydrate microarrays *(1–3)*. This approach makes use of existing cDNA microarray system, including spotter and scanner, for carbohydrate array production. A key technical element of this array platform is the introduction of nitrocellulose-coated microglass slides to immobilize unmodified carbohydrate antigens on the chip surface noncovalently. This technology has achieved the sensitivity to recognize the profiles of human anti-carbohydrate antibodies with as little as a few microliters of serum specimen and reached the chip capacity to include the antigenic preparations of most common pathogens (~20,000 microspots per biochip). We described also an eight-chamber subarray system to produce carbohydrate microarrays of relative smaller scale, which is more frequently applied in our laboratory's routine research activities. Of late, we applied this system to assemble glycan arrays to probe the immunogenic sugar moieties of a recently discovered viral pathogen, SARS-CoV. This research approach is probably applicable for the identification of immunological targets of other microorganisms and for the exploration of complex carbohydrates that are differentially expressed by host cells, including stem cells at various stages of differentiation and human cancers.

References

1. Wang, D., Liu, S., Trummer, B. J., Deng, C., and Wang, A. (2002) Carbohydrate microarrays for the recognition of cross-reactive molecular markers of microbes and host cells. *Nat. Biotechnol.* **20,** 275–281.
2. Wang, D. (2003) Carbohydrate microarrays. *Proteomics* **3,** 2167–2175.
3. Wang, D., and J. Lu. (2004) Glycan arrays lead to the discovery of autoimmunogenic activity of SARS-CoV. *Physiol. Genomics* 2004. **18,** 245–248.
4. Wang, D., and Kabat, E. A. (1996) Carbohydrate antigens (polysaccharides). In: *Structure of Antigens, Vol. 3* (M. H. V. Van Regenmortal, ed.), CRC Press, Boca Raton, FL, pp. 247–276.
5. Brooks, S. A., Dwek, M. V., and Schumacher, U. (2002) *Functional and Molecular Glycobiology.* BIOS Scientific Publishers Ltd., Oxford, United Kingdom.
6. Wang, D. (2004) Carbohydrate antigens. In: *Encyclopedia of Molecular Cell Biology and Molecular Medicine* Vol. 2, (R. A. Meyers, ed.), Wiley-VCH, pp. 277–301.
7. Feizi, T. (1982) The antigens Ii, SSEA-1 and ABH are in interrelated system of carbohydrate differentiation antigens expressed on glycosphingolipids and glycoproteins. *Adv. Exp. Med. Biol.* **152,** 167–177.
8. Hakomori, S. (1985) Aberrant glycosylation in cancer cell membranes as focused on glycolipids: overview and perspectives. *Cancer Res.* **45,** 2405–2414.
9. Focarelli, R., La Sala, G. B., Balasini, M., and Rosati, F. (2001) Carbohydrate-mediated sperm-egg interaction and species specificity: a clue from the *Unio elongatulus* model. *Cells Tissues Organs* **168,** 76–81.
10. Crocker, P. R., and Feizi, T. (1996) Carbohydrate recognition systems: functional triads in cell-cell interactions. *Curr. Opin. Struct. Biol.* **6,** 679–691.
11. Heidelberger, M., and Avery, O. T. (1923) The soluble specific substance of *Pneumococcus. J. Exp. Med.* **38,** 73–80.
12. Dochez, A. R., and Avery, O. T. (1917) The elaboration of specific soluble substance by pneumococcus during growth. *J. Exp. Med.* **26,** 477–493.
13. Ezzell, J. W., Jr., Abshire, T. G., Little, S. F., Lidgerding, B. C., and Brown, C. (1990) Identification of *Bacillus anthracis* by using monoclonal antibody to cell wall galactose-*N*-acetylglucosamine polysaccharide. *J. Clin. Microbiol.* **28,** 223–231.
14. Robbins, J. B., and Schneerson, R. (1990) Polysaccharide-protein conjugates: a new generation of vaccines. *J. Infect. Dis.* **161,** 821–832.
15. Mond, J. J., Lees, A., and Snapper, C. M. (1995) T cell-independent antigens type 2. *Annu. Rev. Immunol.* **13,** 655–692.
16. DeRisi, J., Penland, L., Brown P. O., et al. (1996) Use of a cDNA microarray to analyse gene expression patterns in human cancer. *Nat. Genet.* **14,** 457–460.
17. Brown, P. O., and Botstein, D (1999) Exploring the new world of the genome with DNA microarrays. *Nat. Genet.* **21(Suppl. 1),** 33–37.
18. MacBeath, G., and Schreiber, S. L. (2000) Printing proteins as microarrays for high-throughput function determination (see comments). *Science* **289,** 1760–763.
19. Stoll, D., Templin, M. F., Schrenk, M., Traub, P. C., Vohringer, C. F., and Joos, T. O. (2002) Protein microarray technology. *Front. Biosci.* **7,** C13–C32.
20. Willats, W. G., Rasmussen, S. E., Kristensen, T., Mikkelsen, J. D., and Knox, J. P. (2002) Sugar-coated microarrays: a novel slide surface for the high-throughput analysis of glycans. *Proteomics* **2,** 1666–1671.
21. Fazio, F., Bryan, M. C., Blixt, O., Paulson, J. C., and Wong, C. H. (2002) Synthesis of sugar arrays in microtiter plate. *J. Am. Chem. Soc.* **124,** 14,397–14,402.
22. Fukui, S., Feizi, T., Galustian, C., Lawson, A. M., and Chai, W. (2002) Oligosaccharide microarrays for high-throughput detection and specificity assignments of carbohydrate-protein interactions. *Nat. Biotechnol.* **20,** 1011–1017.

23. Houseman, B. T., and Mrksich, M. (2002) Carbohydrate arrays for the evaluation of protein binding and enzymatic modification. *Chem. Biol.* **9**, 443–454.
24. Park, S., and Shin, I. (2002) Fabrication of carbohydrate chips for studying protein-carbohydrate interactions. *Ang. Chem. Int. Ed. Engl.* **41**, 3180–3182.
25. Kabat, E. A., and Berg, D. (1953) Dextran–an antigen in man. *J. Immunol.* **70**, 514–532.
26. Cisar, J., Kabat, E. A., Dorner, M. M., Liao, J. (1975) Binding properties of immunoglobulin combining sites specific for terminal or nonterminal antigenic determinants in dextran. *J. Exp. Med.* **142**, 435–459.
27. Wang, D., Liao, J., Mitra, D., Akolkar, P. N., Gruezo, F., and Kabat, E. A. (1991) The repertoire of antibodies to a single antigenic determinant. *Mol. Immunol.* **28**, 1387–1397.
28. Wang, D., Chen, H. T., Liao, J., et al. (1990) Two families of monoclonal antibodies to $\alpha(1,6)$dextran, $V_H19.1.2$ and $V_H9.14.7$, show distinct patterns of J_k and J_H minigene usage and amino acid substitutions in CDR3. *J. Immunol.* **145**, 3002–3010.
29. Matsuda, T., and Kabat, E. A. (1989) Variable region cDNA sequences and antigen binding specificity of mouse monoclonal antibodies to isomaltosyl oligosaccharides coupled to proteins T-dependent analogues of $\alpha(1,6)$dextran. *J. Immunol.* **142**, 863–870.
30. Oehler, S., Alex, R., and Barker, A. (1999) Is nitrocellulose filter binding really a universal assay for protein-DNA interactions? *Anal. Biochem.* **268**, 330–336.
31. Van Oss, C. J., Good, R. J., and Chaudhury, M. K. (1987) Mechanism of DNA (Southern) and protein (Western) blotting on cellulose nitrate and other membranes. *J. Chromatogr.* **391**, 53–65.
32. Fouchier, R. A., Kuiken, T., Schutten, M., et al.(2003) Aetiology: Koch's postulates fulfilled for SARS virus. *Nature* **423**, 240.
33. Ksiazek, T. G., Erdman, D, Goldsmith, C. S., et al. (2003) A novel coronavirus associated with severe acute respiratory syndrome. *N. Engl. J. Med.* **348**, 1953–1966.
34. Rota, P. A., Oberste, M. S., Monroe, S. S., et al. (2003) Characterization of a novel coronavirus associated with severe acute respiratory syndrome. *Science* **300**, 1377–1380.
35. Cummings, R. D., and Kornfeld, S. (1984) The distribution of repeating (Gal beta 1,4GlcNAc beta 1,3) sequences in asparagine-linked oligosaccharides of the mouse lymphoma cell lines BW5147 and PHAR 2.1. *J. Biol. Chem.* **259**, 6253–6260.
36. Pacifico, F., Montuori, N., Mellone, S., et al. (2003) The RHL-1 subunit of the asialoglycoprotein receptor of thyroid cells: cellular localization and its role in thyroglobulin endocytosis. *Mol. Cell. Endocrinol.* **208**, 51–59.
37. Calarese, D. A., Scanlan, C. N., Zwick, M. B., et al. (2003) Antibody domain exchange is an immunological solution to carbohydrate cluster recognition. *Science* **300**, 2065–2071.
38. Adams, E. W., Ratner, D. M., Bokesch, H. R., McMahon, J. B., O'Keefe, B. R., and Seeberger, P. H. (2004) Oligosaccharide and glycoprotein microarrays as tools in HIV glycobiology; glycan-dependent gp120/protein interactions. *Chem. Biol.* **11**, 875–881.

PART IV

EMERGING TECHNOLOGIES IN DIAGNOSTICS

Krishnarao Appasani

How microarrays can be used in a robust way for blood diagnostics, prenatal genetic screening, and "lab-on-a-chip" methods is the subject of Part IV. Microplate hemagglutination is the most frequently used detection method for blood donation testing in clinical laboratories; however, it has lower throughput patient testing. To overcome this obstacle, Chapter 15, by Petrik and Robb, describes the successful adoption of a microarray platform to screen blood donations by identifying a set of blood group markers. In addition, their chapter summarizes the details of testing blood-borne pathogens by using rolling circle and polymerase chain amplification methods. In Chapter 16, Gambari et al. describe how they devised lab-on-a-chip to generate cellular arrays based on dielectrophoresis. Laboratory-on-chip technology could represent a very appealing approach, because it involves the miniaturization of several complex chemical and/or physical procedures in a single microchip-based platform, leading to the possibility of manipulating large numbers of single cells or cell populations. This new methodology may open novel opportunities in the field of pharmaceutical science and tumor immunology.

Rapid and highly sensitive diagnostic tests are a necessity for any pathological laboratory to treat and manage various forms of human diseases. Another important emerging scenario in the clinics is how efficiently the genetic disorders can be diagnosed. In the final chapter, Chandak and Hemavathi address an important problem of genetic disorders in developing countries (such as India): these disorders remain largely hidden owing to insufficient diagnostic laboratories. With the aid of prenatal test results and diagnoses, clinicians can provide genetic counseling (with the help of social psychologists), so that parents can make appropriate decisions. However, these tests prompt a host of social and ethical concerns arising from the culture, religion, and politics of any country.

15

Microarrays and Blood Diagnostics

Juraj Petrik and Janine Scott Robb

Summary

Blood donation testing consists of a series of distinct assays determining compatibility of blood and blood products between donor and recipient as well as detecting potential contamination by life-threatening blood-borne pathogens. Current testing algorithms in developed countries provide extremely safe blood supply, although they are rather complex. Microarray technology has the potential to simplify the testing of algorithms by providing a single multiplex testing platform.

Key Words: Blood testing; microarrays; protein arrays; blood grouping.

1. Introduction

Microarray technology has the potential to alter many diagnostic applications, especially high-throughput multiparameter assays. In this chapter, we discuss potential impact of microarrays on blood donation screening algorithms by providing a single testing platform for microbiology and blood group markers. An initial stage of such development is demonstrated on examples of successful blood grouping and antibody screen for some rare alloantibodies.

2. Current Blood Testing Methods

Blood tests can reveal a variety of changes taking place during disease processes as well as genotype and phenotype characteristics necessary for further medical interventions. There are a huge number of tests and instruments in use, reflecting the needs of specialized hospital laboratories as well as individual point of care devices. Microarrays could play an important role predominantly in high-throughput applications. In this chapter, we focus on the potential role of microarrays in future blood donation screening techniques.

2.1. Unique Features of Blood Donation Screening

Blood transfusion occupies rather an unusual place among medical procedures in the level of expected safety of both the procedures and the outcome. Risk assessments for very few other medical procedures would be comparable with figures for residual risk of transfusion-transmitted viruses, currently in the range of one in several million of donations (1). A combination of donor selection methods, testing procedures, and, in some cases, inactivation techniques guarantees an extremely safe blood supply (2).

From: *Bioarrays: From Basics to Diagnostics*
Edited by: K. Appasani © Humana Press Inc., Totowa, NJ

Blood donation screening is a unique process for several reasons:

1. The majority of donations produce a negative result for blood-borne pathogens because blood donors represent a self-selected healthy population.
2. Each sample positive for blood-borne pathogens must, however, be detected, because blood, blood products, and components are destined for recipients. The sensitivity of the screening test needs to be extremely high, even at the expense of slightly lower specificity. Repeat testing is a routine component of testing algorithms.
3. For the same reasons, clinically relevant blood groups and the presence of blood group alloantibodies have to be reliably determined.
4. Screening procedures have to be able to satisfy throughput needs between several hundreds and a few thousand donations per day (for an average size blood center);
5. Turnaround time has to be short owing to the shelf life of some of the components (e.g., 5-day shelf life for platelets) and flexibility of stock management.

2.2. Testing Targets

Each donation has to undergo testing for a set of markers for selected blood-borne pathogens and to determine blood group compatibility. **Table 1** lists the testing targets and parameters of currently used testing procedures.

2.2.1. Testing for Blood-Borne Pathogens

Highly sensitive immunoassays form the basis for the majority of microbiology assays. They usually represent various modifications of sandwich ELISAs. Ideally, the screening would be based on direct antigen detection, confirming the presence of an infectious agent. Often, however, the antigen concentration or titer of the infectious agent in circulating blood is too low to be detected in this way. One exception is hepatitis B surface antigen (HBsAg) produced at quantities sufficient for robust antigen detection. The best HBsAg assays can detect approx 1 pg/mL, which is approx 8.35×10^6 molecules for a 24-kDa protein.

Generally, we are measuring antibodies developed as a response to infection. The appearance of measurable antibodies may take a significant amount of time creating a potentially long window period. For immunoassays, the antibodies (polyclonal or monoclonal) or antigens (recombinant proteins and peptides) are attached to a solid phase, most frequently represented by the surface of a bead (microsphere) or a microplate well. One- or two-step sandwich detection is usually used, with conjugated enzymes converting substrate into color or (chemi)luminescent signal.

Owing to the sensitivity issues with many antigen-detecting immunoassays, much hope was placed in the application of nucleic acid amplification technologies (NATs), most frequently represented by PCR. This technology indeed has an exquisite sensitivity (**Table 1**) and detects the component of the infectious agent rather than antibodies. The period for viruses for which PCR was introduced as a routine screening method has been shortened significantly, and residual risk of transfusion-transmitted events are extremely low. It would seem to be an ideal screening procedure. However, the cost efficiency of this type of testing is being increasingly questioned, especially because the number of NAT-only positive donations is lower than originally expected (*see* **Subheading 2.3.**).

2.2.2. Blood Typing and Detection of Alloantibodies

The traditional method of monitoring blood group serology reactions is hemagglutination. It is an inexpensive technique with easily identifiable end point detection. There

Table 1
Blood Screening Assays: Examples of Assays Used in SNBTS

Detected target	Detection method	Instrumental platform	Signal readout	Detection limit	Throughput (No of samples)
Pathogen Testing					
HBsAg	Sandwich immunoassay	Abbott Prism	Chemiluminiscence	1-10 pg/mL [a]	200/h
Anti-HCV	Sandwich immunoassay	Abbott Prism	Chemiluminiscence	NA [b]	200/h
Anti-HIV1,2	Sandwich immunoassay	Abbott Prism	Chemiluminiscence	NA [b]	200/h
Anti-treponema	Heamagglutination	Olympus	CCD	NA [b]	240/h
HCV RNA	Real time PCR	In-house	Fluorescence	29 geq/mL [c]	30 –50/6 h [d]
HIV1 RNA	Real time PCR	In-house	Fluorescence	34 geq/mL [c]	30 – 50/6 h [d]
Blood Group Serology					
ABO antigens	Hemagglutination	Olympus	CCD	1,400 Ag sites/RBC [e]	240/h
RhD antigen	Hamagglutination	Olympus	CCD	200 Ag sites/RBC [e]	240/h
Antibody screen	Hemagglutination	Olympus	CCD	0.5 IU/mL Anti-D	240/h

[a] Estimate.
[b] Not applicable.
[c] geq: genome equivalent.
[d] Minipools of 95 samples.
[e] Minimum level of antigen on certain phenotypes, may be below detectable limits of the test system.
CCD, charge-coupled device; IU, Iuternational unit; SNTBS, Scottish National Blood Transfusion Service.

are two basic types of assays, both included in routine testing procedures. In "forward blood typing," the red blood cell (RBC) antigens of clinically relevant blood group systems are determined using specific polyclonal (minority) or monoclonal antibodies. IgM monoclonal antibodies are preferred to IgGs, because they cause hemagglutination in one step owing to their structure, providing 10 antigen binding sites per molecule compared with two per IgG molecule. Microplate assay format is most frequently used for donation blood typing (**Table 1**). "Reverse typing" is designed to detect antibodies against blood group antigens. The reason for development of alloantibodies (or isoantibodies) is usually abnormal pregnancy, blood transfusion, or some immunizations. Clinically relevant antibodies need to be detected in donors, and especially in recipients, because they could cause life-threatening reactions. The microplate hemagglutination is the most frequently used detection method for donation testing, whereas lower-throughput patient testing is most often done using gel cards.

2.3. Advantages and Disadvantages of Current Testing Methods

As mentioned in **Subheading 2.2.1.**, current preventative screening and testing procedures used in developed countries provide an extremely safe blood supply. Introduction of a conceptually different method in the form of PCR into a routine testing algorithm for certain blood-borne pathogens ended an era completely dominated by immunoassays. It makes it easier for blood centers to accommodate improved and new technologies as they emerge.

PCR would be a very attractive general testing platform owing to its sensitivity and precision. PCR and other target amplification techniques have been used not only to detect pathogens but also for determination of RBC and platelet antigens (*3–6*). Unfortunately, there is a high cost attached to PCR screening, both direct (e.g., contamination prevention measures, reagents, instrumentation, and specialized staff) and indirect (e.g., significant license fees). The only viable way to achieve satisfactory cost efficiency would be extensive multiplexing. Although some multiplex assays were successfully developed, the number of targets is generally low. This low number is mostly because of interference of primers and probes, which is difficult to predict even with the help of oligonucleotide design software. Future improvements may, however increase the multiplexing potential of target amplification methods. At the same time, there is a continuous improvement in the performance of immunoassays, and some antigen or combi (antigen and antibody) assays are getting closer to NAT assays in respect of closing the window period (*7*). It does not seem probable at present that one type of assay (e.g., nucleic acid based or immunoassays) would be able to replace the other type entirely. It remains to be seen how successfully they can be codeveloped in the future.

We have an extremely safe but also complex testing algorithm in place. Microarray technology could address this complexity by providing a single donation testing instrumental platform for a complete set of required markers (*8*; *see* **Subheading 3.**).

3. Microarrays as the Potential Next Generation Testing Platform

First experiments to further miniaturise the existing immunoassays date back to late 1980s and early 1990s (*9*). In parallel, several groups were developing techniques for simultaneous monitoring of large groups of genes. These techniques required high probe density facilitated by the use of new, nonporous materials such as glass, silicon or plastic, allowing for the reduction in the quantities of deposited probes and reaction volumes. Robotic spotting techniques *(10,11)* and *in situ* oligonucleotide synthesis *(12)* speeded up gene expression studies with information content provided by large sequencing projects. Significant efforts were dedicated to developing new surface chemistries as well as alternative microarray formats, including encoded beads and exploiting microfluidics and additional features such as electronic probe and sample addressing *(13–15)*. Although driven mainly by genomic research, the potential of microarrays for diagnostics has been quickly identified. Indeed, diagnostics may well become the main microarray application in the future.

It soon became clear that some of the techniques used for preparation of DNA arrays could be adapted to protein microarrays *(16,17)* despite the more complex character of protein interactions compared with rules for complementary nucleic acid strands reannealing. Alternative systems describing the use of polyacrylamide patches and subarrays within microplate wells also were reported *(18,19)*. For mass screening applications such as blood donation testing, the cost efficiency is one of the main parameters to consider, together with high sensitivity and specificity, robustness, and a high level of automation, limiting or eliminating the operator-introduced errors. This need will inevitably affect the complexity of future microarray-based testing platforms. An open plan blood typing microarray, with no physical barriers between the probes, has the potential to develop into a highly cost-effective, high-throughput system for blood screening if it can sustain the specificity of reactions and achieve the required sensitivity.

Most protein arrays described to date investigate the interaction of free protein molecules with the defined partner (probe) immobilized on the solid surface. Probes are represented predominantly by antibodies *(18–21)*, in some cases by antigens *(22,23)*. Blood group serology adds another dimension to these interactions, because the interaction involves blood group antigens on the surface of RBCs. Antibody–cell interactions on planar microarrays have been described for applications such as leucocyte typing *(24)* but still are rather rare. Quinn et al. *(25)* described solid phase blood grouping by using the Biacore platform, but this is an analytical rather than high-throughput application platform.

3.1. Potential Benefits of Microarray Testing

In the currently used testing algorithms, several aliquots of donated blood are taken to be used on between three and six instrumental platforms to test for a complete set of required markers (*see* **Subheadings 2.2.1.** and **2.2.2.**). Time required for testing on different instruments varies and sets of results become available at different times, the slowest holding up the release of blood and blood components, which cannot be issued without final data reconciliation. A proportion of samples will undergo repeat testing, because the sensitive tests necessary for initial screening produce some false-positive samples in addition to true positives. **Table 2** provides a comparison of current algorithms with potential changes, which might be possible, for future microarray-based testing platform. The most significant change would be transition from testing many samples in parallel for one or few markers on multiple instrumental platforms, to simultaneous testing of each sample for the required set of markers on one or two chips. Data reconciliation would be simplified by data provided simultaneously. One platform

Table 2
Comparison of Current Testing Algorithm
With Proposed Microarray Based Testing Algorithm

	Testing algorithms	
Parameter	Current	Microarray-Based
No. of instrumental platforms	3–6	1
No. of markers per assay	1–4	n × 10
No. of probes per target in screening assay	up to 5	Multiple
Individual detection of probes	No	Yes
Need for repeat testing	Essential	Non-essential
Result acquisition	Sequential	Simultaneous
IT result reconciliation before products/components issue	Essential	Not necessary
Reagent (probe) consumption	n × uL	n × pl - nL

would not require multiple sets of reagents for various instruments or staff dedicated to individual platforms, as is the case at present. Inclusion of multiple probes per target is easier in microarrays because the incremental cost of additional probes is low. This platform could perhaps eliminate the need for repeat testing, because the reactivity pattern on multiple probes should provide a clear answer on sample reactivity.

3.2. Options to Increase Microarray Sensitivity

Sensitivity is probably the major problem for wider use of microarrays for some diagnostic purposes. This problem is often being addressed by combining target amplification (e.g., PCR) by using degenerate primers or nonspecific amplification, with subsequent microarray analysis *(26–28)*. Such combination would add another extra step to existing blood donation testing algorithms, increasing the complexity of testing as well as costs. Fortunately, there are multiple alternative approaches currently being developed to increase the detection limits.

Ideally, the screening assays would approach the PCR sensitivity at lower cost, effectively through the higher multiplexing power, and perhaps the true multianalyte character of the procedures. Current, PCR techniques allow detection of less than 10 genome equivalents per reaction, amplifying the target approximately a billion times. The best ELISA-based assays can detect 10^6–10^7 molecules per milliliter. How can this gap of approx 6 orders of magnitude be closed in the absence of target amplification methods for proteins? Certain advantage is offered by the presence of hundreds to thousands of copies of the viral antigens compared with one or two copies of genomic RNA or DNA per virion. In addition, there are some alternative methods of signal enhancement as listed in **Table 3**. Signal amplification methods seem to provide the biggest gain in sensitivity in comparison with usual methods using few fluorophore molecules conjugated to primary or secondary reagents in sandwich assays. To apply any amplification method simultaneously to all sites on a planar microarray, the amplified signal needs to be localised, not diffuse. One primer variant of rolling circle amplification (RCA; *29*) produces a long product physically attached to the site and can provide signal amplification exceeding 8×10^3 *(30)*. Another amplification method is tyramide signal amplification (TSA) based on the catalytic activity of horseradish peroxidase generating high-density labeling of a target claiming subpicogram per milliliter sensitivity. Elimination of the need for physical separation of individual sites necessary for target amplification methods makes these techniques less expensive. Other advantages include use of RCA and TSA for both DNA and protein arrays *(31,32)* and isothermal reaction.

A smaller, but significant increase in sensitivity is provided by new detection techniques. Resonance light scattering uses a white-light source scanner to detect monochromatic, scattered light signals of metal (e.g., gold, silver) particles. The material, size, and shape of the particles determine the signal. When resonance light scattering was directly compared with fluorescent detection of bacterial pathogens, the authors observed even 50 times more intense signal *(33)*.

Planar waveguide technology developed by Zeptosens uses a special coating of T_2O_5 or TiO_2 to induce evanescence field for efficient fluorescence detection *(34)*, at least 10 times more sensitive than the conventional fluorescence if using the company's chips and a reader. New materials such as quantum dots (or semiconductor nanoparticles) provide brighter fluorescence, narrower emission spectra, and higher resistance to photobleaching compared with the conventional small molecule dyes. Signal also can be increased, however, by using small molecule dyes enclosing thousands of dye molecules within a derivatized nanoparticle and enhancing the signal many times *(35)*.

Instead of using a linear probe, usually labeled with a single fluorophore, dendrimers or multiply branched molecules provide the opportunity to introduce many dye molecules and to increase generated signal. When used for DNA microarray human herpes virus diagnosis, the signal has been enhanced at least 30 times *(36)*. Some of the described approaches can be further combined, leading to very sensitive assays, potentially approaching the sensitivity of nucleic acid amplification techniques. Combined with easier multiplexing such assays could be well suited for high-throughput blood screening of the future.

4. Development of Microarray Blood Testing Format

As outlined in **Table 2**, a microarray-based blood screening platform could provide advantages over existing testing algorithms. An ultimate testing platform would combine pathogen testing with blood group serology. As the first step in this development, we have investigated the applicability of blood grouping in a microarray format.

4.1. Blood Group Serology

The majority of blood serological methods are based on the ability of erythrocytes to agglutinate. Reactions in the liquid phase using tubes were replaced by slide and tile reactions and later by microplate format. Solid phase assays *(37,38)* were more suitable for predispensed, dried reagents and automation, including operator-independent readout. An alternative solid phase assay is the gel card system containing microtubes with antibodies suspended within a gel or glass microbeads *(39,40)*.

Table 3
Methods, Techniques and Materials Increasing the Microarray Sensitivity

Signal amplification	Label/signal readout	Gain[a]	Reference/website
RCA	Fluorescence/various	10^3–10^4	*29,30*; www1.qiagen.com/molecularstaging.aspx
TSA	Fluorescence/various	n × 100–1000	*31*
Quantum dots	Semiconductor nanoparticles/ fluorescence	n × 10	www.qdots.com
Planar waveguide	Fluorescence	10	*34*; www.zeptosens.com
Dye-doped nanoparticles	Fluorescence	n × 10	*35*
Resonance light scattering	Metal nanoparticles	50	*33*; www.geniconsciences.com
Dendrimers	Fluorescence/various	30	*36*; www.genisphere.com

* In relation to confocal scanning of conventional small molecule dye-labeled reagents.

Agglutination techniques on solid phase have been exploited with lesser success in the reverse format to test for clinically relevant alloantibodies against blood group antigens. It proved more difficult to reproducibly immobilize and store erythrocytes or erythrocyte ghosts than immobilized antibodies *(41,42)*, and despite various systems being developed, gel cards are still the most widely used format.

The experiments described below aimed at developing an array platform for the determination of the antigens of main blood group systems by using a set of well-defined proprietary antibodies routinely used in current agglutination-based assays.

4.1.1. Blood Grouping

Microarray blood grouping could replace an agglutination reaction with fluorescent or other quantifiable readout signals. Affinity of immobilized antibodies needs to be sufficiently high to keep specifically bound RBCs attached during the incubation and washing steps. In addition, the reaction conditions must guarantee preserving the integrity of the RBCs necessary for binding of the labeled RBCs or fluorophore-conjugated secondary reagent. We have investigated multiple parameters of these interactions on a variety of slide surfaces under various printing regimes, reaction conditions, and so on. An extensive set of well functionally characterized proprietary antibodies, used in a variety of current blood grouping assays, has been exploited in these studies *(43,44)*. We have shown that a reproducible ABO grouping can be achieved and successful typing can be extended to Rhesus (D, C, c, E, e), Kell, and other clinically relevant blood group systems. **Table 4** lists the most clinically relevant of the 25 blood group systems and shows the basis of their characteristics. Even a selection of most important of blood group systems reveals the structural variability of blood group antigens as well as large differences in their abundance. When adding the functional heterogeneity and various modes of attachment to RBC membrane affecting the access to antigens, the character of various interactions involving antigens of blood group systems could be studied by microarrays, apart from purely diagnostic application.

Figure 1 depicts a typical blood grouping experiment conducted on gold slides. Monoclonal antibodies specific for A, B, A(B), Rhesus D, and K blood group antigens were immobilized on gold slides, and their reactivity was investigated separately with RBCs of different phenotypes of the aforementioned blood group antigens. In **Fig. 1A**, RBC carrying blood group A antigen bound to both anti-A and anti-A(B) antibodies and RBC carrying both, blood group A and B antigens bound in addition to anti-B antibody, as expected. RBC carrying blood group antigen B bound to anti-B antibody. They could be expected to react as well with anti-A(B) antibody, but reaction with this unique, single available antibody of this specificity is known to be weak, even in normal hemagglutination assays *(48)*. RBCs of blood group O specificity do not react with any of the antibodies, because they do not carry A or B antigens and serve at the same time as a negative control. Various levels of reactivity can be seen in **Fig. 1B** with RBCs carrying two (DD) one (Dd) or no (dd) copy of the RhD antigen. Again, RBCs of dd phenotype serve at the same time as a negative control in this assay. **Figure 1C** shows a dose-dependent signal produced by binding homozygous (KK) and heterozygous (Kk) cells to the immobilized K specific antibody. All other four RBC preparations are K negative (kk) and show no or minimal reactivity. We were assaying two different types of blood group antigens in this experiment: carbohydrate A and B blood group antigens and protein Rhesus D and K antigens **(Table 4)**.

Table 4
Characteristics of Most Important Blood Group Antigens (Compiled From 45–47)

Antigen	Blood group system	Antigen	Determinants	Size	No. of antigenic sites/erythrocyte	Clinical significance[a]
H (groupO)	ABO	A, B Precursor carried on glyco-shingolipid or glycoprotein	Fucose determinant	Variable(branched); up to 60 carbohydrate residues	1.5–2 million	H
A	ABO	Sugar attached to H precursor oligosaccharide chain	N-Acetylgalactosamine determinant (GalNAc)	Variable(branched); up to 60 carbohydrateresidues	A1: 8.1–1.17×10^5 A2: 1.6–4.4×10^5	H
B	ABO	Sugar attached to H precursor oligosaccharide chain	Galactose determinant (Gal)	Variable(branched); up to 60 carbohydrate residues	6.1–8.3×10^5	H
D	Rhesus[b]	Multi-pass membrane protein Protein/lipid aggregates	Eight substitutions on extracellular loops	30–32 k Da 417 Aa	Common: 1–3×10^3 Weak: 2×10^2–10^4 High: $7.5 \times 10^4 - 2 \times 10^5$	H
C, c	Rhesus[b]	Multi-pass membrane protein Protein/lipid aggregates.	Differences at positions 60, 68, 103,16 of RhCE polypeptide	30–32 kDa 417 aa	$\sim4 \times 10^4$	M to H
E, e	Rhesus[b]	Multi-pass membrane protein Protein/lipid aggregates.	Difference at pos. 226 of RhCE polypeptide	30–32 kDa 417 aa	$\sim2 \times 10^4$	M to H
Cw	Rhesus[b]	Multi-pass membrane protein Protein/lipid aggregates.	Difference at pos. 41 of RhCE polypeptide	30–32 kDa 417 aa	$2.15 - 4 \times 10^4$	L to M
K, k	Kell	Single-pass membrane glycoprotein	Difference at pos. 193	93 kDa 732 aa	$3.5 \times 10^3 - 1.8 \times 10^4$	M to H
Fya, Fyb	Duffy	Multi-pass membrane glycoprotein	Difference at pos. 42	35–43 kDa 336 aa	1.35×10^4	M to H (Fya) L (Fyb)
Jka, Jkb	Kidd	Multi-pass membrane protein	Difference at pos. 280	45 kDa 391 aa	1.4×10^4	M to H
S, s	MNS	Single pass membrane protein GPB	Difference at pos. 29	24 kDa 72 aa	2×10^5	M
M, N	MNS	GPA	Difference at pos. 1, 5 (O-glycans at pos. 2, 3 &4)	37 kDa 131 aa	8×10^5	L
P1	P	Globotetraosylceramide	Terminal-Galactose	Variable	5×10^5	L
Le^a, Le^b	Lewis	40-100 carbohydrate chains per 300 kDa carrier molecule	Fucose (Le^a) or 2 Fucose (Le^b) to precursor	300 kDa incl.carrier (average);heterodisperse	Variable (attached by passive adsorption)	L

[a] H, high; M, moderate; L low .
[b] RhCE and RhD proteins are 92% identical with only 35 amino acid substitutions variance.

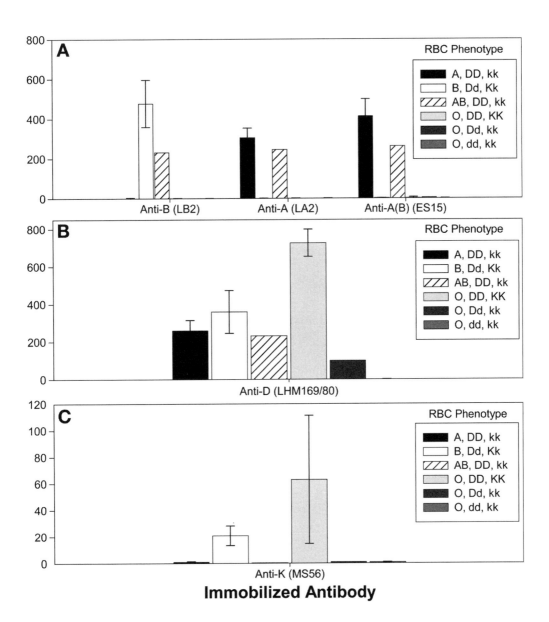

Fig. 1. Microarray red cell grouping. Monoclonal antibodies specific for blood group antigens A and B **(A)**, Rhesus D **(B)**, or K antigen **(C)** were printed using MicroGrid II Arrayer (BioRobotics) on gold slides (Erie Scientific), blocked, and incubated for 1 h at room temperature with 1% suspension of fluorescently labeled erythrocytes of six different phenotypes: ABO, cells carrying A, B, both (AB), or no (O) antigen. DD cells carry two copies of D antigen, Dd one and dd none. Similarly, KK cells carry two copies of K antigen, Kk one copy, and kk no copies. Slides were washed twice with phosphate-buffered saline briefly spun and scanned on Packard Bioscience ScanArray 5000 by using QuantArray® Microarray Analysis Software (GSI Lumonics). Reactivity is expressed as S/N, with signal from phosphate-buffered saline spots providing noise values. Values represent median of at least three replicates.

4.1.2. Antibody Screen

Another part of blood group serology is the determination of alloantibodies against blood group antigens, which may cause significant, sometime life-threatening situations if undetected in donors and especially recipients. These assays are more difficult to adapt to a solid phase, because they require immobilization of red cells, although alternative systems can be developed.

During the initial stage of microarray antibody screen development, we immobilized RBCs of various specificities on planar microarray slides modified with different surface treatments. **Figure 2A** shows that signals produced by specific binding of monoclonal anti-A and anti-B antibodies to gold slides-immobilized RBCs carrying corresponding antigens are several times higher than those caused by nonspecific binding, crossreactivity, or both and can be clearly distinguished. **Figure 2** shows an initial series experiment under conditions not fully optimized yet. The reaction parameters are being continuously improved to reduce nonspecific signal. However, the background will always be higher with immobilized whole cells than with isolated antigens. It can be documented by using the synthetic blood group B antigen (last column), producing lower signal-to-noise ratio (S/N) for specific (anti-B) reaction but no crossreactivity or nonspecific signal. Much lower S/N values were obtained for anti-D monoclonal antibody binding to immobilised RhD+ RBCs **(Fig. 2B)**. Again, the signals on RhD– cells are detectable but significantly lower than on RhD+ cells. As in **Fig. 1**, in this case we were also measuring the reactivity of two compositionally and structurally different blood group antigens. Carbohydrate A and B antigens protrude from the surface of RBCs and are easily accessible, whereas access to protein Rhesus D antigen with epitopes close to the membrane surface is more difficult. As in other types of blood group assay, this accessibility could explain the differences in the obtained S/N values. Another important factor is the number of antigenic sites per RBC **(Table 4)**.

Fig. 2 In addition to higher background on immobilized whole cells, the stability of such reagents is a cause for concern. As an alternative, recombinant blood group antigens could be used, and this approach is a subject of intense development. However, some of the antigens are multipass transmembrane proteins, making it difficult to adopt proper configuration preserving the conformational epitopes. Linear epitopes, in contrast, can often be mimicked by synthetic peptides. Immobilization of these probes instead of red cells provides an alternative way for antibody screening in the future.

5. Future Developments

A complete set of currently used reagent panels for blood grouping and antibody screening needs to be evaluated on real samples to confirm applicability of a planar microarray to this type of assay. We are working on adapting mandatory blood-borne pathogen tests to microarray platform. Although blood group serology usually does not suffer from sensitivity problem, it can be an issue with some antigens present in low numbers on the surface of RBCs. It is a much more pronounced problem for pathogen detection, and we envisage exploiting signal-enhancing methods to overcome insufficient sensitivity in some cases. Signal amplification methods could be used for both weak blood grouping and pathogen detection. In addition, these techniques could be applied to both DNA and protein arrays, should both types be necessary for complete

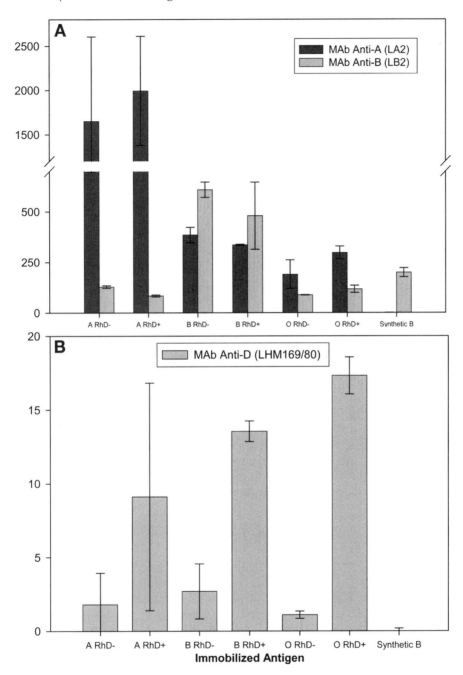

Fig. 2. Microarray blood group antibody detection. Group A, B, or O red blood cells, positive (RhD+) or negative (RhD–) for Rhesus D antigen were spotted on gold slides (Erie Scientific), alongside the synthetic blood group B antigen (Dextra Laboratories) by using manual spotter (V&P Scientific) with solid pins. Slides were incubated for 1 h at room temperature with monoclonal antibodies for blood group antigens A and B **(A)** or Rhesus D antigen **(B)**. After repeated washing, the slides were incubated with fluorescein isothiocyanate anti-mouse IgM (Sigma) at 33.3 μg/mL **(A)** or Cy3-conjugated anti human IgG (Sigma) at same concentration **(B)**, repeatedly washed , briefly spun, and scanned using Axon 4100A scanner and Axon GenePix Pro 4.1. S/N values are as in **Fig. 1**.

blood donation screening. Although we have focused on protein microarrays in this chapter, there is an extensive development work going on for blood group genotyping. It is very possible that no single type of assay, DNA based or protein based, will be able to completely replace the other type, owing to some genotype–phenotype discrepancies as well as existence of certain protein-only agents such as prions. Microarray technology has, however, potential to accommodate all required assay formats on one testing platform.

Acknowledgments

We thank Prof. Peter Ghazal, Dr. Colin Campbell, and Alan Ross (Centre for Genomic Technology and Informatics) for printing slides for and help with the experiment described in **Fig. 1**.

References

1. Soldan, K., Barbara, J. A. J., Ramsay, M. and Hall, A. J. (2003) Estimation of the risk of hepatitis B virus, hepatitis C virus and human immunodeficiency virus infectious donations entering the blood supply in England, 1993–2001. *Vox Sang.* **84,** 274–286.
2. Barbara, J. A. (1998) Prevention of infections transmissible by blood derivatives. *Transfus. Sci.* **19,** 3–7.
3. Bugert, P., McBride, S., Smith, G., et al. (2005) Microarray-based genotyping for blood groups: comparison of gene array and 5'-nuclease assay techniques with human platelet antigen as a model. *Transfusion* **45,** 654–659.
4. Denomme, G. A., and Van Oene, M. (2005) High-throughput multiplex single-nucleotide polymorphism analysis for red cell and platelet antigen genotypes. *Transfusion* **45,** 660–666.
5. Beiboer, S. H. W., Wieringa-Jelsma, T., Maaskant-Van Wijk, P. A., et al. (2005) Rapid genotyping of blood group antigens by multiplex polymerase chain reaction and DNA microarray hybridisation. *Transfusion* **45,** 667–679.
6. Hashmi, G., Shariff, T., Seul, M., et al. (2005) A flexible array format for large-scale, rapid blood group DNA typing. *Transfusion* **45,** 680–688.
7. Weber, B., Gurtler, L., Thorstensson, R., et al. (2002) Multicenter evaluation of a new automated fourth-generation human immunodeficiency virus screening assay with a sensitive antigen detection module and high specificity. *J. Clin. Microbiol.* **40,** 1938–1946.
8. Petrik, J. (2001) Microarray technology: the future of blood testing? *Vox Sang.* **80,** 1–11.
9. Ekins, R. P., and Chu, F. W. (1991) Multianalyte microspot immunoassay—microanalytical "compact disk" of the future. *Clin. Chem.* **37,** 1955–1967.
10. Schena, M., Shalon, D., Davis, R. W., and Brown, P. O. (1995) Quantitative monitoring of gene expression patterns with a complementary DNA microarray. *Science* **270,** 467–470.
11. Shalon, D., Smith, S .J., and Brown, P. O. (1996) A DNA microarray system for analyzing complex DNA samples using two-color fluorescent probe hybridization. *Genome Methods* **6,** 301–306.
12. Fodor, S. P. A., Read, J. L., Pirrung, M. C., Stryer, L., Lu, A. T. and Solas, D. (1991) Light-directed, spatially addressable parallel chemical synthesis. *Science* **251,** 767–773.
13. Michael, K. L., Taylor, L. C., Schultz, S. L., and Walt, D. R. (1998) Randomly ordered addressable high-density optical sensor arrays. *Anal. Chem.* **70,** 1242–1248.
14. Cheng, J., Sheldon, E. L., Wu, L., et al. (1998) Preparation and hybridization analysis of DNA/RNA from *E. coli* on microfabricated bioelectronic chips. *Nat. Biotechnol.* **16,** 541–546.

15. Vignali, D. A. (2000) Multiplexed particle-based flow cytometric assays. *J. Immunol. Methods* **243,** 243–255.

16. MacBeath, G., and Schrieber, S. L. (2000) Printing proteins as microarrays for high-throughput function determination. *Science* **289,** 1760–1763.

17. Haab, B. B., Dunham, M. J. and Brown, P. O. (2001) Protein microarrays for highly parallel detection and quantitation of specific proteins and antibodies in complex solutions. *Genome Biol.* **2,** Research0004.

18. Mendoza, L. G., McQuary, P., Mongan, A., Gangadharan, R., Brignac, S., and Eggers, M. (1999) High-throughput microarray-based enzyme-linked immunosorbent assay (ELISA). *Biotechniques* **27,** 782–786.

19. Arenkov, P., Kukhtin, A., Gemmell, A., Voloshchuk, S., Chupeeva, V., and Mirzabekov, A. (2000) Protein microchips: use for immunoassay and enzymatic reactions. *Anal. Biochem.* **278,** 123–131.

20. Borrebaeck, C. A, Ekstrom, S., Hager, A. C., Nilsson, J., Laurell, T., and Marko-Varga, G. (2001) Protein chips based on recombinant antibody fragments: a highly sensitive approach as detected by mass spectrometry. *BioTechniques* **30,** 1126–1132.

21. de Wildt, R. M. T., Mundy, C. R., Gorick, B. D., and Tomlinson, I. M. (2000) Antibody arrays for high-throughput screening of antigen-antibody interactions. *Nat. Biotechnol.* **18,** 989–994.

22. Bussow, K., Cahill, D., Nietfeld, W., et al. (1998) A method for global protein expression and antibody screening on high-density filters of an arrayed cDNA library. *Nucleic Acids Res.* **26,** 5007–5008.

23. Bacarese-Hamilton, T., Messazoma, L., Ardizzoni, A., Bistoni, F., and Crisanti, A. (2004) Serodiagnosis of infectious diseases with antigen microarrays. *J. Appl. Microbiol.* **96,** 10–17.

24. Belov, L., de la Vega, O., dos Remedios, C. G., Mulligan, S. P., and Christopherson, R. I. (2001) Immunophenotyping of leukemias using a cluster of differentiation antibody microarray. *Cancer Res.* **61,** 4483–4489.

25. Quinn, J. G., O'Kennedy, R., Smyth, M., Moulds, J., and Frame, T. (1997) Detection of blood group antigens utilising immobilised antibodies and surface plasmon resonance. *J. Immunol. Methods* **206,** 87–96.

26. Lin, B., Vora, G. J., Thach, D., et al. (2004) Use of oligonucleotide microarrays for rapid detection and serotyping of acute respiratory disease-associated adenoviruses. *J. Clin. Microbiol.* **42,** 3232–3239.

27. Foldes-Papp, Z., Egerer, R., Birch-Hirschfeld, E., et al. (2004) Detection of multiple human herpes viruses by DNA microarray technology. *Mol. Diagn.* **8,** 1–9.

28. Klaassen, C. H. W., Prinsen, C. F. M., de Valk, H. A., Horrevorts, A. M., Jeunink, M. A. F., and Thunnissen, F.B.J.M. (2004) DNA microarray format for detection and subtyping of human Papillomavirus. *J. Clin. Microbiol.* **42,** 2152–2160.

29. Schweitzer, B., Wiltshire, S., Lambert, J., et al. (2000) Immunoassays with rolling circle DNA amplification: a versatile platform for ultrasensitive antigen detection. *Proc. Natl. Acad. Sci. USA* **97,** 10,113–10,119.

30. Nallur, G., Luo, C., Fang, L., et al. (2001) Signal amplification by rolling circle amplification on DNA microarrays. *Nucleic Acids Res.* **29,** e118.

31. Karsten, S. L., Van Deerlin, V. M. D., Sabatti, C., Gill, L. H., and Geschwind, D. H. (2002) An evaluation of tyramide signal amplification and archived fixed and frozen tissue in microarray gene expression analysis. *Nucleic Acids Res.* **30,** e4.

32. Varnum, S. M., Woodburry, R. L., and Zangar, R. C. (2004) A protein microarray ELISA for screening biological fluids. *Methods Mol. Biol.* **264,** 161–172.

33. Francois, P., Bento, M., Vaudaux, P., and Schrenzel, J. (2003) Comparison of fluorescence and resonance light scattering for highly sensitive microarray detection of bacterial pathogens. *J. Microbiol. Methods* **55,** 755–762.

34. Pawlak, M., Schick, E., Bopp, M. A., Schneider, M. J., Oroszlan, P., and Ehrat, M. (2002) Zeptosens' protein microarrays: a novel high performance microarray platform for low abundance protein analysis. *Proteomics* **2,** 383–393.

35. Lian, W., Litherland, S. A., Badrane, H., et al. (2004) Ultrasensitive detection of biomolecules with fluorescent dye-doped nanoparticles. *Anal. Biochem.* **334,** 135–144.

36. Striebel, H. M., Birch-Hirschfeld, E., Egerer, R., Foldes-Papp, Z., Tilz, G. P., and Stelzner, A. (2004) Enhancing sensitivity of human herpes virus diagnosis with DNA microarrays using dendrimers. *Exp. Mol. Pathol.* **77,** 89–97.

37. Sinor, L. T., Rachel, J. M., Beck, M. L., Bayer, W. L., Coenen, W. M., and Plapp, F. V. (1985) Solid phase ABO grouping and Rh typing. *Transfusion* **25,** 21–23.

38. Beck, M. L., Rachel, J. M., Sinor, L. T., and Plapp, F. V. (1984) Semi automated solid phase adherence assays for pre-transfusion testing. *Med. Lab. Sci.* **41,** 374–381.

39. Lapierre, Y., Rigal, D., Adam, J., et al. (1990) The gel test: a new way to detect red cell antigen-antibody reactions. *Transfusion* **30,** 109–113.

40. Reis, K. J., Lachowski, R., Cupido, A., Davies, A., Hackway, J., and Setcavage, T. M. (1993) Column agglutination technology: the antiglobulin test. *Transfusion* **33,** 639–643.

41. Scott, M. L. (1991) The principles and applications of solid-phase blood grouping serology. *Transfus. Med. Rev.* **5,** 60–72.

42. Knight, R. C., and de Silva, M. (1996) New technologies for red-cell serology. *Blood Rev.* **10,** 101–110.

43. Robb, J. S., Roy, D. J., Ghazal, P., Allan, J., and Petrik, J. (2006) Development of non-agglutination microarray blood grouping. *Transfus. Med.* **16,** 119–129.

44. Campbell, C. J., O'Looney, N., Chong Kwan, M., et al. (2006) Cell interaction microarray for blood phenotyping. *Anal Chem.* **78,** 1930–1938.

45. Reid, M. E., and Lomas-Francis, C. (1997). *The Blood Group Antigen Facts Book. Academic Press*, Harcourt Brace & Company, CA.

46. Daniels, G. (1995) *Human Blood Groups.* Blackwell Science Limited, Oxford, United Kingdom.

47. Schenkel-Brunner, H. (2000) *Human Blood Groups: Chemical and Biochemical Basis of Antigen Specificity*, Springer, New York.

48. Moore, S., Chirnside, A., Micklem, L. R., McClleland, D. B. L., and James.K. (1984) A mouse monoclonal antibody with anti-A,(B) specificity which agglutinates Ax cells. *Vox Sang.* **47,** 427–434.

16

"Lab-on-a-Chip" Devices for Cellular Arrays Based on Dielectrophoresis

Roberto Gambari, Monica Borgatti, Enrica Fabbri, Riccardo Gavioli, Cinzia Fortini, Claudio Nastruzzi, Luigi Altomare, Melanie Abonnenc, Nicolò Manaresi, Gianni Medoro, Aldo Romani, Marco Tartagni, and Roberto Guerrieri

Summary

Dielectrophoresis (DEP)-based lab-on-a-chip devices represent a very appealing approach for cell manipulation and will enable laboratory testing to move from laboratories into nonlaboratory settings. DEP-based lab-on-a-chip platforms involve the miniaturization of several complex chemical and physical procedures in a single microchip-based device, allowing the identification and isolation of cell populations or single cells, separation of cells exhibiting different DEP properties, isolation of infected from uninfected cells, and viable from nonviable cells. In addition, DEP-arrays allow cellomics procedures based on the parallel manipulation of thousands of cells. Arrayed lab-on-a-chip platforms are also suitable to study cell–cell interactions and cell targeting with programmed numbers of microspheres.

Key Words: Cellular arrays; delivery; dielectrophoresis; lab-on-a-chip; microspheres.

1. Introduction

The postgenomic era is characterized by the development of approaches enabling researchers to study the expression of several hundreds of genes at the mRNA (gene expression profiling by using microarray technology) or protein (proteomic strategies) levels (1–3). Accordingly, several studies have provided strong evidence that thousands of transcripts can be quantified using microarrays exposing cDNA or oligonucleotide probes (4,5). At the same time, several hundreds of proteins can be quantified using arrayed monoclonal antibodies (6).

Although the number of studies on gene expression profiling and proteomics has been dramatically increased over the past 5 yr, the number of reports on the possibility to analyze complex cell populations is much lower. However, the possibility to manipulate large numbers of single cells is very interesting, and several groups are currently involved in projects aimed at analysis at the single cell level (7).

For analysis at the single cell level, the so-called "laboratory-on-chip" (lab-on-a-chip) technology could represent a very appealing approach, because it involves the miniaturization of several complex chemical and physical procedures in a single microchip-

From: *Bioarrays: From Basics to Diagnostics*
Edited by: K. Appasani © Humana Press Inc., Totowa, NJ

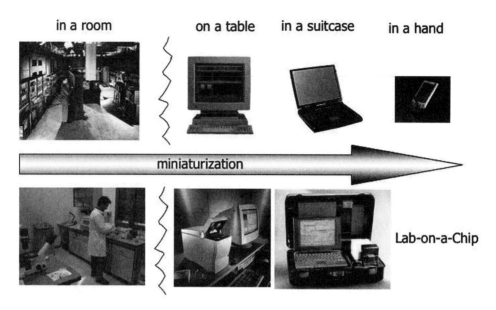

Fig. 1. A parallel analysis of miniaturization of computers (**top**) and laboratory equipment (**bottom**).

based platform *(8–12)*, leading to the possibility of manipulating large numbers of single cells or cell populations. There is a general agreement that lab-on-a-chip technology will enable laboratory testing to move from laboratory into nonlaboratory settings *(12–19)*.

The concept of lab-on-a-chip is depicted in **Fig. 1**, in which a parallelism is shown between technological improvements and miniaturization of computers and lab-on-a-chip platforms. As it occurred for computers (that were located in large rooms, later on tables, and recently within suitcases and even within hand-held devices), laboratory technologies can move from laboratories using complex equipment into miniaturized lab-on-a-chip platforms.

As far as the physical basis of the design, production, and characterization of lab-on-a-chip devices for cell manipulation, dielectrophoresis (DEP) *(20–24)* has been reported as a very valuable approach *(25–28)*. Application of DEP protocols *(23,24)* results in the movement of particles in nonuniform electric fields *(22,23,27,29,30)* in which the particles experience a translational force (DEP force) of magnitude and polarity depending not only on the electrical properties of particles and medium but also on the magnitude and frequency of the applied electric field. Thus, for a given particle type and suspending medium, the particles can experience, at a certain frequency of the electrode-applied voltages, a translational force directed toward regions of high electric field strength. This phenomenon is called positive DEP (pDEP). By simply changing the frequency, they may experience a force that will direct them away from high electric field strength regions. This phenomenon is called negative DEP (nDEP). A scheme outlining the concept of pDEP and nDEP is shown in **Fig. 2A**.

Using DEP-based lab-on-a-chip platforms, isolation of cell populations or single cells, separation of cells exhibiting different DEP properties, and isolation of infected from uninfected cells and viable from nonviable cells is possible. In addition, DEP

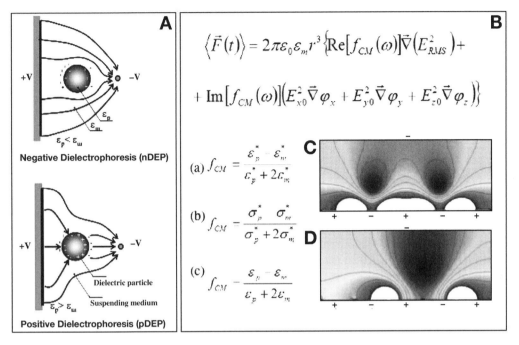

$$\langle \vec{F}(t) \rangle = 2\pi\varepsilon_0 \varepsilon_m r^3 \left\{ \mathrm{Re}[f_{CM}(\omega)]\vec{\nabla}\left(E_{RMS}^2\right) + \right.$$

$$\left. + \mathrm{Im}[f_{CM}(\omega)]\left(E_{x0}^2\vec{\nabla}\varphi_x + E_{y0}^2\vec{\nabla}\varphi_y + E_{z0}^2\vec{\nabla}\varphi_z\right)\right\}$$

(a) $f_{CM} = \dfrac{\varepsilon_p^* - \varepsilon_m^*}{\varepsilon_p^* + 2\varepsilon_m^*}$

(b) $f_{CM} = \dfrac{\sigma_p^* - \sigma_m^*}{\sigma_p^* + 2\sigma_m^*}$

(c) $f_{CM} = \dfrac{\varepsilon_p - \varepsilon_m}{\varepsilon_p + 2\varepsilon_m}$

Fig. 2. (A) Scheme outlining the concept of nDEP and pDEP. (B) Algorithms on which dielectrophoresis is based. (C,D) Scheme showing two DEP-cages (C) that are forced to move one against the other, originating a single cage (D).

arrays allow cellomics procedures based on the parallel manipulation of thousands of cells. Arrayed lab-on-a-chip platforms are suitable to study cell–cell interactions and cell targeting with programmed numbers of microspheres.

2. Theory Supporting DEP-Based Levitation and Movement of Biological and Physical Objects

For cells, the DEP properties largely depend on several biological parameters, including membrane capacitance (determined for living cells by the membrane dielectric permittivity and composition, thickness, and area) and conductance. For spherical geometries, a first order approximation of the DEP force can be expressed as the general equation shown in **Fig. 2B**, where e_0 is the vacuum dielectric constant, R is the particle radius, E_{a0} and j_a (a = x, y, z) are the magnitude and phase of each component of the electric field in a Cartesian coordinate frame, $ERMS$ is the root mean square value of the electric field, $\mathrm{Re}(\Delta'CM)$ and $\mathrm{Im}(\Delta'CM)$ are the real (in-phase) and imaginary (out-of-phase) components of the Clausius–Mossotti factor (**Eq. a of Fig. 2B**), which is a function of the complex permittivities of the particle e^*_p and the medium e^*_m, defined as $e^* = e_0 + s/jw$, with e_0 the permittivity of vacuum, e the relative dielectric permittivity, and s the conductivity.

For nDEP, it should be **Re(f_{CM})<0**. At low frequency ($\omega \ll \sigma/\varepsilon$), equation a can be approximated by **Eq. b of Fig. 2B**, whereas at high frequency ($\omega \gg \sigma/\varepsilon$), **Eq. c of Fig. 2B** is obtained. Thus, particle levitation by nDEP is possible at low frequency if $\sigma_p < \sigma_m$, and at high frequency if $\varepsilon_p < e_m$.

Figure 2C,D shows a numerical simulation of the effects of the general working principle of DEP. By applying suitable potentials to the electrodes, it is possible to generate time-dependent electric fields in the liquid. These fields can then generate DEP fields acting on the particles in the fluid. A DEP force is then generated owing to the differences in the dielectric permittivities of the different materials. An important point of this approach is that the overall system can be designed to force the DEP fields to create closed cages (DEP-cages) that can trap inside particles in a stable way. In **Fig. 2C**, two DEP-cages are simulated that can entrap, when suitable electric potentials are applied to the electrodes, microspheres, target cells, or both. By looking at **Fig. 2C**, it is possible to see a local minimum of electric fields associated with the presence of the two DEP-cages. Because these electric potentials can be applied under software control, it is possible to change how particles are moved by modifying the settings on a computer. In addition, it is possible to change in time the location of these closed DEP-cages. After changing the potentials applied to the electrodes, the location of the two cage changes (**Fig. 2D**), and all the microparticles entrapped in the two starting DEP-cages are now concentrated within a single DEP-cage. DEP-based lab-on-a-chip devices are of great interest to manipulate single cells, as published previously *(31–34)*.

3. Description of Lab-on-a-Chip Platforms for Cell Manipulation

The general setup for lab-on-a-chip platforms is described in **Fig. 3** and includes a microscope (**Fig. 3A**) connected with a charge-coupled device-camera and computer, and platforms suitable for inclusion of the lab-on-a-chip devices. In **Fig. 4B,C** are shown two lab-on-a-chip platforms, described in detail by Medoro et al. *(35)* and Manaresi et al. *(34)* constituted by paralleled (the SmartSlide depicted in **Fig. 4B**) or arrayed (the DEP array depicted in **Fig. 4C**) electrodes.

Several DEP-based lab-on-a-chip devices were recently described and found to be suitable for biotechnological applications in isolation of single cell populations as well as manipulation of single biological objects, including cells and microspheres.

3.1. Lab-on-a-Chip With Spiral Electrodes

Examples of biological applications of lab-on-a-chip platforms with spiral electrodes have been reported by Wang et al. *(36)* and by Gascoyne et al. *(37)*. By using a microelectrode array consisting of four parallel spiral electrode elements energized with phase-quadrature signals of frequencies between 100 Hz and 100 MHz, Wang et al. *(36)* demonstrated that MDA-MB231 breast cancer cells can be concentrated at the center of the chip, as shown in the scheme depicted in **Fig. 5A**. MDA-MB231 cells were studied in suspensions of conductivities 18, 56, and 160 mS/m *(36)*. At low frequencies, cells were levitated and transported toward or away from the center of the spiral array, whereas at high frequencies cells were trapped at electrode edges. These results suggest that spiral electrode arrays could be applied to the isolation of cells of clinical relevance *(36)*.

3.2. Lab-on-a-Chip With Parallel Electrodes

Two examples of lab-on-a-chip devices with parallel electrodes are shown in **Fig. 5B,C**. The first example shown in **Fig. 5B** *(38)* combines DEP forces with fluid flow; in this case, the separation is obtained by the combination of fluid and DEP forces, separating particles differently attracted by the microelectrodes *(38)*.

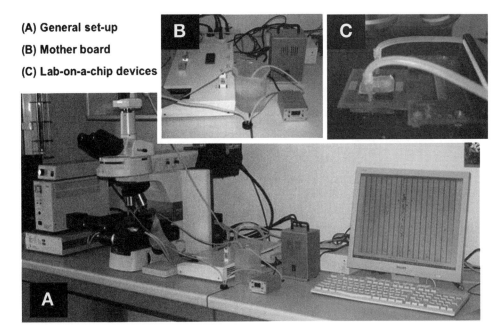

Fig. 3. (A) Setup of the lab-on-a-chip assembly, consisting of a microscope, a mother board **(B)**, a personal computer, and a micropump for temperature control **(C)**.

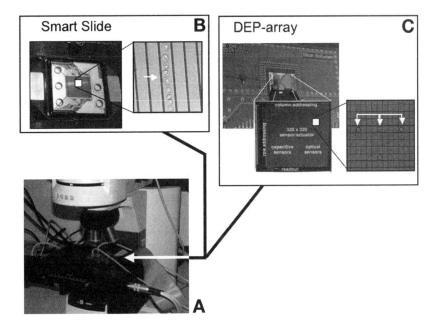

Fig. 4. (A) Detail of the location of the lab-on-a-chip devices under a microscope and structures of the SmartSlide **(B)** and of the DEP array **(C)**. In the insets, separation of cell populations **(B)** and single cells **(C)** is shown within the SmartSlide (containing paralleled electrodes) and the DEP-chip (containing arrayed electrodes). Scheme concerning the SmartSlide are from Medoro et al. *(35,42)* schemes concerning the DEP array are from Medoro et al. *(33,42)* and Manaresi et al. *(34)*.

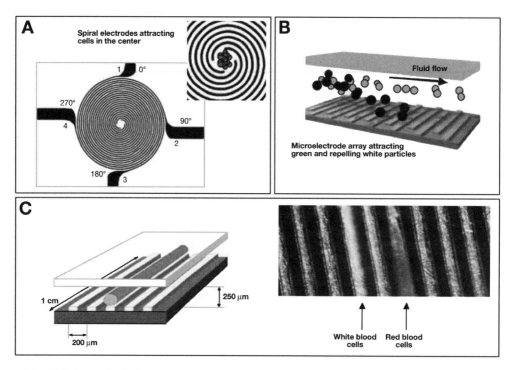

Fig. 5. Scheme depicting devices based on spiral (**A**) and parallel (**B,C**) electrodes. The inset in A shows concentration of cell populations in the center of the device.

The second example, shown in **Fig. 5C**, was reported in a study published by Altomare et al. *(39)* and reviewed by Gambari et al. *(40)*. This device is mad up of a microchamber, delimited on the top by a conductive and transparent lid (which is itself an electrode and is electrically connected to the device by means of a conductive glue) and on the bottom by a support. A spacer (constituted of optic fibers) determines the chamber height, whereas a silicon elastomer gasket delimits and seals the microchamber on the sides.

A mother-board is used to generate and distribute to each electrode in the device the proper phases needed to create and move the DEP-cages and to perform the sensing operations, whereas a software tool allows to control the actuation and sensing operations flows. By changing the electrode programming, each DEP-cage can be independently moved from electrode to electrode along the whole microchamber, dragging with it the trapped elements. This device allows the separation, concentration, or both, of cells and microspheres in a fully electronic system without the need for fluid flow control. This device, with 39 parallel electrodes by which it is possible to generate from 0 to 13 cylinder-shaped DEP-cages, was applied to separate white blood cells from red blood cells *(39,40)*. The same system was recently used to demonstrate that these cells are suitable for genomic studies *(41)*.

More recently, the design, technical parameters, building approach, and manufacturing of the SmartSlide, with 193 parallel electrodes by which it is possible to generate from 0 to 63 cylinder-shaped DEP-cages, has been described *(35,42)*. Set up of this device and use for separating cells are shown in **Fig. 4B**.

3.3. Lab-on-a-Chip With Low-Density Arrayed Electrodes

Cells belonging to the U937 human monocytic *(43)*, HTB glioma *(43)*, SH-SY5Y neuroblastoma *(43)*, and HeLa cervical carcinoma *(44)* cell lines have been moved throughout lab-on-a-chips carrying 5 × 5 electrode arrays. The cells are moved as clusters of few cells and cell separation and isolation performed.

3.4. Lab-on-a-Chip With High-Density Arrayed Electrodes

Design, technical parameters, building approach, and manufacturing of a DEP-based arrayed device have been described by Medoro et al. *(33,42)* and Manaresi et al. *(34)*. This DEP-array **(Fig. 4B)** is constituted by a microchamber defined by the chip surface and a conductive-glass lid. The chip surface implements a two-dimensional array of microsites, each consisting of a superficial electrode, embedded sensors and logic. The electrode array is implemented with complementary metal oxide semiconductor top-metal and protected from the liquid by the standard complementary metal oxide semiconductor passivation. In this system, a closed DEP-cage in the spatial region above a microsite can be created by connecting the associated electrode and the microchamber lid to a counterphase sinusoidal voltage, whereas the electrode of the neighboring microsites is connected to an in-phase sinusoidal voltage. A field minimum is thus created in the liquid, corresponding to a DEP-cage in which, depending on its size, one or more particles can be trapped and levitated. By changing, under software control, the pattern of voltages applied to the electrodes, spherical DEP-cages can be independently moved around the device plane, thus grabbing and dragging cells, microbeads, or both across the chip. Particles in the sample can be detected by the changes in optical radiation impinging on the photodiode associated with each microsite. Because of the small pitch of the electrodes, single cells can be individually trapped in separate cages **(Fig. 4C, inset)** and independently moved on the device. Particle position is digitally controlled step by step in a deterministic way, by applying corresponding pattern of voltages to the array that set the position of the DEP-cages.

4. Biological Applications of Lab-on-a-Chip Platforms

4.1. Isolation and Concentrations of Cell Populations

The possibility to isolate or concentrate cell population by the aid of DEP-lased lab-on-a-chip platforms has been reported previously *(24,36,40,45,46)*. For example, Yu et al. *(46)* recently reported the use of an efficient method for trapping neurons and constructing ordered neuronal networks on bioelectronic chips by using arrayed nDEP forces. A special bioelectronic chip with well-defined positioning electrode arrays was designed and fabricated on silicon substrate. When a high frequency ac signal was applied, the cell positioning bioelectronic chip is able to provide a well-defined non-uniform electric field and thus generate nDEP forces *(46)*.

Interestingly, when a neuronal suspension was added onto the energized bioelectronic chip, the neurons were immediately trapped and quickly formed the predetermined pattern. According with the protocol developed by Yu et al. *(46)*, neurons may adhere and then be cultured directly on the cell positioning bioelectronic chip and show good neuron viability and neurite development *(46)*, suggesting that this approach could be used to characterize functional activities of neuronal networks. It should be pointed out that the development of novel technologies aimed to pattern neurons into regular networks is of great scientific interest in neurological research.

4.2. Separation of Cell Populations Exhibiting Different DEP Properties

Several examples have been reported on separation of cell populations exhibiting different DEP properties *(27,29,30,33)*. For example, Huang et al. *(27)* reported a dielectrophoretic field-flow-fractionation method and used it to purge human breast cancer MDA-435 cells from hematopoietic CD34[+] stem cells. In the case reported by these authors, an array of interdigitated microelectrodes lining the bottom surface of a thin chamber was used to generate dielectrophoretic forces that levitated the cell mixture in a fluid flow profile. CD34[+] stem cells were levitated higher, were carried faster by the fluid flow, and exited the separation chamber earlier than the cancer cells. The same group was able to demonstrate high separation performance in separating T- (or B)-lymphocytes from monocytes, T- (or B)-lymphocytes from granulocytes, and monocytes from granulocytes *(27)*.

Another interesting application of DEP-based lab-on-a-chip is the possibility to separate viable from nonviable yeast by using DEP *(47,48)*. A further example is represented by the possibility of using DEP-based lab-on-a-chip for the isolation of infected cells from uninfected counterparts. This possibility was shown by recent articles demonstrating that lab-on-a-chip platforms carrying spiral electrodes can be used for isolating malaria-infected cells from blood *(49)*. Therefore, DEP-based protocols offer great potential in cell discrimination and isolation.

4.3. Forced Interactions between Cells and Microspheres

The software-guided interactions between microspheres and target cells can be achieved with the use of DEP-based devices. For example, DEP-based devices carrying paralleled electrodes can be used to force the interactions between microspheres and K562 cells *(50)*, after moving them at the central electrode within the corresponding DEP-cage.

A second example is shown in **Fig. 6** and was performed with the DEP-array to determine whether this system would be suitable for directing single microspheres *(51)* to a single identified target cell **(Fig. 6B)**. In the experiment shown in **Fig. 6**, three cationic microspheres (M1, M2, and M3) and two K562 cells are shown, entrapped in five independent spherical DEP-cages. Only one of the two cells, named T-K562, was designed to be the cellular "target" of the three microspheres. We first moved microsphere M1, generating the cell–microsphere complex shown in **Fig. 6C**, and then microsphere M2, generating the T-K562/M1M2 complex shown in **Fig. 6D**. Finally, we moved the M3 microsphere for a further targeting of the T-K562 cell, obtaining the T-K562/M1M2M3 complex shown in **Fig. 6E**. The details of this experiment are described in Borgatti et al. *(52)*.

4.4. Forced Interactions Between Cells

Forced interaction between cells is of great relevance in several research fields in which cell–cell interaction is the major experimental step. Among these projects are studies on interactions between target tumor cells and cytotoxic T-lymphocytes or natural killer cells. From this point of view, the DEP-based arrayed device described by

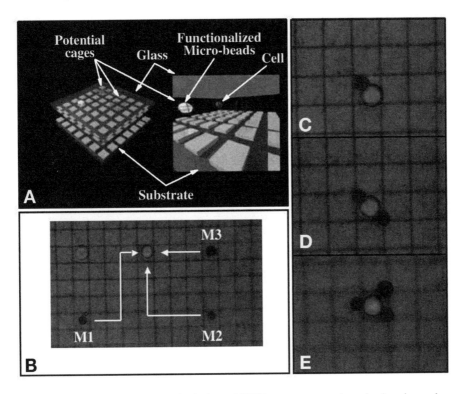

Fig. 6. (A) Scheme showing spherical-shaped DEP-cages entrapping single microspheres and cells on the DEP array. (B–E) Programmed sequential interactions between three cationic microspheres (M1, M2, and M3) and one target K562 cell. The microsphere M1 was moved first to obtain the T-K562/M1 complex shown in C. Then, the M2 and M3 microspheres were moved to obtain the T-K562/M1M2 and T-K562/M1M2M3 complexes shown in D and E, respectively. The scheme reported in A was from Medoro et al. *(33,42)* and Manaresi et al. *(34)*.

Medoro et al. *(33,42)* and Manaresi et al. *(34)* could be of great interest, as shown in the application described in **Fig. 7**, in which software-guided interaction between a cytotoxic T-lymphocyte and a single target K562 cell are shown (Borgatti et al., manuscript in preparation).

5. Conclusions

Manipulation of single cells and microspheres is one of the frontiers of lab-on-a-chip technology *(21,28,33,37)*. With respect to the possibility of easy, reproducible, software-guided levitation, movement, and targeting of single cells, few examples are present in the literature *(37)*. However, DEP-based movement and separation of cells by using a platform consisting of spiral, parallel, and arrayed electrodes has been reported in several studies *(42–51)*. Accordingly, DEP-based manipulation of microparticles was described previously *(52)*. The results herein demonstrate that software-guided manipulation of a very high number of target cells, microspheres, or both can be operated in DEP-based lab-on-a-chip platforms able to generate cellular arrays. These DEP-based approaches enable, in our opinion, novel possibilities in pharmaceutical science, technology, and tumor immunology.

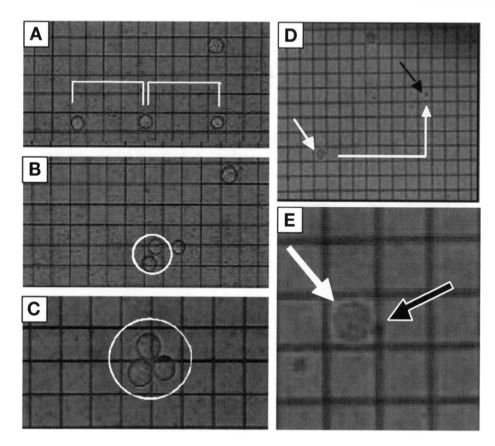

Fig. 7. (A–C) Forced interactions between three single K562 cells independently moved to originate the complex shown within circles of B and C. **(D,E)** Forced interactions between a single K562 cell (white arrow) and a single cytotoxic T-lymphocyte (black arrow).

Technologically, the use of an active substrate implemented with microelectronic semiconductor technology allows integration of an array of optical sensors to detect the position and possibly the status of all particles inside the device. This information can be used to provide meaningful feedback on device operations. Thus, the device has the potential to be used without bulky and expensive external microscopes and cameras. This approach will be important, in perspective, for portable lab-on-a-chip platforms. A final point is that these systems are suitable for cellomics approaches because thousands of cells to be concurrently and independently moved and detected owing to numerous electrodes.

Acknowledgments

This work was supported by MIUR-COFIN-2000 and UE MEDICS Project (IST-2001-32437) to R.Gu. and MURST-COFIN-2002 (Applications of a dielectrophoresis-based lab-on-a-chip to diagnosis and drug research and development) to R. Ga. and by FIRB-2001 (Development of a lab-on-a-chip based on microelectronic technologies and its biotechnological validation) to R. Gu. and R. Ga.. This research also was supported by Fondazione Italiana Ricerca sulla Fibrosi Cistica, Associazione Italiana Ricerca sul Cancro, Associazione Veneta per la Lotta alla Talassemia, Rovigo, and Fondazione Cassa di Risparmio di Padova e Rovigo.

References

1. Rhodes, D. R., and Chinnaiyan, A. M. (2002) DNA microarrays: implications for clinical medicine. *J. Investig. Surg.* **15**, 275–279.
2. Lian, Z., Kluger, Y., Greenbaum, D. S., et al. (2002) Genomic and proteomic analysis of the myeloid differentiation program: global analysis of gene expression during induced differentiation in the MPRO cell line. *Blood* **100**, 3209–3220.
3. Mendez, E., Cheng, C., Farwell, D. G., et al. (2002) Transcriptional expression profiles of oral squamous cell carcinomas. *Cancer* **95**, 1482–1494.
4. Schuppe-Koistinen, I., Frisk, A. L., and Janzon, L. (2002) Molecular profiling of hepatotoxicity induced by a aminoguanidine carboxylate in the rat: gene expression profiling. *Toxicology* **179**, 197–219.
5. Moos, P. J., Raetz, E. A., Carlson, M. A., et al. (2002) Identification of gene expression profiles that segregate patients with childhood leukemia. *Clin. Cancer Res.* **8**, 3118–3130.
6. Tamiya, E., Zhi, Z. L., Morita, Y., and Hasan, Q. (2005) Nanosystems for biosensing: multianalyte immunoassay on a protein chip. *Methods Mol. Biol.* **300**, 369–381.
7. Muller, T., Pfennig, A., Klein, P., Gradl, G., Jager, M. and Schnelle, T. (2003) The potential of dielectrophoresis for single-cell experiments. *IEEE Eng. Med. Biol. Mag.* **22**, 51–61.
8. Kricka, L. J. (2001) Microchips, microarrays, biochips and nanochips: personal laboratories for the 21st century. *Clin. Chim. Acta* **307**, 219–223.
9. Jain, K. K. (2002) Post-genomic applications of lab-on-a-chip and microarrays. *Trends Biotechnol.* **20**, 184–185.
10. Chovan, T., and Guttman, A. (2002) Microfabricated devices in biotechnology and biochemical processing. *Trends Biotechnol.* **20**, 116–122.
11. Jain, K. K. (2001) Cambridge Healthtech Institute's Third Annual Conference on Lab-on-a-Chip and Microarrays. *Pharmacogenomics* **2**, 73–77.
12. Mouradian, S. (2002) Lab-on-a-chip: applications in proteomics. *Curr. Opin. Chem. Biol.* **6**, 51–56.
13. Polla, D. L., Erdman, A. G., Robbins, W. P., et al. (2000) Microdevices in medicine. *Annu. Rev. Biomed. Eng.* **2**, 551–576.
14. Hasselbrink, E. F., Shepodd, T. J., and Rehm, J. E. (2002) High-pressure microfluidic control in lab-on-a-chip devices using mobile polymer monoliths. *Anal. Chem.* **74**, 4913–4918.
15. Figeys, D. (2002) Adapting arrays and lab-on-a-chip technology for proteomics. *Proteomics* **2**, 373–382.
16. Weigl, B. H., and Hedine, K. (2002) Lab-on-a-chip-based separation and detection technology for clinical diagnostics. *Am. Clin. Lab.* **21**, 8–13.
17. Wang, J., Pumera, M., Chatrathi, M. P., et al. (2002) Towards disposable lab-on-a-chip: poly(methylmethacrylate) microchip electrophoresis device with electrochemical detection. *Electrophoresis* **23**, 596–601.
18. Mouradian, S. (2002) Lab-on-a-chip: applications in proteomics. *Curr. Opin. Chem. Biol.* **6**, 51–56.
19. Gawad, S., Schild, L., and Renaud, P. H. (2001) Micromachined impedance spettroscopy flow cytometer for cell analysis and particle sizing. *Lab Chip* **1**, 76–82.
20. Pohl, H. A., and Crane, J. S. (1972) Dielectrophoretic force. *J. Theor. Biol.* **37**, 1–13.
21. Crane, J. S., and Pohl, H. A. (1972) Theoretical models of cellular dielectrophoresis. *J. Theor. Biol.* **37**, 15–41.

22. Voldman, J., Braff, R. A, Toner, M., Gray, M. L. and Schmidt, M. A. (2001) Holding forces of single-particle dielectrophoretic traps. *Biophys. J.* **80,** 531–541.
23. Gascoyne, P. R., and Vykoukal. J. (2002) Particle separation by dielectrophoresis. *Electrophoresis* **23**, 1973–1983.
24. Fiedler, S., Shirley, S. G., Schnelle, T., and Fuhr, G. (1998) Dielectrophoretic sorting of particles and cells in a microsystem. *Anal. Chem.* **70,** 1909–1915.
25. Morgan, H., Hughes, M. P., and Green, N. G. (1999) Separation of submicron bioparticles by dielectrophoresis. *Biophys. J.* **77,** 516–525.
26. Gascoyne, P. R., Vykoukal, J. V., Schwartz, J. A., et al. (2004) Dielectrophoresis-based programmable fluidic processors. *Lab Chip* **4**, 299–309.
27. Huang, Y., Yang, J., Wang, X. B., Becker, F. F. and Gascoyne, P. R. (1999) The removal of human breast cancer cells from hematopoietic CD34+ stem cells by dielectrophoretic field-flow-fractionation. *J. Hematother. Stem Cell Res.* **8**, 481–490.
28. Yang, J., Huang, Y., Wang, X. B., Becker, F. F., and Gascoyne, P. R. (2000) Differential analysis of human leukocytes by dielectrophoretic field-flow-fractionation. *Biophys. J.* **78,** 2680–2689.
29. Wang, X. B., Yang, J., Huang, Y., Vykoukal, J., Becker, F. F., and Gascoyne, P. R. (2000) Cell separation by dielectrophoretic field-flow-fractionation. *Anal. Chem.* **72,** 832–839.
30. Yang, J., Huang, Y., Wang, X. B., Becker, F. F., and Gascoyne, P. R. (1999) Cell separation on microfabricated electrodes using dielectrophoretic/gravitational field-flow fractionation. *Anal. Chem.* **71,** 911–918.
31. Ogata, S., Yasukawa, T., and Matsue, T. (2001) Dielectrophoretic manipulation of a single chlorella cell with dual-microdisk electrode. *Bioelectrochemistry* **54,** 33–37.
32. Muller, T., Pfennig, A., Klein, P., Gradl, G., Jager, M., and Schnelle, T. (2003) The potential of dielectrophoresis for single-cell experiments. *IEEE Eng. Med. Biol. Mag.* **22,** 51–61
33. Medoro, G., Manaresi, N., Tartagni, M., Altomare, L., Leonardi, A. and Guerrieri, R. (2002) A lab-on-a-chip for cell separation based on the moving-cages approach. In: *Proceedings of Eurosensors XVI.* Prague, Czech Republic.
34. Manaresi, N., Romani, A., Medoro, G., et al. (2003) A CMOS chip for individual cell manipulation and detection. *IEEE J. Solid-State Circuits* **38,** 2297–2304.
35. Medoro, G., Manaresi, N., Tartagni, M., and Guerrieri, R. (2000) CMOS-only sensor and manipulator for microorganisms. *IEEE International Electron Devices Meeting (IEDM) Technical Digest*, pp. 415–418.
36. Wang, X. B., Huang, Y., Wang, X., Becker, F. F., and Gascoyne, P. R. Dielectrophoretic manipulation of cells with spiral electrodes. *Biophys. J.* **72,** 1887–1899.
37. Gascoyne, P., Mahidol, C., Ruchirawat, M., Satayavivad, J., Watcharasit, P., and Becker, F.F. (2002) Microsample preparation by dielectrophoresis: isolation of malaria. *Lab Chip* **2,** 70–75.
38. Ermolina, I., and Morgan, H. (2005) The electrokinetic properties of latext particles: comparison of electrophoresis and dielectrophoresis. *J. Colloid Interface Sci.* **285,** 419–428
39. Altomare, L., Borgatti, M., Medoro, G., et al. (2003) Levitation and movement of human tumor cells using a printed circuit board device based on software-controlled dielectrophoresis. *Biotechnol. Bioeng.* **82,** 474–479.
40. Gambari, R., Borgatti, M., Altomare, L., et al. (2003) Applications to cancer research of "lab-on-a-chip" devices based on dielectrophoresis (DEP). *Technol. Cancer Res. Treat.* **2,** 31–40.
41. Borgatti, M., Altomare, L., Baruffa, M., et al. (2005) Separation of white blood cells from erythrocytes on a dielectrophoresis (DEP) based 'lab-on-a-chip' device. *Int. J. Mol. Med.* **15,** 913–920.

42. Medoro, G., Manaresi, N., Leonardi, A., Altomare, L., Tartagni, M., and Guerrieri, R. (2002) A lab-on-a-chip for cell detection and manipulation. In: *Proceedings of IEEE Sensors Conference.*

43. Huang, Y., Joo, S., Duhon, M., Heller, M., Wallace, B., and Xu, X. (2002) Dielectrophoretic cell separation and gene expression profiling on microelectronic chip arrays. *Anal. Chem.* **74,** 3362–3371.

44. Cheng, J., Sheldon, E. L., Wu, L., Heller, M. J., and O'Connell, J. P. (1998) Isolation of cultured cervical carcinoma cells mixed with peripheral blood cells on a bioelectronic chip. *Anal. Chem.* **70,** 2321–2326.

45. Xu, C., Wang, Y., Cao, M., and Lu, Z. (1999) Dielectrophoresis of human red cells in microchips. *Electrophoresis* **20,** 1829–1831.

46. Yu, Z., Xiang, G., Pan, L., et al. (2004) Negative dielectrophoretic force assisted construction of ordered neuronal networks on cell positioning bioelectronic chips. *Biomed. Microdevices* **6,** 311–324.

47. Markx, G. H., Talary, M. S., and Pethig, R. (1994) Separation of viable and non-viable yeast using dielectrophoresis. *J. Biotechnol.* **32,** 29–37.

48. Huang, Y., Holzel, R., Pethig, R., and Wang, X. B. (1992) Differences in the AC electrodynamics of viable and non-viable yeast cells determined through combined dielectrophoresis and electrorotation studies. *Phys. Med. Biol.* **37,** 1499–1517.

49. Gascoyne, P., Satayavivad, J., and Ruchirawat, M. (2004) Microfluidic approaches to malaria detection. *Acta Trop.* **89,** 357–369.

50. Bianchi, N., Chiarabelli, C., Borgatti, M., Mischiati, C., Fibach, E., and Gambari, R. (2001) Accumulation of gamma-globin mRNA and induction of erythroid differentiation after treatment of human leukaemic K562 cells with tallimustine. *Br. J. Haematol.* **113,** 951–961.

51. Mischiati, C., Sereni, A., Finotti, A., et al. (2004) Complexation to cationic microspheres of double-stranded peptide nucleic acid-DNA chimeras exhibiting decoy activity. *J. Biomed. Sci.* **11,** 697–704.

52. Borgatti, M., Altomare, L., Abonnec, M., et al. (2005) Dielectrophoresis (DEP) based 'lab-on-a-chip'' devices for efficient and programmable binding of microspheres to target cells. *Int. J. Oncol.* **27,** 1559–1566.

17

Genetic Disorders and Approaches to Their Prevention

Giriraj Ratan Chandak and Hemavathi Jayaram

Summary

The new genetics or—as it is often called—genomics is one of the most exciting and challenging areas of science today. It promises to revolutionize medicine and healthcare in the twenty-first century. We are now only just beginning to understand the genetic basis of common diseases; but the number of patients with the potential to be affected by genetic susceptibility and the capability to identify them using newer advances in genetics is rising rapidly. The simple rules of inheritance proposed by Gregor Mendel are no longer adequate to explain complex patterns of inheritance, reflecting intriguing phenomena such as mitochondrial inheritance and genomic imprinting. The distinction blurs between the genetic factors and the environment in the expression of the clinical phenotype. Genes are no longer thought to operate in isolation but to interact with each other and with the environmental milieu, both external and internal.

Mapping of the human genome, one of the greatest scientific developments, places researchers at the edge of a new frontier that is already yielding medical breakthroughs and shows promise for many others. The resultant discoveries have come to herald a new era wherein these new scientific advances are fast becoming an integral part of modern medicine and have led to an increased understanding of diseases. Contemporary aspects, such as molecular diagnostic testing for various single gene disorders, bacterial and viral identification, chromosome analysis in congenital anomalies, and DNA analysis for forensic and parental identification, have come to be regarded as representative and crucial facets of present standards of health care.

This chapter aims to create an interface among the subject, its students, and its practitioners at the field level, encompassing diverse walks of life, ranging from diagnostics to industrial and pharmaceutical applications, as well keeping pace with the current emerging developments and the challenges involved. It is fondly hoped that this will serve as a platform for further scientific inquiries and pave the way for the refinement of the discipline in the times ahead.

Key Words: Genetic disorders; inheritance; genetic analysis; prenatal diagnosis; genetic counselling.

1. Introduction

The association of human beings and disease is as old as humanity. The major concern has always been the infectious diseases that are common and associated with immense suffering. However, another class of diseases that not only cause physical misery to the patients but also psychological upset is genetic disorders, characterized by transmission down the generations. Even worse are the late-onset genetic disorders, which affect individuals when their family is almost complete. The worst, however, is the ugly truth that there is no available cure for these diseases; hence, prevention of inheritance of such

From: *Bioarrays: From Basics to Diagnostics*
Edited by: K. Appasani © Humana Press Inc., Totowa, NJ

diseases becomes a necessity. The burden of genetic disorders in the society is becoming increasingly evident as infectious diseases are being managed better, owing to the awareness about immunization, better understanding of the pathogenesis of the diseases, and the ever-growing development of medical science. More than 5000 human diseases are currently being classified as resulting from the action of a single mutant gene. However, the genetic defect in many of these diseases is yet to be determined. The past two decades have seen the emergence of modern cytogenetics and molecular genetics along with the development of medical genetics from a purely academic discipline into a clinical specialty of relevance. The role of genetic counselling, prenatal diagnosis, carrier detection, and other forms of genetic screening in the prevention of genetic diseases is now well established and is reflected in the increasing provision of genetic services throughout the world. It is indeed unfortunate that only after there is an affected child in the family that the parents or relatives get concerned of the genetic nature of the disease. Prospective approaches are required for combating disorders of high prevalence and with known genetic basis, such as for thalassemias, cystic fibrosis, and so on. The classical examples may include near eradication of thalassemia in Sardinia by adopting mandatory screening and genetic counselling for the whole population, and routine screening for cystic fibrosis in many countries.

The new genetics—often called genomics—is one of the most exciting and challenging areas of science today. It holds the promise to revolutionize medicine and healthcare in 21st century. We are now only just beginning to understand the genetic basis of common diseases, but the number of patients with the potential to be affected owing to genetic susceptibility and the need to identify those individuals by using newer advances in genetics is rising rapidly. The simple rules of inheritance proposed by Gregor Mendel are no longer adequate to explain complex patterns of inheritance reflecting intriguing phenomena such as mitochondrial inheritance and genomic imprinting. The distinction blurs between the genetic factors and the environment in the expression of the clinical phenotype. Genes are no more thought to operate in isolation but rather to interact with each other and with the environmental milieu, both external and internal. Recently, more and more genes causing "complex diseases" such as cancer, asthma, diabetes mellitus, and heart disease are being identified, and analysis of such disorders with particular reference to the role of environment is very challenging. The concept of susceptibility and predisposition to various diseases, based on the genetic constitution of an individual is also taking shape. Mapping of the human genome, one of the greatest scientific developments, places researchers at the edge of a new frontier that is already yielding medical breakthroughs and shows promise for many others. The resultant discoveries have come to herald a new era wherein these new scientific advances are fast becoming an integral part of modern medicine and have led to an increased understanding of disease.

With the completion of Human Genome Project, mapping the associated genes for such disorders and understanding the genetic mechanism will be simpler. The research on gene therapy is in a very interesting stage with a recent study reporting steady state of expression of factor IX gene in mice for a period of 6 mo. Many other therapeutic strategies also are being envisaged with specific stress on stem cells transplantation. The contemporary aspects such as molecular diagnostic testing for various single gene disorders, bacterial and viral identification, chromosome analysis in congenital anomalies, and DNA analysis for forensic and parental identification have come to be regarded as representative and crucial facets of present standards of healthcare. The advent of microarray has added another dimension to the functional genomics. However, it will take some more time before the concept can be put to routine use.

The treatment for most of the genetic disorders is rehabilitation, except for a few metabolic errors, such as phenylketonuria, where modification of diet may lead to substantial improvement of the phenotype. In the absence of specific treatment and gene therapy also being a long-cherished goal, it is very important to persist with the concept of molecular diagnosis, carrier detection, genetic counseling, prepregnancy monitoring, preimplantation genetic diagnosis, and prenatal diagnosis. The identification of molecular defect in the proband can be used as a handle to track the inheritance of defective gene in the fetus by performing fetal sampling procedure at relevant gestational age. Overall, the scenario for DNA diagnosis has immense potential with particular stress on prospective screening and counseling for common single-gene disorders such as thalassemias, muscular dystrophies, and hemophilias, and so on.

The problem of genetic disorders in developing countries such as India remains largely hidden, comparable with the tip of an iceberg. It is unfortunate that although countries such as India contribute to almost one-sixth of the world population and that there are more than 10 million births annually, very few laboratories are available to offer competent diagnostic services for chromosomal abnormalities or metabolic errors, and fewer still offer DNA-level diagnosis. Most of the laboratories that have competence in DNA-based diagnosis are housed in scientific research institutions without functional linkages to hospitals and the problems of patients care. Although some laboratories are engaged in the carrier detection and prenatal diagnosis of some common genetic diseases, the basic data regarding the incidence and molecular nature of common genetic disorders is lacking. This discrepancy can be partly attributed to the failure to have registry for individual patients and partly to lack of awareness about the genetic diseases both in patients as well as in the clinicians. The same situation also exists elsewhere, including many European countries. Fortunately, the scenario has moved toward a change, and the knowledge about such diseases and their impact on the society is now being appreciated. In this chapter, we present information on some of the commonly occurring diseases, various approaches to analyze them, and how their prevalence can be minimized by identification and selective termination of defective fetuses.

This chapter is aimed at creating an interface between the subject, its students, and practitioners at the field level by encompassing diagnostics and industrial and pharmaceutical applications as well as current emerging developments and the challenges involved. It is hoped that this chapter can serve as a platform for further scientific inquiries and pave the way for the refinement of the discipline.

2. Genetic Diseases and Their Patterns of Inheritance

Genetic disorders are conditions associated with genetic mutations that may be inherited or may arise spontaneously. Single gene disorders arise as a result of mutation in one or both alleles of a gene located in an autosome, a sex chromosome, or a mitochondrial gene. They generally refer to conditions inherited in patterns that follow the rules originally proposed by Mendel. In contrast, polygenic disorders are a result of involvement of two or more genes, whereas multifactorial disease arises owing to complex interplay of genetic and environmental factors.

2.1. Dominant and Recessive Disorders

Dominant disorders need only one copy of the defective gene to express the clinical phenotype, whereas recessive disorders traditionally involve diseases where individuals who are clinically affected have both copies defective (double dose). However, the final phenotypic expression of even single gene disorders is subject to many regulations; thus, it is important to understand the pattern of inheritance and peculiarity, if any, especially with autosomal dominant disorders. Recently, co-dominant inheritance has been described, where each allele at a particular locus can have distinct phenotypic expression *(1)*. There is an extra wrinkle in this pattern with regard to the genes on the sex-determining chromosomes, X and Y. Not everyone has two copies of the genes on the sex chromosome. Men are XY and women are XX. Because of this anomaly, genetic diseases are also classed as autosomal and sex-linked (**Table 1**).

2.1.1. Autosomal Dominant

Disorders with autosomal dominant pattern of inheritance account for almost one-half of all single gene disorders. They are usually inherited and only the affected can pass the disease to the next generation, showing characteristic vertical pattern on a pedigree (**Fig. 1**). Transmission of the trait from father to son is a key feature of this type of disorders. The recognition of the characteristic pattern may be complicated by variable expression and incomplete penetrance of the trait; there may be different features in the individuals of same family and the manifestation of the disease may vary, despite having the same mutant gene. An affected individual has 50% chance to pass the disease to next generation. Many of these conditions, such as Marfan syndrome, have a high new mutation rate that maintains the prevalence of these diseases.

2.1.2. Autosomal Recessive

Autosomal recessive inheritance typically means that parents, although clinically normal, are obligate carriers. Consanguinity is an important clue to identify this pattern of inheritance. These disorders are known to skip generations and show the characteristic pattern of horizontal inheritance on a pedigree. Such parents will have a 25% chance of having an affected son or daughter in each pregnancy. These disorders are usually very severe, and majority of inborn errors of metabolism such as phenylketonuria, hemoglobinopathies, and so on belong to this category.

2.1.3. X-linked Dominant

X-linked Dominant disorders such as vitamin D-resistant rickets can affect both males and females. The condition in males is uniformly severe, but females are more variably affected because of X-chromosome inactivation. This pedigree resembles that of an autosomal dominant trait except for the lack of male-to-male transmission.

2.1.4. X-Linked Recessive

These pedigrees show a marked discrepancy in the sex ratio with only affected males often in more than one generation. Because females have two X-chromosomes, daughters are unaffected but are usually carriers and transmit the condition to the next generation, resulting in a "knight's move" pedigree pattern of affected males. Rarely, the disease can affect females owing to X-chromosome inactivation. Carrier women have a 50%

Table 1
Features in the Pedigree to Identify the Pattern of Inheritance

Inheritance pattern	Features
Autosomal dominant	• Vertical inheritance • No gender bias—both males and females are affected • Male to male transmission • Affected individuals in multiple successive generations • Unaffected do not transmit the condition • Incomplete penetrance and variable expression
Autosomal recessive	• Horizontal inheritance • No gender bias—both males and females are affected • Male to male transmission • Affected individuals usually in one generation only • Parental consanguinity • Parents are considered obligate carriers
X-linked dominant	• Vertical inheritance • No male-to-male transmission • Males affected • Females have milder phenotype
X-linked recessive	• Knight's move • Only males affected • No male to male transmission • Females are usually carriers, may rarely be affected
Mitochondrial	• Maternal transmission • No transmission from father • No gender bias—both males and females are affected • Affected individuals in multiple successive generations • Highly variable clinical expressivity owing to heteroplasmy and threshold effect
Multifactorial	• No gender bias—both males and females are affected • Does not follow Mendelian pattern of inheritance • Skips generations • Only few family members affected

chance of having an affected son and an equal chance of having a carrier daughter. An unaffected male never transmits the condition, but all his daughters are obligate carriers. New mutations are known to account for about one-third of all cases. Diseases such as Duchenne muscular dystrophy, hemophilias, and so on follow this pattern of inheritance.

2.1.5 Mitochondrial Disorders

Mitochondrial disorders are characterized by a peculiar pedigree pattern—both males and females are affected, but disease is transmitted through females only. However, expression of the clinical phenotype is highly variable, because only a fraction of cellular mitochondrial DNA may be mutated (heteroplasmy). A "threshold effect" also determines the final clinical picture, because a certain proportion of mutant mitochondria are tolerated within a cell. These disorders are mostly associated with central nervous system, heart, skeletal muscle, and so on.

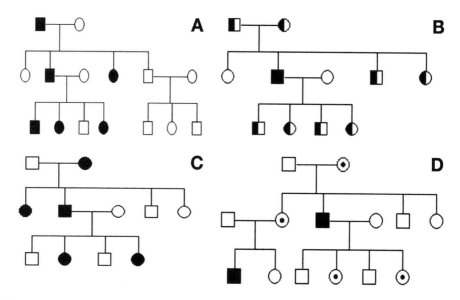

Fig. 1. Patterns of Mendelian inheritance. **(A)** Autosomal dominant inheritance. **(B)** Autosomal recessive inheritance. **(C)** X-Linked dominant inheritance. **(D)** X-Linked recessive inheritance.

2.1.6. Multifactorial Disorders

In contrast to single gene disorders, multifactorial disorders or complex diseases have both environmental and genetic components; the genetic component is determined by multiple genes. Variations in each gene in the cascade also influence disease phenotype. There is no gender bias in the pedigree and usually there are no or few affected family members. It is difficult to calculate the recurrence risk for these disorders, which is mostly derived from the empirical data from the populations. Commonly occurring diseases such as diabetes and coronary artery disease are examples of complex diseases.

2.1.7. Disorders Associated With Dynamic Mutations

Most genes are transmitted unaltered (stable transmittance) through each generation of a family. These disorders are characterized by unstable triplet repeats such as CAG, CGG with a propensity to expand beyond a critical size threshold in successive generations. An intermediate size range between normal range and disease range exists and is described as premutation. Premutation is also the explanation for the phenomenon of anticipation where the clinical picture worsens over successive generations. The age of onset is inversely correlated and the severity directly associated with the number of repeats. Huntington's disease (HD) and myotonic dystrophies are the examples of this type of disorder.

3. Approach for Analysing Genetic Disorders

The approach for genetic analysis for inherited disorders can be summarised in five important stages *(2)*.

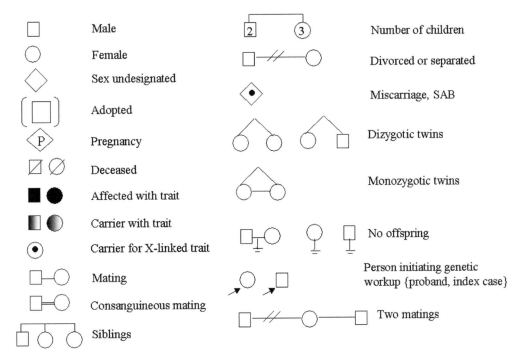

Fig. 2. Commonly described pedigree symbols.

3.1. Clinical History and Pedigree Charting

The gateway to recognizing inherited disorders is a thorough medical family history recorded in the form of a pedigree. The word "pedigree" first appeared in the English language in the 15th century and has its origin in the French term *pie de grue* or "crane's foot," describing curved lines, resembling a bird's claw that were used to connect an individual with their offspring. Pedigree for at least three generations is constructed using standardized set of symbols **(Fig. 2)**. A standard medical history is required for the proband (affected individual who draws attention to the family) and for other affected persons in the family. Direct questions need to be asked for similarly affected individuals, miscarriages, early deaths, consanguinity, major and minor malformations, age of presentation, and so on. The pedigree assists in distinguishing genetic from other risk factors, establishing diagnosis, and in deciding on testing strategies *(3)*. It is an invalubale tool in understanding the pattern of inheritance, calculating recurrence risks, determining reproductive options apart from making decisions on medical management and surveillance, and proper patient counselling.

The importance of a complete physical examination and relevent anthropometric measurements to understand the clinical phenotype cannot be underestimated. A careful clinical examination and investigation of parents and family members for mild manifestations can be of great use in genetic counseling, because incomplete penetrance or variable expression is characteristic of many autosomal dominant conditions having no sex bias with vertical pattern of inheritance, coupled with genetic anticipation (increased severity of the disease or decreased age of onset).

3.2. Genetic Testing

3.2.1. Identification of Genetic Defect by Molecular Methods

Genetic tests use a variety of laboratory techniques to determine whether a person has a genetic condition or disease or is likely to get the disease. Individuals may want to get tested if there is a family history of one specific disease, they show symptoms of a genetic disorder, or they are concerned about passing on a genetic problem to their children. These tests include techniques to examine genes or markers near the genes. **Direct testing** for diseases such as cystic fibrosis and sickle cell anemia comes from analysis of an individual's specific genes and identification of the defect therein responsible for the phenotype. A technique called linkage analysis, or **indirect testing**, is used when the gene cannot be directly identified but can be located within a specific region of a chromosome. Indirect testing is usually performed with the help of DNA markers, which are located close to the gene (extragenic), if not in the gene itself (intragenic). This testing requires a positive family history and additional DNA from affected family members for comparison *(2)*.

Genetic testing is usually **diagnostic testing** and involves diagnosis of a disease or condition in an affected individual. It also may be useful to help predict the course of a disease and determine the choice of treatment. **Carrier testing** is performed to determine whether an individual carries one copy of an altered gene for a particular recessive disease. Finally, **predictive (presymptomatic) testing** determines the probability that a healthy individual with or without a family history of a certain disease might develop that disease. Historically, the term presymptomatic testing has been used when testing for diseases or cycling conditions such as HD, in which the likelihood of developing the condition is very high in people with a positive test result. This type of testing is going to become very popular in the perspective of susceptibility screening for various complex diseases and modification of the risk for these diseases.

Genetic testing is a complex process, and the results depend both on reliable laboratory procedures and accurate interpretation of results. Tests also vary in sensitivity, that is, their ability to detect mutations or to detect all patients who have or will get the disease. Interpretation of test results is often complex even for trained physicians and other healthcare specialists. When interpreting the results of any genetic test, the probability of false-positive or false-negative test results must be considered. Special training is required to be able to analyze and convey information about genetic testing to affected individuals and their families. Persons in high-risk families live with troubling uncertainties about their own future as well as that of their children. A negative test, especially a test that is strongly predictive, can create a tremendous sense of relief. Simultaneously, a positive test also can produce benefits by relieving uncertainty and allowing a person to make informed decisions about his or her future. Under the best of circumstances, a positive test creates an excellent opportunity for counseling and interventions to reduce risk.

3.2.2. Carrier Analysis

The main aim of carrier testing is to identify carriers to assess the risk of a couple having an affected child to provide information on the options available to avoid such an eventuality. It is usually done in the context of reproductive planning or prenatal diagnosis and should ideally be done before a pregnancy occurs. Identification of carriers forms the cornerstone toward eradication of the recessive diseases, such as thalassemias. There are several possible strategies for screening, depending on factors such as frequency of the disease; heteogeneity of the genetic defects; and social, cultural, and religious factors. Screening for carrier status is ideally "prospective," i.e., when carriers are identified before having an affected child. This screeening is usually done in view of the family history or biochemical parameters or a high prevalence of specific genetic disorder such as hemoglobinopathies or cystic fibrosis. Unfortunately, often, it is "retrospective," i.e., when couples already have an affected child and are planning for future pregnancy.

3.3. Prenatal Diagnosis

The pinnacle of any genetic diagnostic testing is prenatal diagnosis, which simply means diagnosing a genetic disease or condition in an unborn fetus. Without the knowledge gained by prenatal diagnosis, there could be an untoward outcome for the fetus, the mother, or both, because congenital anomalies account for 20 to 25% of perinatal deaths. Specifically, prenatal diagnosis is helpful for managing the remaining weeks of the pregnancy, determining the outcome of the pregnancy, planning for possible problems that may occur in the newborn infant, deciding whether to continue the pregnancy, and finding cycling conditions that may affect future pregnancies.

3.3.1. Prenatal Sampling and Stage of Gestation

There are a variety of noninvasive and invasive sampling procedures available for prenatal diagnosis. Each of them can be applied only during specific time periods during the pregnancy for greatest utility. The prenatal procedures used for such diagnosis are as follows.

3.3.1.1. INVASIVE PROCEDURES

3.3.1.1.1. Chorion Villus Sampling. Chorion villus sampling is a specialized test that can be performed beginning at the 10th and up to the12th week of pregnancy. It involves removing a small amount of the tissue called the chorionic villi, which is located on the outside of the fetal gestational sac. The chorion is a fetal tissue and shares its genetic makeup with the fetus, not the mother. The chorion has many small, finger-like projections on its outer surface, and a few of these projections may easily be removed under ultrasonographic guidance without disturbing the pregnancy. However, a possibility of maternal tissue getting sampled along with the fetal tissue also remains; hence, the removed samples need to be processed *(4)*. Although the process needs special equipment, and technical expertise, it is fairly safe in experienced hands. However, women undergoing chorion villus sampling may sometimes experience a miscarriage (0.3%).

3.3.1.1.2. Amniocentesis. Amniocentesis involves guiding a thin needle through the mother's abdomen into the amniotic sac and removing a small amount of the fluid that is surrounding the fetus. Ultrasound is used to determine the location of the fetus and the best place to withdraw the fluid. The amniotic fluid contains cells that have been shed by the fetus during normal development. Amniocentesis is the safest prenatal sampling procedure, but in rare cases, there may be fluid leakage or an infection. Any of these complications may cause a miscarriage (0.1%). Amniocentesis is most commonly performed between the 14th and 20th week of pregnancy.

3.3.1.1.3. Fetal Cord Blood Sampling. Fetal cord blood sampling is also known as percutaneous umbilical cord sampling or cordocentesis and collects fetal blood directly from the umbilical cord or fetus. This procedure is technically most challenging and is associated with higher risk of miscarriage (1–2%). However, if properly performed, it is the best procedure because it gives pure material in good quantity. The procedure is usually performed between 18 and 20 weeks of gestation.

3.3.1.2. NONINVASIVE PROCEDURES

3.3.1.2.1. Preimplantation Genetic Diagnosis (PGD). PGD is a recently developed technique that allows screening and identification of specific genetic defects in early embryos created through in vitro fertilization (IVF) before transferring them into the uterus. Because only unaffected embryos are transferred to the uterus for implantation, PGD provides a good alternative to current prenatal sampling procedures such as amniocentesis or chorionic villus sampling, which are frequently followed by pregnancy termination if results are unfavourable *(5)*. The technique involves making a small hole in the zona pellucida of the eight-cell embryo and removing a single cell from the embryo. Removing one cell does not harm the embryo at this stage, because the cell is identical to the rest of the embryo. Couples wishing to have PGD have to undergo conventional IVF or intracytoplasmic sperm injection even if they are able to conceive pregnancies naturally, because there is a risk of getting contamination by any extraneous sperm. Although PGD has many advantages, a misdiagnosis could occur from this technique. Conventional prenatal diagnosis methods provide large amounts of material available for analysis; therefore, errors are considered extremely rare. Because PGD analysis is performed on a single cell, the diagnosis often is very difficult because one or more of the genes may fail to amplify, a direct consequence of analyzing such a minute quantity of material.

3.3.1.2.2. Fetal Cells in Maternal Circulation. Fetal cells are historically known to circulate in the blood of pregnant women. More recently, molecular analysis of plasma DNA during pregnancy led to the discovery that maternal plasma contains both free fetal and maternal DNA *(6,7)*. Fetal cells in maternal circulation were thought to be good source of noninvasive prenatal diagnosis, but fetal cells from a 9-yr-old pregnancy have been shown to ciruclate in the maternal blood. Of three types of fetal cells sought as a source for fetal DNA, including erythroblasts, lymphocytes, and trophoblasts, erythroblasts are best suited, because they are highly differentiated and do not persist after the delivery. The detection of fetal DNA in maternal plasma is technically much simpler and more robust than the detection of fetal nucleated cells in maternal blood. Fetal DNA may be detected in a small amount of maternal plasma. The detection and analysis of fetal DNA is based on the exclusion of the paternally derived pathological allele, assuming that it was different to the mother's mutation. The major limitation is its inability to distinguish the maternally derived allele, thus making the diagnosis only 50% effective.

3.3.2. Postsampling Analysis

It is important that the specific mutation responsible for the genetic disorder in the family is known before deciding about the sampling. Genomic DNA can be isolated from the aforementioned samples directly except from chorionic villus sample, which needs processing under the microscope to remove maternal tissue. Because the amount of DNA available is very little, specific experiments are performed to assess the status of the fetus and confirmed by repeating by another individual. Finally, 10-loci microsatellite analysis is done to rule out the maternal contamination. The sex of the fetus is never analyzed and reported.

3.4. Genetic Counseling

Gentic counseling may be defined as the "process" by which patients or relatives at risk of a disorder that may be hereditary are advised of the probability of developing and transmitting the disorder, its consequences, and the ways in which it may be prevented or ameliorated. It is a multistep process and involves clinical expertise similar to any other medical speciality. Counseling must be nonjudgemental and nondirective and hence needs to include all aspects of the condition *(8)*. The aim is to deliver a balanced version of the facts; hence, the depth of explanation should be matched to the educational background of the couple. Natural course of the disease and treatment (sometimes only supportive) needs to be highlighted. The risk of recurrence in future pregnancy is explained, if necessary with the help of diagrams. It also is made clear that there may be two or three or more consecutively affected children in a family as chance does not have memory. It is often useful to compare this recurrence risk against the general population risk for the condition and for common birth defects. It is stressed that any family can have children affected with genetic diseases or congenital malformations and parenting such a child or carrier status for genetic disease is not a stigma nor should be considered a discriminating factor. A common misconception about heredity also may need to be dispelled. The last and the most important part of the counseling is about reproductive options. Depending upon the parents' judgement about the risk of recurrence, they may go for contraception, adoption, in vitro fertilization, or further pregnancy with or without prenatal diagnosis **(Fig. 3)**.

3.5. Follow-Up

The most important but unfortunately the least rigorously pursued step is the follow-up. It is a good practice to follow the counseling session with a summary of the information provided and suggesting interaction, whenever required. The consultands should be contacted and any new or improved test explained, whenever it becomes available.

4. Illustrative Examples of Individual Diseases

A few commonly occurring diseases are briefly explained as a prototype for each type of genetic disorder.

4.1. Huntington's Disease

HD is an autosomal dominant progressive neurodegenerative disorder characterized by uncontrolled movements, general motor impairment, psychiatric disturbances, dementia, and associated with selective neuronal death, primarily in the cortex and the striatum. Onset of the disease is usually in the third or fourth decade of life, and symptoms progressively worsen over next 15–20 yr before the patient succumbs. The disease is very severe if it manifests before the age 20 yr and is termed juvenile HD. The gene for

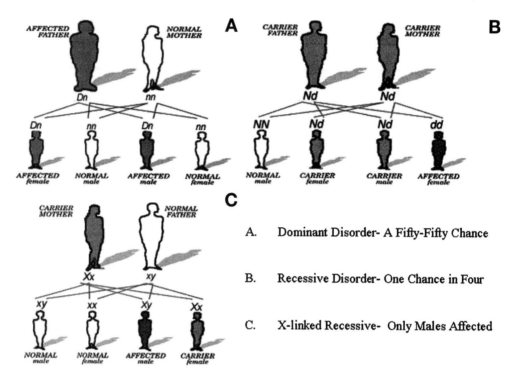

Fig. 3. Genetic counseling for various disorders with specific patterns of inheritance. (**A**) Autosomal dominant inheritance. (**B**) Autosomal recessive inheritance. (**C**) X-Linked recessive inheritance.

this disorder, IT15, is located on chromosome-4p16.3 (*9*). The underlying mutation has been identified as triplet repeat CAG/polyglutamine expansion in the first exon of the gene. The CAG tract has been shown to range up to 34 copies on normal chromosomes, whereas it is expanded above 39 in affected individuals. The range between 34 and 39 repeats is considered as the premutation range, suggesting susceptibility for amplification in the next generation. The degree of repeat expansion is inversely associated with age of onset, and a change in the length of CAG tract with transmission from parent to child represents the rule rather than the exception. This explanation is also the biological explanation for the phenomenon of anticipation, which is related to the increasing disease severity or decreasing age of onset in successive generations. Anticipation occurs more commonly in paternal transmission of the mutated allele. The disease pathology has been linked to abnormal protein–protein interaction related to the elongated polyglutamine tract (*10*).

Apart from CAG repeat expansion, very rare cases of other mutations in IT15 gene have been reported. The molecular diagnosis focuses on the identification of number of CAG repeats in the gene and is usually done by PCR by using fluorescent primers. PCR-based methods can detect alleles with size up to about 115 CAG repeats.

4.2. β-*Thalassemia*

Thalassemias are inherited autosomal recessive disorders that affect the production of hemoglobin and cause anemia. Thalassemias originate when there is an imbalance in the final α-globin and β-globin chain ratios. An equal amount of α and β globins is required to form a functionally efficient haemoglobin molecule. The incidence of β-thalassemia, which is associated with defective β-globin chain production is very high and >30 million people are thought to carry the defective gene. Carrier frequency varies from 3 to 17% in different populations. The disease is of variable severity and correlates with the amount of β-globin chain produced *(11)*. Thalassemia major patients have severe microcytic hypochromic anemia owing to severely impaired β-globin chains production, because both β-globin genes are mutated, whereas thalassemia minor patients are usually heterozygous for $β^0$ or $β^+$ mutation and have mild-to-nonexistent symptoms. Certain individuals have symptoms of severity in between the two spectra and are known as thalassemia intermedia. The most characteristic haematological marker associated with the carrier stage is an elevated level of HbA_2 (>4.5%).

β-Thalassemias are very heterogeneous at molecular level with > 200 different mutations of variable severity described in the literature (http://globin.cse.psu.edu). Despite this marked molecular heterogeneity, the prevalent molecular defects are limited in each at-risk population. For example, five mutations in HBB gene account for approx 95% of all the patients in India. These mutations are classified as either $β^+$ (reduced expression of β-globin gene) or $β^0$ (complete absence of β-globin) type *(12)*.

Mutations in the HBB gene can be detected by a number of PCR-based procedures such as allele-specific oligonucleotide screening, reverse dot blot hybridization, and amplification refractory mutation system (ARMS). The most commonly used methods are reverse dot blot analysis with a set of probes or primer-specific amplification, or primers complementary to the most common mutations in the population **(Fig. 4)**. The usual approach is to screen the proband for the eight common mutations by using PCR coupled with either allele-specific oligonucleotide or reverse dot blot hybridization approach *(13)*. In the situation of failure to identify the mutation (~5–7%), complete β-globin gene can be sequenced to identify the mutation. Usually, the parents are obligate carriers in this situation, but it is important to confirm the mutation in the parents by performing ARMS-PCR for identified mutation(s) in the proband. Very rarely, one of the mutations may originate for the first time in either of the parents and thus the recurrence risk will be comparable with the general risk. These facts are important for genetic counseling and planning for future pregnancy and genetic testing. Once the mutation is identified, the couple can be advised to proceed for appropriate sampling procedure depending on the age of gestation followed by prenatal diagnosis.

4.3. Duchenne/Becker Muscular Dystrophy

Duchenne muscular dystrophy (DMD) is a severe, progressive, and lethal form of muscular dystrophy occurring in up to 1 in 3500 males, whereas Becker muscular dystrophy (BMD) is a related disorder with milder clinical course. DMD is one of the most frequent muscle diseases in children with the age of onset usually before 3 yr. The first symptoms are walking problems owing to symmetrical weakness of the hip and lower proximal limb muscles, which slowly spreads to upper limbs, neck, and respiratory muscles. Calf hypertrophy is also present in 95% of DMD patients and they have difficulties in climbing stairs as well as in getting up from the floor (Gower's sign). BMD shows a variable phenotype from a little less severe DMD-like condition to very mild

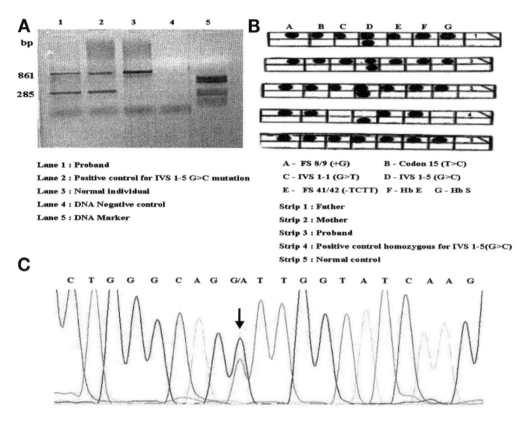

Lane 1 : Proband
Lane 2 : Positive control for IVS 1-5 G>C mutation
Lane 3 : Normal individual
Lane 4 : DNA Negative control
Lane 5 : DNA Marker

A - FS 8/9 (+G)   B - Codon 15 (T>C)
C - IVS 1-1 (G>T)   D - IVS 1-5 (G>C)
E - FS 41/42 (-TCTT)   F - Hb E   G - Hb S

Strip 1 : Father
Strip 2 : Mother
Strip 3 : Proband
Strip 4 : Positive control homozygous for IVS 1-5(G>C)
Strip 5 : Normal control

Fig. 4. Genetic analysis of β-thalassemia. (**A**) ARMS-PCR analysis showing the proband to be positive for IVS 1-5 G>C mutation (mutant allele). (**B**) Screening of eight common mutations in β-globin gene by reverse dot blot hybridization, showing the proband to be homozygous for IVS 1-5 G>C mutation. The mutation has been inherited from the heterozygous parents. (**C**) Electropherogram showing a heterozygous mutation (IVS1-1 G>A) in the β-globin gene

in patients that remain ambulant throughout their lives. Serum creatinine phosphokinase levels are elevated in both the conditions initially, but later get fixed at certain levels as the muscles become completely atrophic.

Both DMD and BMD follow X-linked recessive pattern of inheritance and are caused by mutations in the dystrophin gene on Xp21. The dystrophin gene, the largest human gene known to date spans 2.4 megabases in Xp21. Deletions are the commonest mutations and responsible for two-thirds of patients, whereas duplications and point mutations account for the remaining. Deletions, which alter the open reading frame of the gene generally, cause DMD, whereas in BMD, the reading frame is usually intact and this frame shift model complies with the phenotype in 92% of cases *(14)*. In both DMD and BMD, partial deletions and duplications cluster in two recombination hot spots, one hot spot proximal and one hot spot more distal, but the site and size of these mutations are very heterogeneous. The mutations are usually inherited but as is true with all lethal X-linked disorders, DMD also has a high rate of new mutations. Moreover, approximately one-third of sporadic cases are owing to a new mutation and have an associated risk of approx 20% for gonadal mosaicism in the parent of origin of the X-chromosome carrying the mutation.

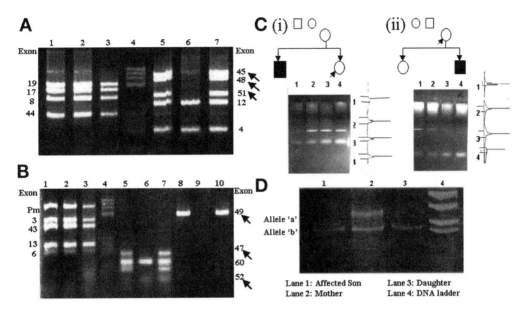

Fig. 5. Genetic analysis of Duchenne muscular dystrophy. **(A,B)** Deletion analysis of various exons in DMD gene by multiplex PCR. Arrowheads indicate the deleted exons. **(C)** Carrier analysis using quantitative PCR. **(i)** Exclusion of the proband as carrier since the band intensity and peak height is same as the normal female. **(ii)** Inclusion of the proband as carrier because the band intensity and the peak height is same as the normal male. **(D)** Carrier analysis using intragenic markers shows the daughter as the carrier since she shares the same allele with his affected brother.

The diagnosis of DMD and BMD is usually clinical and confirmed by immunostaining the muscle biopsy with dystrophin antibody. Molecular genetic analysis initially begins with deletion analysis using multiplex PCR for a number of exons **(Fig. 5)**. In total, 25 exons accounting for approx 70% of all deletions can be performed in parallel. In the remaining cases, linkage analysis using intragenic microsatellite markers is applied to identify the "at risk" or "affected" X-haplotype. Thus, a positive family history and DNA from additional family members is essential. Once a marker (either direct or indirect) has been identified, identification of the carrier status is next important step.

Pedigree analysis is never more informative than in X-linked recessive diseases such as DMD. It is possible to identify the obligate carriers if the detailed family history is available. Although a rough idea can be obtained by elevated serum creatinine phosphokinase levels, by muscle histology, or by dystrophin immunoreactivity, DNA-based analysis accurately detects carriers in >95% of cases. In families where a deletion is identified, quantitative PCR can be performed for detection of the copy number of DMD gene. Linkage analysis using the specific microsatellite marker can be used to trace the inheritance of affected X-haplotype and establish the carrier status.

4.4. Multifactorial Disorders (Factor V Leiden)

Deep venous thrombosis is a thrombophilic state commonly affecting the veins in the legs. Factor V Leiden is a genetically acquired trait that can result in a hypercoaguable state owing to the phenomenon of activated protein C resistance

(APCR). APCR was first described in 1993 followed by factor V Leiden in 1994, and >95% of patients with APCR have factor V Leiden. Factor V Leiden thrombophilia is characterized by a poor anticoagulant response to activated protein C and an increased risk of venous thromboembolism. Presence of the factor V Leiden allele also has been reported in individuals with deep vein thrombosis, central retinal vein occlusion, cerebral sinus thrombosis, and hepatic vein thrombosis.

The term "factor V Leiden" refers to the specific G-to-A substitution at nucleotide 1691 in the gene for factor V that predicts a single amino acid replacement (Arg506Gln) at one of three activated protein C cleavage sites in the factor Va molecule, which plays an important role in blood clotting *(15)*. Individuals heterozygous for the factor V Leiden mutation have a slightly increased risk for venous thrombosis, whereas homozygous individuals have a much greater thrombotic risk. The G>A mutation at position 1691 in factor V gene leads to the loss of site for MnlI enzyme and can be identified by PCR amplification followed by restriction digestion (PCR-restriction fragment length polymorphism). This analysis can be used for susceptibility screening for thrombophilic condition states and appropriate genetic counselling to modify the risk associated with it.

5. Future Thoughts

Currently available technologies can help eliminate some single gene disorders in the future such as haemoglobinopathies, cystic fibrosis, X-linked dystrophies, and so on. Complete cure for many genetic diseases are not likely to be found soon; therefore, preventing the disease is preferable to waiting for a possible cure. Furthermore, available treatments often have multiple adverse effects. Prolonging the life span of affected patients could cause them to develop diseases not previously known to be associated with the particular genetic conditions. For example, as improved treatment prolongs life for individuals with cystic fibrosis, other manifestations of pancreatic insufficiency and nutritional malabsorption associated with the disease, such as diabetes mellitus and osteoporosis, begin to emerge. Pharmacogenetics and genotype based therapy are also gaining momentum, and it is not far away that prescriptions will become completely individualized. Additionally, newer technologies are being developed that will help in providing the diagnostic testing faster and at more reasonable rates. Thus, the future holds great promise for genetic testing with an ever-widening horizon for single gene disorders and additional scope for susceptibility screening for various complex diseases.

References

1. Vogel, F., and Motulsky, A. G. (1996) *Human Genetics: Problems and Approaches*, 3rd ed. Springer, Berlin, Germany.
2. Emery, A. E. H., and Rimoin, D. L. (1990) *The Principles and Practice of Medical Genetics*, 3rd ed. Churchill Livingstone, Edinburgh, Scotland, UK.
3. Bennet, R. L. (1999) *The Practical Guide to the Genetic Family History*, 1st ed. Wiley-Liss, New York.
4. Old, J. M. (1986) Foetal DNA analysis, in *Genetic Analysis of the Human Disease: A Practical Approach* (K. E. Davies, ed.), IRL Press, Oxford, UK.
5. Kanavakis, E., and Traeger-Synodinos, J. (2002) Preimplantation genetic diagnosis in clinical diagnosis. *J. Med. Genet.* **39,** 6–11.

6. Simpson, J. L., and Elias, S. (1993) Isolating fetal cells from maternal blood: Advances in prenatal diagnosis through molecular technology. *J. Am. Med. Assoc.* **270,** 2357–2361.

7. Lo, Y. M. (2000) Fetal DNA in maternal plasma. *Ann. NY Acad. Sci.* **906,** 141–147.

8. Harper, P. S. (1988) *Practical Genetics Counselling,* 3rd ed. John Wright, Bristol, United Kingdom.

9. Huntington's Disease Collaborative Research Group (1993) A novel gene containing a trinucleotide repeat that is expanded and unstable on Huntington's disease chromosomes. *Cell* **729,** 71–83.

10. Andrew, S. E., Goldberg, Y. P., Kremmer, B., et al. (1993) The relationship between trinucleotide (CAG) repeat length and clinical features of Huntington's disease. *Nat. Genet.* **4,** 398–403.

11. Weatherall, D. J., and Clegg, J. B. (2001) *The Thalassemia Syndromes.* Blackwell Science, Oxford, UK.

12. Varawalla, N. Y., Old, J. M., Sarkar, R., Venkatesan, R., and Weatherall, D. J. (1991) The spectrum of β-thalassemia mutations on the Indian subcontinent: the basis of prenatal diagnosis. *Br. J. Haemat.* **78,** 242–247.

13. Maggio, A., Giambona, A., Cai, S. P., et al (1993) Rapid and simultaneous typing of hemoglobin S, hemoglobin C and seven Mediterranean β-thalassemia mutations by covalent reverse dot-blot analysis: Applications to prenatal diagnosis in Sicily. *Blood* **81,** 239–242.

14. Emery, A. E. H. (1997) Duchenne and other X-linked muscular dystrophies, in *Emory and Rimoin's Principles of Medical Genetics,* 3rd ed. (D. L. Rimoin, J. M. Connor, and R. E. Pyeritz, eds.), Churchill Livingstone, New York, pp. 2337–2354.

15. Grody, W. W., Griffin, J. H., Taylor, A. K., Korf, B. R., and Heit, J. A. (2001) Americal College of Medical Genetics consensus statement on Factor V Leiden mutation testing. *Genet. Med.* **3,** 139–148.

Index

263

Printed in the United States of America.